RIGOROUS RESULTS IN QUANTUM DYNAMICS

RIGOROUS RESULTS IN QUANTUM DYNAMICS

Liblice, Czechoslovakia 10 – 15 June 1990

Edited by

J. Dittrich and P. Exner

Nuclear Physics Institute
Czechoslovak Academy of Sciences

World Scientific
Singapore • New Jersey • London • Hong Kong

Published by

World Scientific Publishing Co. Pte. Ltd.
5 Toh Tuck Link, Singapore 596224
USA office: 27 Warren Street, Suite 401-402, Hackensack, NJ 07601
UK office: 57 Shelton Street, Covent Garden, London WC2H 9HE

British Library Cataloguing-in-Publication Data
A catalogue record for this book is available from the British Library.

RIGOROUS RESULTS IN QUANTUM DYNAMICS

ISBN-13 978-981-02-0561-4
ISBN-10 981-02-0561-9

Preface

The depth of rigor in theoretical physics is always a relative matter, and it is probably not a good idea to try to draw a borderline between rigorous and non-rigorous results. One can, of course, always find advocates of the claim that people who pay attention to whether the symbols they use make sense are good just for "cleaning the battlefield", however, everybody who understands the complex and intricate ways in which the science develops can quote examples against such a claim.

You need not be an insider to observe the rapid development of mathematical physics in the past two decades as well as the fact that it brought not only existence proofs but also some quite practical results ; let us just mention the purely mathematical trick of J.Aguilar and J.-M.Combes which has developed into one of the most efficient methods of investigating resonances in atomic and molecular physics. It is also clear that a large part of this boom is in one or another way connected with dynamics of quantum systems. This trend is likely to continue since quantum theory becomes the basis of most of the modern applied physics from which it gets in turn a lot of inspiration.

The core of quantum mechanics is traditionally the spectral and scattering theory of Schrödinger and Dirac operators. Even after seven decades there are still many open problems for wide classes of these operators to investigate : precise bounds on the eigenvalue numbers, nodal structure of eigenfunctions, criteria for presence or absence of the singularly continuous spectrum *etc.* In some situations the mathematical problem is directly related to an important physical question like in the case of molecule stability. In the first chapter we collect contributions containing some fresh results in this field. They are followed by two lectures aimed at description of new methods of solving the Schrödinger equation, and a chapter devoted to what might be called non-standard Schrödinger operators, *i.e.*, with interaction supported by a zero-measure set or restricted to a region in configuration space, or with a stratification.

An independent chapter collects contributions dealing with Hamiltonians dependent on time. This field is less developed than its time-independent counterpart, and new results appearing here concern both the general properties like smoothing effect of propagators, as well as particular physically interesting phenomena like the change in phase gained by a system during a periodic motion, or quantum chaos.

There is, however, more in quantum dynamics than the Schrödinger equation. Several contributions deal with the problems of spin systems which are the simplest framework to investigate infinite number of degrees of freedom, and at the same time they are related to basic physical phenomena like magnetic properties of matter. The present volume covers also some new results concerning stochastic analysis, non-linear equations, coherent states and others topics.

The Liblice meeting belonged to the series of conferences which started at Dubna in 1987, and which was aimed, in particular, at opening ways to East-West contacts. The world has changed a lot since then, and one can hope that East and West will stay from now on primarily geographical and not political categories. Hence the aim of the meetings like this should be now to help to intensify collaboration among mathematical physicists in this newly open but still weakly connected world.

Prague, February 1991

Jaroslav Dittrich
Pavel Exner

Jiří Blank (17.5.1939 – 22.2.1990)

IN MEMORIAM

Before the conference began, we lost our colleague Jiří Blank, senior research fellow of the Charles University in Prague, who had helped to organize this meeting.

He was a dedicated scientist, one of those who use all their strength to move the vehicle of science forward. He was also an excellent teacher whom we are indebted for showing to bright young people way to mathematical physics.

As a man of firm civic attitudes he had to encounter many troubles in the past two decades, and it was sad to lose him just at the time when the nightmares of the past began to scatter.

Above all, he was a man of warm personality and a good friend to many of us. We dedicate this volume to his memory.

The editors

Contents

Schrödinger Equation: Methods of Solution

Waveguides and Contact Interactions

Time-dependent Systems, Berry Phase, Quantum Chaos

Miscellaneous

Schrödinger and Dirac Operators

Commutator Methods and Schrödinger Operators

ARNE JENSEN
Department of Mathematics and Computer Science
Institute for Electronic Systems, Aalborg University
Fredrik Bajers Vej 7, DK-9220 Aalborg Ø, Denmark

Abstract
An abstract short range scattering theory is presented. Its application to Schrödinger operators is based on commutator methods. Applications to ordinary Schrödinger operators and to Stark Hamiltonians are given. Finally the concepts of a short range and a long range potential are discussed, and a discrepancy between classical mechanics and quantum mechanics is observed in the case of Stark Hamiltonians.

1 Introduction

In this paper an abstract short range scattering theory with applications to both ordinary Schrödinger operators and Stark Hamiltonians is presented. There are two basic assumptions in this abstract theory. One assumption is a pair of propagation estimates for the unperturbed operator H_0 involving a second selfadjoint operator A, and the other assumption is an abstract short range condition on the potential V defined in terms of A.

The verification of the assumptions in the abstract theory is often based on commutator computations. We illustrate this by discussing applications to both ordinary Schrödinger operators and to Stark Hamiltonians. We also briefly discuss an abstract approach to propagation estimates. In the last section of the paper we discuss briefly the concept of a short range potential and a long range potential. For the ordinary Schrödinger operator it is well known that the borderline between short range and long range potentials is given by the Coulomb potential $1/|x|$, both in classical and in quantum scattering theory. For Stark Hamiltonians the situation is different, and a discrepancy between classical mechanics and quantum mechanics is observed.

2 An Abstract Short Range Scattering Theory

We assume familiarity with the basic concepts in scattering theory, and refer to standard textbooks and monographs[2, 10, 16] for these results. There are several abstract versions of scattering theory. They include the trace class scattering theory[10] and the abstract theory developed by Kato and Kuroda[11]. The results presented here are distinguished by the introduction af an abstract short range condition. The proofs are time-dependent. These results were first given a paper with Mourre and Perry[8], without proof. Since the proofs seem to be of interest, a version is presented here.

We consider the following framework: Let $\mathcal{H}$ be a separable Hilbert space and H_0 a selfadjoint operator in $\mathcal{H}$. The domain of H_0 is denoted $\mathcal{D}(H_0)$. We consider a perturbation V of H_0, called the potential, and the operator H, which is a selfadjoint extension of $H_0 + V$, defined suitably. Here we need not be concerned with such technicalities and assume that V is a symmetric H_0-bounded operator with relative bound less than one. Then $H = H_0 + V$ is the operator sum, and H is selfadjoint on $\mathcal{D}(H) = \mathcal{D}(H_0)$. The spectral family for H is denoted E_H. The subspace of absolute continuity is denoted $\mathcal{H}_{ac}(H)$ and the singular subspace $\mathcal{H}_s(H)$. The corresponding projections are denoted $P_{ac}(H)$ and $P_s(H)$, respectively. The spectrum of H is denoted $\sigma(H)$, the absolutely continuous part $\sigma_{ac}(H)$, and so on. Let $I_0 \subseteq \mathbf{R}$ be an open interval. We consider the scattering theory for the pair (H, H_0), localized to the interval I_0. The two basic questions are:

(i) Existence of the wave operators

$$W_\pm(H, H_0; I_0) = \mathrm{s} - \lim_{t\to\pm\infty} e^{itH} e^{-itH_0} E_{H_0}(I_0) P_{ac}(H_0). \tag{1}$$

(ii) Asymptotic completeness of the wave operators, i.e. the two results

$$\mathrm{Ran}(W_\pm(H, H_0; I_0)) = E_H(I_0) P_{ac}(H)\mathcal{H}, \tag{2}$$

and

$$\sigma_s(H) \cap I_0 \quad \text{is discrete in } I_0. \tag{3}$$

The statement (ii) above means that there is no singular continuous spectrum of H in I_0 and the point spectrum of H in I_0 consists at most of discrete eigenvalues with finite multiplicity. The only possible accumulation points of $\sigma_p(H) \cap I_0$ are the end points of I_0. The abstract theory is based on the following two definitions:

Definition 2.1 *Let A be a selfadjoint operator on $\mathcal{H}$. We say that H_0 satisfies propagation estimates with respect to A on I_0, if there exist real numbers $s > s' > 1$ such that for all $g \in C_0^\infty(I_0)$ the following two estimates hold:*

$$\|(1 + A^2)^{-s/2} e^{-itH_0} g(H_0)(1 + A^2)^{-s/2}\| \le c(1 + |t|)^{-s'} \quad \textit{for all } t \in \mathbf{R}, \tag{4}$$

$$\|(1 + A^2)^{-s/2} e^{-itH_0} g(H_0) P_A^\pm\| \le c(1 + |t|)^{-s'} \quad \textit{for all } \pm t > 0. \tag{5}$$

Here $P_A^+ = E_A((0, \infty))$ and $P_A^- = I - P_A^+$.

Definition 2.2 *Let A be a selfadjoint operator. The potential V is said to be a short range perturbation of H_0 with respect to A, if*

$$(H+i)^{-1}-(H_0+i)^{-1} \quad \textit{is a compact operator on } \mathcal{H}, \tag{6}$$

and if there exist a real number $\mu > 1$ and integers $j,\, k \geq 0$ such that the operator

$$(H+i)^{-j}V(H_0+i)^{-k}(1+A^2)^{\mu/2} \tag{7}$$

extends to a bounded operator on $\mathcal{H}$.

The abstract theorem given here is a version of the Enss method[5, 12, 15, 17].

Theorem 2.3 *Let H_0, V, I_0 and H be as above. Assume that there exists a selfadjoint operator A such that H_0 satisfies the propagation estimates with respect to A and such that the potential V is a short range perturbation of H_0 with respect to A. Then the wave operators $W_\pm(H, H_0; I_0)$ exist and are asymptotically complete.*

Proof: The assumption Eq. 6 implies that $g(H)-g(H_0)$ is compact for all $g \in C_0^\infty(I_0)$. Let us note that Eq. 4 implies $I_0 \cap \sigma_s(H_0) = \emptyset$, hence the projection $P_{ac}(H_0)$ can be omitted in the computations. We consider only the +-case, and start by proving existence of

$$\widetilde{W} = \mathrm{s}-\lim_{t\to\infty}(H+i)^{-j}e^{itH}e^{-itH_0}E_{H_0}(I_0). \tag{8}$$

By the standard Cook-Kuroda argument it suffices to prove

$$\int_0^\infty \|(H+i)^{-j}Ve^{-itH_0}g(H_0)\psi\|\,dt < \infty \tag{9}$$

for a dense set of $\psi \in \mathcal{H}$, and all $g \in C_0^\infty(I_0)$. Without loss of generality we can assume that $s = \mu$, with s from Definition 2.1 and μ from Definition 2.2. Fix $g \in C_0^\infty(I_0)$ and take $\psi = (1+A^2)^{-s/2}\phi$, $\phi \in \mathcal{H}$ arbitrary. Using Eq. 4 and Eq. 7 we get

$$\begin{aligned}
\|(H+i)^{-j}Ve^{-itH_0}g(H_0)\psi\| &\leq \|(H+i)^{-j}V(H_0+i)^{-k}(1+A^2)^{s/2}\| \\
&\quad \cdot \|(1+A^2)^{-s/2}e^{-itH_0}(H_0+i)^k g(H_0)(1+A^2)^{-s/2}\| \cdot \|\phi\| \\
&\leq c(1+|t|)^{-s'},
\end{aligned}$$

which is integrable, since $s' > 1$. To remove the factor $(H+i)^{-j}$ in Eq. 8, we note (with an obvious notation)

$$\begin{aligned}
e^{itH}e^{-itH_0}g(H_0) &= e^{itH}e^{-itH_0}(H_0+i)^{-j}\tilde{g}(H_0) \\
&= e^{itH}((H_0+i)^{-j}-(H+i)^{-j})e^{-itH_0}\tilde{g}(H_0) \\
&\quad + (H+i)^{-j}e^{itH}e^{-itH_0}\tilde{g}(H_0).
\end{aligned}$$

Since $(H_0+i)^{-j}-(H+i)^{-j}$ is compact by Eq. 6 and $e^{-itH_0}\tilde{g}(H_0)$ tends to zero weakly as $t\to\infty$, the existence of $W_+(H,H_0;I_0)$ follows.

We write $W_+ = W_+(H,H_0;I_0)$ to simplify the notation. Let $g\in C_0^\infty(I_0)$. We prove that the operators

$$(W_\pm - I)g(H_0)P_A^\pm \tag{10}$$

and

$$g(H)(W_\pm - I)P_A^\pm \tag{11}$$

are compact. This is done by first proving that the operators

$$g_1(H)(W_\pm - I)g_2(H_0)P_A^\pm \tag{12}$$

are compact for all $g_1, g_2\in C_0^\infty(I_0)$. We have

$$\begin{aligned} g_1(H)(e^{itH}e^{-itH_0} - I)g_2(H_0)P_A^\pm \\ = \int_0^t g_1(H)e^{i\tau H}iVe^{-i\tau H_0}g_2(H_0)P_A^\pm\, d\tau, \end{aligned} \tag{13}$$

where the integrand is norm continuous in τ and compact, due to Eq. 6. We estimate for $\tau > 0$

$$\begin{aligned} &\|g_1(H)e^{i\tau H}iVe^{-i\tau H_0}g_2(H_0)P_A^+\| \\ &\quad\le \|g_1(H)V(H_0+i)^{-k}(1+A^2)^{s/2}\|\cdot\|(1+A^2)^{-s/2}e^{-\tau H_0}\tilde{g}_2(H_0)P_A^+\| \\ &\quad\le c(1+|t|)^{-s'}. \end{aligned}$$

Thus the limit exists in norm as $t\to\infty$. An analogous result holds for $t\to-\infty$. Hence the operators in Eq. 12 are compact. The extra g-factors can be removed using the interwining relation for wave operators and the compactness of $g(H)-g(H_0)$. Thus compactness of the operators in Eq. 10 and Eq. 11 has been proved.

As the first step towards completeness we show that $\sigma_s(H)\cap I_0$ is discrete in I_0. Let $J\subset I_0$ be a relatively compact open interval and take $g\in C_0^\infty(I_0)$, $g(\lambda)=1$ for all $\lambda\in J$. A general property of wave operators is

$$\operatorname{Ran}(W_\pm)\subseteq \mathcal{H}_{ac}(H) = \mathcal{H}_s(H)^\perp,$$

hence $P_s(H)W_\pm = 0$.

$$\begin{aligned} P_s(H)E_H(J) &= P_s(H)E_H(J)g(H) \\ &= P_s(H)E_H(J)g(H)(P_A^+ + P_A^-) \\ &= P_s(H)E_H(J)g(H)(I-W_+)P_A^+ \\ &\quad + P_s(H)E_H(J)g(H)(I-W_-)P_A^-. \end{aligned}$$

By the compactness of Eq. 11 we conclude that this operator is compact, hence of finite rank, which proves the result in Eq. 3.

The final step in the proof of completeness consists in proving

$$\mathrm{Ran}(W_\pm) = E_H(I_0)\mathcal{H}_{ac}(H).$$

Since $I_0\backslash\sigma_p(H)$ is open by the first part of the proof, we can take $g \in C_0^\infty(I_0\backslash\sigma_p(H))$. By general results on wave operators[10, 16] it suffices to prove

$$\mathrm{s}-\lim_{t\to\pm\infty} e^{itH_0}e^{-itH}g(H) = W_\pm^* g(H). \tag{14}$$

Take $\psi \in \mathcal{H}_{ac}(H)$ with $\psi = g(H)\psi$ and compute as follows in the +-case:

$$\begin{aligned}
&\|e^{itH_0}e^{-itH}g(H)\psi - W_+^* g(H)\psi\| \\
&\quad = \|e^{-itH}g(H)\psi - e^{-itH_0}W_+^* g(H)\psi\| \\
&\quad = \|(P_A^+ + P_A^-)(e^{-itH}g(H)\psi - e^{-itH_0}W_+^* g(H)\psi)\| \\
&\quad \le A_+(t) + A_-(t)
\end{aligned}$$

with an obvious notation.

$$\begin{aligned}
A_+(t) &= \|P_A^+(e^{-itH}g(H)\psi - e^{-itH_0}W_+^* g(H)\psi)\| \\
&= \|P_A^+(I - W_+^*)g(H)e^{-itH}\psi\|,
\end{aligned}$$

where we used the intertwining relation. By Eq. 11 the operator $P_A^+(I - W_+^*)g(H)$ is compact, and $e^{-itH}\psi \to 0$ as $t \to \infty$ weakly, so $A_+(t) \to 0$ as $t \to \infty$.

$$A_-(t) \le \|P_A^- e^{-itH}g(H)\psi\| + \|P_A^- e^{-itH_0}W_+^* g(H)\psi\|. \tag{15}$$

The estimate in Eq. 5 implies

$$\mathrm{s}-\lim_{t\to\infty} P_A^- e^{-itH_0}g(H_0) = 0.$$

Since $W_+^* g(H)\psi = g(H_0)W_+^*\psi$, the second term in Eq. 15 tends to zero.

$$\begin{aligned}
\|P_A^- e^{-itH}g(H)\psi\| &\le \|P_A^- W_+^* e^{-itH}g(H)\psi\| \\
&\quad + \|P_A^-(I - W_+^*)e^{-itH}g(H)\psi\|.
\end{aligned}$$

These two terms tend to zero, as above. This concludes the proof of Theorem 2.3.

Remark 2.4 (i) Note that we need the existence of both W_+ and W_- to prove the completeness of W_+.

(ii) One can state the result in a form that does not introduce the potential. Instead, one assumes the existence of two selfadjoint operators H_0 and H satisfying Eq. 6. The assumption Eq. 7 is reformulated as: For some integer $k \ge 0$ and a real number $\mu > 1$ the operator

$$E_H(I_0)(H - H_0)(H_0 + i)^{-k}(1 + A^2)^{\mu/2}$$

extends to a bounded operator on $\mathcal{H}$.

3 Propagation Estimates I

In this section we show how to derive the essential propagation estimates in Definition 2.1 in two cases where the results are obtained using straighforward commutator calculations.

3.1 The Case $H_0 = -\Delta$

We consider first the Laplace operator $H_0 = -\Delta$ on $\mathcal{H} = L^2(\mathbf{R}^n)$. Let $A = \frac{1}{2i}(x \cdot \nabla + \nabla \cdot x)$ denote the generator of dilations. We state the result in the following form:

Theorem 3.1 *The operators H_0 and A above satisfy the estimates Eq. 4 and Eq. 5 with $I_0 = (0, \infty)$ for all $s > s' > 0$*

Proof: Define $K_0 = \log H_0$ using the functional calculus. A straightforward calculation proves

$$i[K_0, A] = 2I.$$

Therefore we have

$$e^{itK_0} A e^{-itK_0} = A + 2tI$$

which implies three propagation estimates for K_0:

$$\|(1 + A^2)^{-s/2} e^{-itK_0} (1 + A^2)^{-s/2}\| \leq c(1 + |t|)^{-s} \quad \text{for all } t \in \mathbf{R}. \tag{16}$$

$$\|(1 + A^2)^{-s/2} e^{-itK_0} P_A^{\pm}\| \leq c(1 + |t|)^{-s} \quad \text{for all } \pm t > 0. \tag{17}$$

$$P_A^{\mp} e^{-itK_0} P_A^{\pm} = 0 \quad \text{for all } \pm t > 0. \tag{18}$$

By the functional calculus

$$e^{-itK_0} = H_0^{-it}. \tag{19}$$

The propagation estimates for H_0 are now derived using these results and the one-dimensional Mellin transform in momentum space. Let us recall the definitions: $\mathcal{M} : L^2(\mathbf{R}_+) \to L^2(\mathbf{R})$ is given by

$$(\mathcal{M} f)(\lambda) = \int_0^\infty f(p) \, p^{-i\lambda - 1} \, dp$$

with inverse

$$(\mathcal{M}^{-1} g)(p) = \frac{1}{2\pi} \int_{-\infty}^\infty g(\lambda) \, p^{i\lambda} \, d\lambda.$$

Let $g \in C_0^\infty(\mathbf{R}_+)$. In momentum space $e^{-itH_0} g(H_0)$ is multiplication by the function $e^{-it\rho} g(\rho)$, $\rho = p^2$. Using the Mellin transform (shifting a 2π factor) we write

$$e^{-it\rho} g(\rho) = \int_{-\infty}^\infty G_t(\lambda) \, \rho^{i\lambda} \, d\lambda.$$

with

$$G_t(\lambda) = \frac{1}{2\pi} \int_0^\infty e^{-it\rho} g(\rho) \, \rho^{-i\lambda - 1} \, d\rho.$$

Lemma 3.2 *The function $G_t(\lambda)$ satisfies the following estimates:*

$$|G_t(\lambda)| \le C_N\,|t|^{-N}(1+|\lambda|)^N \quad \textit{for all } N \ge 1 \tag{20}$$

and for $t \cdot \lambda > 0$

$$|G_t(\lambda)| \le C_N\,(1+|t+\lambda|)^{-N} \quad \textit{for all } N \ge 1 \tag{21}$$

Proof: The results follow from straightforward (non-)stationary phase computations.

We can now prove the propagation estimates for H_0. By the functional calculus we have

$$e^{-itH_0}g(H_0) = \int_{-\infty}^{\infty} G_t(\lambda)\, {H_0}^{i\lambda}\, d\lambda.$$

Using Eq. 16, Eq. 19, and Eq. 20 we obtain for $s > 1$

$$\begin{aligned}\|(1+A^2)^{-s/2}e^{-itH_0}g(H_0)(1+A^2)^{-s/2}\| &\le \int_{-\infty}^{\infty} |G_t(\lambda)|(1+|\lambda|)^{-s}\, d\lambda \\ &\le C_{N,s}\,|t|^{-N},\end{aligned}$$

if N is an integer satisfying $N < s-1$. Thus the estimate holds for all s and N satisfying this condition, and the general result Eq. 4 follows by complex interpolation.

We prove Eq. 5 in the $t > 0$ case. Write

$$(1+A^2)^{-s/2}e^{-itH_0}g(H_0)P_A^+ = \int_{-\infty}^{\infty} G_t(\lambda)(1+A^2)^{-s/2}{H_0}^{i\lambda}P_A^+\, d\lambda.$$

Then the integral over negative λ is estimated using Eq. 17 and Eq. 20 for $N < s-1$

$$\begin{aligned}\|\int_{-\infty}^{0} G_t(\lambda)(1+A^2)^{-s/2}{H_0}^{i\lambda}P_A^+\, d\lambda\| &\le c\int_{-\infty}^{0} |G_t(\lambda)|(1+\lambda)^{-s}\, d\lambda \\ &\le C_{N,s}\,|t|^{-N}.\end{aligned}$$

For the integral over positive λ we use the estimate Eq. 21 to get

$$\begin{aligned}\|\int_{0}^{\infty} G_t(\lambda)(1+A^2)^{-s/2}{H_0}^{i\lambda}P_A^+\, d\lambda\| &\le c\int_{0}^{\infty} |G_t(\lambda)|\, d\lambda \\ &\le c\int_{0}^{\infty} (1+t+\lambda)^{-m}\, d\lambda \\ &\le c\,t^{-m+1}\end{aligned}$$

for any $m > 1$. The proof is now completed using complex interpolation.

Remark 3.3 The Mellin transform was used by Perry[14] in a different manner to derive the basic propagation estimate used in the Enss method.

3.2 The Case $H_0 = -\Delta + Fx_1$

As our second example we take the free Stark Hamiltonian

$$H_0 = -\Delta + Fx_1, \quad F \neq 0, \tag{22}$$

on $\mathcal{H} = L^2(\mathbf{R}^n)$. Here we choose

$$A = \frac{i}{F}\frac{\partial}{\partial x_1}. \tag{23}$$

The spectrum of H_0 is equal to $\mathbf{R}$ and is purely absolutely continuous. A straightforward commutator computation yields

$$i[H_0,\, A] = I.$$

Thus the estimates given in Eq. 16 and Eq. 17 hold in this case too, and we have proved the following theorem:

Theorem 3.4 *The operators H_0 from Eq. 22 and A from Eq. 23 satisfy the estimates Eq. 4 and Eq. 5 with $I_0 = (-\infty,\, \infty)$ for all $s = s' > 0$.*

4 Propagation Estimates II

The two explicit examples above were based on the canonical commutation relation. There is a far reaching generalization of this approach based on some ideas due to Mourre[8, 13]. We state the result[6, 8] in the following form:

Theorem 4.1 *Let H_0 and A be selfadjoint operators on a Hilbert space $\mathcal{H}$. Let $\lambda_0 \in \mathbf{R}$. We make the following assumptions:*

(a) *$\mathcal{D}(A) \cap \mathcal{D}(H_0)$ is a core for H_0.*

(b) *$e^{i\theta A}\mathcal{D}(H_0) \subseteq \mathcal{D}(H_0)$ and for each $\psi \in \mathcal{D}(H_0)$ we have*

$$\sup_{|\theta|\leq 1} \|H_0 e^{i\theta A}\psi\| < \infty.$$

(c) *The commutator $i[H_0,\, A]$, defined as a form on $\mathcal{D}(A)\cap\mathcal{D}(H_0)$, is bounded below and closable. The selfadjoint operator associated with its closure is denoted iB_1. Assume $\mathcal{D}(B_1) \supseteq \mathcal{D}(H_0)$. Assume inductively for $j = 2, 3, \ldots$ that the form $i[iB_{j-1},\, A]$ is bounded below and closable. The associated operator is denoted iB_j. Assume $\mathcal{D}(B_j) \supseteq \mathcal{D}(H_0)$.*

(d) *There exist $\alpha > 0$, $\delta > 0$, and a compact operator K such that with $J = (\lambda_0 - \delta,\, \lambda_0 + \delta)$ the Mourre estimate*

$$E_{H_0}(J) iB_1 E_{H_0}(J) \geq \alpha E_{H_0}(J) + K \tag{24}$$

holds.

Then we have

(i) $\sigma_s(H_0) \cap J$ *is discrete in* J.

(ii) *The estimates in Eq. 4 and Eq. 5 hold with* $I_0 = J \setminus \sigma_s(H_0)$ *for all* $s > s' > 0$.

We refer to the original papers for the proofs[8, 13]. The essential point is that the exact canonical commutation relation $i[H_0, A] = I$ can be replaced by the estimate Eq. 24.

5 Short Range Potentials

To apply the abstract short range scattering theory it is useful to have criteria to determine whether a given potential V is short range with respect to a given pair of operators H_0 and A, for which the propagation estimates have been established. We shall discuss the two examples from Section 3.

5.1 The Case $H_0 = -\Delta$

In this case the well known usual short range condition implies the short range condition in Definition 2.2. In rough asymptotic form the short range condition is $V(x) = O(|x|^{-1-\varepsilon})$, $\varepsilon > 0$, as $|x| \to \infty$ for a multiplicative potential. A precise statement for a general potential is given in the following proposition:

Proposition 5.1 *Let* V *be a symmetric* H_0*-compact operator such that for some* $\mu > 1$ *the operator* $V(H_0+i)^{-1}(1+x^2)^{\mu/2}$ *extends to a bounded operator on* $L^2(\mathbf{R}^n)$. *Then* V *is a short range perturbation of* $H_0 = -\Delta$ *with respect to* $A = 1/(2i)(\nabla \cdot x + x \cdot \nabla)$.

Proof: The assumptions imply Eq. 6 immediately, and a simple complex interpolation argument shows that $V(H_0+i)^{-2}(1+A^2)^{\mu/2}$ extends to a bounded operator on $L^2(\mathbf{R}^n)$. Thus the conditions in Definition 2.2 are verified.

More interesting is to note that some classes of oscillatory potentials also can be shown to fit into the framework given here. There is already a considerable literature[3, 15, 18] on this subject, so we only give one representative result, which seems to be new, since we only require $\beta > 0$, whereas previous results[18] required $\beta > 1/2$. Note that our technique allows us to treat general non-central oscillatory potentials.

Proposition 5.2 *Let* $\alpha > 1$, $0 < \beta \leq 1$, $b, c \in \mathbf{R}$ *and define for* $x \in \mathbf{R}^n$, $x \neq 0$,

$$V(x) = c\frac{\sin(b|x|^\alpha)}{|x|^\beta}.$$

Then V *is a short range perturbation of* $H_0 = -\Delta$ *with respect to* $A = 1/(2i)(\nabla \cdot x + x \cdot \nabla)$.

Proof: The proof is based on the following lemma:

Lemma 5.3 *Let $g : \mathbf{R}^n \to \mathbf{R}^n$ be a real-valued continuously differentiable function. Let U and W be bounded real-valued functions such that for some integer k and real number $\mu > 1$ the operators*

$$g_j(x)U(x)(H_0 + i)^{-k}(1 + A^2)^{\mu/2}, \quad j = 1, 2, \ldots, n,$$

$$(\nabla \cdot g(x))U(x)(H_0 + i)^{-k}(1 + A^2)^{\mu/2},$$

and

$$W(H_0 + i)^{-k}(1 + A^2)^{\mu/2}$$

extend to bounded operators on $L^2(\mathbf{R}^n)$. Let

$$V(x) = [g(x) \cdot \nabla,\, U(x)] + W(x).$$

Then the operator

$$(H_0 + i)^{-1}V(H_0 + i)^{-k-1}(1 + A^2)^{\mu/2}$$

extends to a bounded operator on $L^2(\mathbf{R}^n)$.

Proof: We compute:

$$\begin{aligned}
&(H_0 + i)^{-1}V(H_0 + i)^{-k-1}(1 + A^2)^{\mu/2} \\
&\quad = \Big((H_0 + i)^{-1}\nabla\Big) \cdot (gU)(H_0 + i)^{-k-1}(1 + A^2)^{\mu/2} \\
&\quad + (H_0 + i)^{-1}(\nabla \cdot g)U(H_0 + i)^{-k-1}(1 + A^2)^{\mu/2} \\
&\quad - (H_0 + i)^{-1}(gU)(H_0 + i)^{-k} \cdot (\nabla(H_0 + i)^{-1})(1 + A^2)^{\mu/2}.
\end{aligned}$$

Since

$$(1 + A^2)^{-\mu/2}(H_0 + i)^{-1}(1 + A^2)^{\mu/2}$$

and

$$(1 + A^2)^{-\mu/2}\nabla(H_0 + i)^{-1}(1 + A^2)^{\mu/2}$$

extend to bounded operators, the result follows.

Using this lemma we can prove the proposition. It suffices to consider the case $b = c = 1$. Furthermore, using a smooth cutoff function, it suffices to consider $r = |x| \geq 1$. Assume $\alpha > 1$ and $\beta > 0$. Compute the commutator

$$\left[\frac{1}{\alpha r^{\alpha-1}}\frac{\partial}{\partial r}, \frac{\sin(r^\alpha)}{r^\beta}\right] = \frac{\cos(r^\alpha)}{r^\beta} - \frac{\beta}{\alpha}\frac{\sin(r^\alpha)}{r^{\alpha+\beta}}. \tag{25}$$

Assume $\alpha + \beta > 2$. Using the lemma, we conclude that

$$\frac{\cos(r^\alpha)}{r^\beta}$$

is short range, and similarly for

$$\frac{\sin(r^\alpha)}{r^\beta}.$$

Now use these results in Eq. 25 to conclude that

$$\frac{\cos(r^\alpha)}{r^\beta}$$

is short range for $2\alpha + \beta > 3$. By induction, we get that these two perturbations are short range for $(\alpha - 1)m + \beta > 1$, m an integer. This concludes the proof.

5.2 The Case $H_0 = -\Delta + Fx_1$

For Stark Hamiltonians we recover the results due to Avron-Herbst[1] and Jensen[7]. The following conditions imply the short range conditions in Definition 2.2. Consider the case $F > 0$. We assume that V is a symmetric H_0-compact operator such that the operator $V(H_0+i)^{-1}(1+\chi_+(x_1)x_1^2)^{\mu/2}$ extends to a bounded operator on $\mathcal{H} = L^2(\mathbf{R}^n)$ for some $\mu > 1/2$. Here χ_+ denotes the characteristic function for the set $(0, \infty)$. In rough asymptotic form for a multiplicative potential the conditions are, with the notation $x = (x_1, x_\perp) \in \mathbf{R}^1 \times \mathbf{R}^{n-1}$,

$$V(x_1, x_\perp) = \begin{cases} O(|x_1|^{-1/2-\epsilon}) & \text{as } x_1 \to -\infty, \\ o(|x_1|) & \text{as } x_1 \to \infty, \\ o(1) & \text{as } |x_\perp| \to \infty. \end{cases}$$

For the one-dimensional case one can find a class of short range potentials with oscillatory behavior. We state the following sufficient condition[7]: Let $V(x_1) = W''(x_1)$, where W is a bounded function with four bounded derivatives. Then V is a short range perturbation of $H_0 = -d^2/dx_1^2 + Fx_1$ with respect to $A = (i/F)(d/dx_1)$. This class of potentials includes periodic and almost periodic functions. See the paper[7] for examples.

6 Short Range Versus Long Range Potentials

We conclude with a brief discussion of the concept of a short range potential and a long range potential. The theory presented above, and in particular the definition of a short range perturbation given in Definition 2.2, seems to capture most of the cases where the ordinary wave operators exist and are asymptotically complete. In contrast, there seems to be no general theory for long range potentials. Note that the concept of a short range perturbation introduced here differs from the terminology used by other authors. In particular, the type of oscillatory potential in Proposition 5.2 is often described as a long range potential.

For the case of the ordinary Schrödinger operator $H_0 = -\Delta$ it is well known that the borderline between the short range and the long range behavior is given by the Coulomb potential $c/|x|$ in the sense that the usual wave operators do not exist for this potential[4]. In classical scattering the borderline is also given by the Coulomb potential. In contrast, for Stark Hamiltonians it was recently discovered[9] that the borderline is not the same. In quantum scattering the borderline is a behavior $O(|x|^{-1/2})$ as $Fx_1 \to -\infty$, whereas in classical one-dimensional scattering the classical wave operators exist and are asymptotically complete for potentials with a decay rate as slow as $O((\log(x))^{-a})$, $a > 1$.

Acknowledgment. The author wishes to thank the organizers for organizing an interesting conference. The results of this paper have also been presented at lectures at Université de Nantes, Nantes, France, and Gakushuin University, Tokyo, Japan. The author wishes to thank the two universities for their hospitality.

References

[1] J. E. Avron and I. W. Herbst. Spectral and scattering theory for Schrödinger operators related to Stark effect. *Commun. Math. Phys.* **52** (1977), 239–254.

[2] H. Baumgärtel and M. Wollenberg. *Mathematical Scattering Theory*, volume 9 of *Operator Theory: Advances and Applications.* Birkhäuser Verlag, Basel, 1983.

[3] M. Combescure. Spectral and scattering theory for a class of strongly oscillating potentials. *Commun. Math. Phys.* **73** (1980), 43–62.

[4] J. D. Dollard. Asymptotic convergence and the Coulomb interactions. *J. Math. Phys.* **57** (1964), 29–738.

[5] V. Enss. Asymptotic completeness for quantum mechanical potential scattering, I. Short range potentials. *Commun. Math. Phys.* **61** (1978), 285–291.

[6] A. Jensen. Propagation estimates for Schrödinger-type operators. *Trans. Amer. Math. Soc.* **291** (1985), 29–144.

[7] A. Jensen. Asymptotic completeness for a new class of Stark effect Hamiltonians. *Commun. Math. Phys.* **107** (1986), 21–28.

[8] A. Jensen, E. Mourre, and P. Perry. Commutator estimates and resolvent smoothness in quantum scattering theory. *Ann. Inst. H. Poincaré, Sect. A (N.S.)* **41** (1984), 207–225.

[9] A. Jensen and T. Ozawa. Classical and quantum scattering for Stark Hamiltonians with slowly decaying potentials. Preprint RIMS-713, RIMS, Kyoto University, August 1990.

[10] T. Kato. *Perturbation Theory for Linear Operators*, volume 132 of *Die Grundlehren der Mathematischen Wissenschaften*. Springer Verlag, Berlin, Heidelberg, New York, second edition, 1976.

[11] T. Kato and S. T. Kuroda. Theory of simple scattering and eigenfunction expansions. In *Functional Analysis and Related Fields*, pages 99–131. Springer Verlag, 1970.

[12] E. Mourre. Link between the geometrical and the spectral transformation approaches in scattering theory. *Commun. Math. Phys.* **68** (1979), 91–94.

[13] E. Mourre. Absence of singular continuous spectrum for certain self-adjoint operators. *Comm. Math. Phys.* **78** (1981), 391–408.

[14] P. A. Perry. Mellin transforms and scattering theory. I, short-range potentials. *Duke Math. J.* **47** (1980), 187–193.

[15] P. A. Perry. Scattering theory by the Enss method. *Math. Report*, 1, 1983.

[16] M. Reed and B. Simon. *Methods of Modern Mathematical Physics. III: Scattering Theory.* Academic Press, New York, 1979.

[17] B. Simon. Phase space analysis of simple scattering systems: Extensions of some work of Enss. *Duke Math. J.* **46** (1979), 119–168.

[18] D. A. White. Schrödinger operators with rapidly oscillating central potentials. *Trans. Amer. Math. Soc.* **275** (1983), 641–677.

Discrete spectrum in a gap of perturbed periodic operator at large coupling constants

M.Sh.Birman

Department of Mathematical Physics
Leningrad State University
198904 Leningrad, USSR

1. The present communication represents a continuation of the series [1-5]. Let A be the operator in $L_2(\mathbb{R}^d)$, $d \geq 3$, which is associated with the differential expression

$$- \operatorname{div}(a(x) \operatorname{grad} u) + f(x)u , \qquad (1)$$

the positive definite matrix a and the potential f being real -valued and periodic. Without loss of generality we may assume that the period lattice is $\mathbb{Z}^d$. Together with A we consider the perturbed operator $B_\alpha(\pm)$:

$$B_\alpha(\pm)u = Au \mp \alpha v(x)u , \alpha > 0 , v(x) \geq 0 .$$

Let $\Lambda = (\lambda_-, \lambda_+)$ be a gap in the spectrum $\sigma(A)$ of the operator A . We assume the spectrum of $B_\alpha(\pm)$ in Λ to be discrete and finite for any $\alpha > 0$. This assumption is automatically fulfilled under the conditions we impose on v below. For $\chi \in \bar{\Lambda}$ we denote by $N_{\pm}(\alpha, A, v, \chi)$ the number of eigenvalues of the operator $B_t(\pm)$ crossing the point χ while the coupling constant t grows from 0 to $\alpha > 0$ (in the case $\chi = \lambda_{\pm}$ we have in mind the appearance or absorption of the levels, respectively). For $q \geq d/2$ we define the functionals

$$\Delta_q^{(\pm)}(A, v, \chi) = \limsup_{\alpha \to \infty} \alpha^{-q} N_{\pm}(\alpha, A, v, \chi) , \tag{2}$$

$$\delta_q^{(\pm)}(A, v, \chi) = \liminf_{\alpha \to \infty} \alpha^{-q} N_{\pm}(\alpha, A, v, \chi) \tag{3}$$

describing the behaviour of $N_{\pm}(\alpha)$ in the power scale.

Our aim is to study the quantities (2),(3) at the points $\chi = \lambda_{\pm}$. The results differ qualitatively for $2q = d$ and $2q > d$ (we speak about the *regular* and *non-regular* cases - cf.[3]). In the regular case our solution reduces to the usual quasi- classical formula. In the non-regular case the results are given in terms of a model problem. Description of the model includes so-called effective mass tensors. In [4,6,7] the functionals (2), (3) for $\chi \in \Lambda$ are discussed in a more general situation.

2. Let Q be the unit cube in $\mathbb{R}^d$, ω_d be the volume of unit ball in $\mathbb{R}^d$ and $\mathbb{T}^d$ a (flat) torus. By H^{σ} we denote the Sobolev classes. Let A_0 be the operator in $L_2(\mathbb{R}^d)$ associated with $(-\Delta)$ so that $\mathrm{Dom}\, A_0 = H^2(\mathbb{R}^d)$.

Let us recall the structure of spectrum of the operator A. Denote by $A(k)$, $k \in \mathbb{T}^d$ the family of operators in $L_2(Q)$ associated with the expression (1) and quasi-periodic boundary conditions. For the functions $u \in \mathrm{Dom}\, A(k)$ the product $e^{-ikx}u(x)$ can be extended periodically to the whole $\mathbb{R}^d$. The self-adjoint operators $A(k)$ have a discrete spectrum. Let $E_s(k) \le E_{s+1}(k)$, $s = 1,2,\ldots$, be the sequence of eigenvalues and $\psi_s(x,k)$ the corresponding normalized eigenfunctions of $A(k)$. The spectrum $\sigma(A)$ is the union of images of all maps $\mathbb{T}^d \ni k \longrightarrow E_s(k)$, $s = 1,2,\ldots$.

Let us state now the conditions we want to impose on a,f. It is more convenient to present most of them in an implicit form. Let us begin with the smoothness properties.

Condition 1. (a) $a \in C^1(\bar{Q})$, $f \in L_2(Q)$,

(b) $\mathrm{Dom}\, A = \mathrm{Dom}\, A_0 = H^2(\mathbb{R}^d)$,

(c) the eigenfunctions $\psi_s(.,k)$, $k \in \mathbb{T}^d$, belong to $H^{\sigma}(Q)$ for some $\sigma > d/2$.

If $d = 3$ then Condition 1c is obviously contained in 1a,b . The following condition is of spectral nature. It describes structure of the edges $\lambda_{\pm}$ of the gap Λ . It is clear that for some integer l one has

$$\lambda_- = \max_k E_{l-1}(k) , \quad \lambda_+ = \min_k E_l(k) . \tag{4$_\pm$}$$

Condition $2^{(+)}$. (a) $\min_k E_{l+1}(k) > \lambda_+$,

(b) the relation $(4)_+$ is fulfilled at finitely many points $k_n^{(+)} \in \mathbb{T}^d$, $n = 1,\ldots,m_+$, each of them being a non-degenerate minimum point.

Remark 1. If $\lambda_+ = \inf \sigma(A)$, then Condition $2^{(+)}$ is fulfilled automatically, thereby $l = 1$, $m_+ = 1$, $k_1^{(+)} = 0$.

Concerning the point λ_-, the last condition is replaced by

Condition $2^{(-)}$. (a) $\max_k E_{l-2}(k) < \lambda_-$,

(b) the relation $(4)_-$ is fulfilled at finitely many points $k_n^{(-)} \in \mathbb{T}^d$, $n = 1,\dots,m_-$, each of them being a non-degenerate maximum point.

Remark 2. Condition $2^{(+)}$a implies that λ_+ is a *simple* eigenvalue of $A(k)$ at $k = k_n^{(+)}$, $n = 1,\dots,m_+$, so $E_l(k)$ depends analytically on k in a vicinity of $k_n^{(+)}$. Condition $2^{(+)}$b implies

$$E_l(k) - \lambda_+ = p_n^{(+)}(k-k_n^{(+)}) + o(|k-k_n^{(+)}|^2) ,$$

where $p_n^{(+)}$ is a positive definite quadratic form, $n = 1,\dots,m_+$ Similarly, Condition $2^{(-)}$ implies

$$\lambda_- - E_{l-1}(k) = p_n^{(-)}(k-k_n^{(-)}) + o(|k-k_n^{(-)}|^2) ,$$

$n = 1,\dots,m_-$, $p_n^{(-)}$ being a positive definite quadratic form. The so-called effective mass tensors at the points $k_n^{(\pm)}$ are introduced by means of $p_n^{(\pm)}$.

3. *The regular case* is singled out by the condition

$$v \in L_{d/2}(\mathbb{R}^d) . \tag{5}$$

Theorem $1^{(+)}$. Let the Conditions $1, 2^{(+)}$ and (5) be fulfilled, then

$$\Delta^{(+)}_{d/2}(A,v,\lambda_+) = \delta^{(+)}_{d/2}(A,v,\lambda_+)$$

$$= (2\pi)^{-d}\,\omega_d \int \frac{v^{d/2}}{(\det a)^{1/2}}\,dx\ . \tag{6}$$

Remark 3. if $\lambda_+ = \inf \sigma(A)$, then the asymptotics (6) is valid under weaker assumptions on a and f without the condition $v(x) \geq 0$ (cf. [3],[4] ; of course, one has to replace then v on the *rhs* of (6) by v_+).

Theorem 1$^{(-)}$. Let the Conditions $1,2^{(-)}$ and (5) be fulfilled, then

$$\Delta^{(+)}_{d/2}(A,v,\lambda_-) = \delta^{(+)}_{d/2}(A,v,\lambda_-)$$

$$= (2\pi)^{-d}\,\omega_d \int \frac{v^{d/2}}{(\det a)^{1/2}}\,dx\ , \tag{7}$$

$$\Delta^{(-)}_{d/2}(A,v,\lambda_-) = 0\ .$$

Remark 4. Suppose Conditions 1, $2^{(-)}$, $2^{(+)}$ and (5) are fulfilled. Comparing the relations (6) and (7) we see that the number of levels for the operator $B_\alpha(+)$ in the gap Λ behaves as $o(\alpha^{d/2})$ for large $\alpha > 0$. This estimate can be hardly improved without additional assumptions on v . An asymptotics for the number of levels in the power scale of potentials v for the one-dimensional case has been obtained by Sobolev [8].

4. *The non-regular case.* If $v \notin L_{d/2}$, then it is more convenient to formulate conditions on v in terms of the standard operator $A_0-\alpha v$.

Condition 3. For some $q > d/2$, one has

$$\Delta_q^{(+)}(A_0,v,0) < \infty , \quad \Delta_q^{(+)}(A_0+1,v,0) = 0 . \qquad (8)$$

The requirements (8) are related to the problem of negative discrete spectrum of a Schroedinger operator, which has been extensively studied ; hence reduction to it is preferable to any explicit assumptions on the potential v. A more complete discussion of the relations (8) can be found in [3].

Let us describe now *the model operator* at the point $\lambda_\pm$. It acts in the space $L_2(\mathbb{R}^d,\mathbb{C}^{m_\pm})$ of vector-valued functions. We introduce the matrices $V_\pm := v\Psi_\pm$,

$$\Psi_+ = \left\{ \overline{\psi_l(x,k_i^{(+)})}\, \psi_l(x,k_j^{(+)}) \right\}_{i,j=1}^{m_+} ,$$

$$\Psi_- = \left\{ \overline{\psi_{l-1}(x,k_i^{(-)})}\, \psi_{l-1}(x,k_j^{(-)}) \right\}_{i,j=1}^{m_-} .$$

The matrix $\Psi_\pm$ is of rank one and non-negative. Set

$$M_\pm = \mathrm{diag}\left(p_1^{(\pm)}(D), p_2^{(\pm)}(D), \dots, p_{m_\pm}^{(\pm)}(D) \right) .$$

By Condition $2^{(\pm)}$, the differential operator with constant coefficients $M_\pm$ is elliptic and positive. For the model operator $M_\pm - \alpha V_\pm$ the functionals of the form (2),(3), $\Delta_q^{(+)}(M_\pm, V_\pm, 0)$ and $\delta_q^{(+)}(M_\pm, V_\pm, 0)$, characterize the large-coupling-constant behaviour of the function $N_+(\alpha, M_\pm, V_\pm, 0)$ which equals the number of negative eigenvalues. It follows from (8) that

$$\Delta_q^{(+)}(M_\pm, V_\pm, 0) < \infty , \quad \Delta_q^{(+)}(M_\pm + 1, V_\pm, 0) = 0 .$$

The following theorem has the same form for the two signs.

Theorem $2^{(\pm)}$. Let Conditions 1, $2^{(\pm)}$ and 3 be fulfilled, then

$$\Delta_q^{(\pm)}(A, v, \lambda_\pm) = \Delta_q^{(+)}(M_\pm, V_\pm, 0) ,$$
$$\delta_q^{(\pm)}(A, v, \lambda_\pm) = \delta_q^{(+)}(M_\pm, V_\pm, 0) , \tag{9}$$

$$\Delta_q^{(\pm)}(A, v, \lambda_\mp) = 0 . \tag{10}$$

Next we make an additional assumption on v which will allow us to replace the matrix $\Psi_\pm$ by the unit matrix $I_{m_\pm}$ in the model operator. Let $\hat{v}$ be Fourier transform of v.

Condition 4. (a) If $d \geq 4$ or $d = 3$, $q \geq 2$, then $\hat{v} \in L_\infty$ on compact sets separated from the origin.

(b) If $d = 3$, $3 < 2q < 4$, then $\hat{v} \in H^{\sigma}$, $2\sigma > 3$, on compact sets separated from the origin.

Theorem $3^{(\pm)}$. Let Conditions 1, $2^{(\pm)}$, 3 and 4 be fulfilled, then $V_{\pm}$ can be replaced by $V_{\pm}^{0} := vI_{m_{\pm}}$ in the relations (9).

5. Let us comment finally on the distinctions between the case $v \in L_{d/2}(\mathbb{R}^d)$ and the non-regular one :
(a) Theorems $2^{(\pm)}$, $3^{(\pm)}$ are symmetric with respect to exchange of the endpoints $\lambda_{\pm}$.
(b) Comparison of (9) and (10) shows that the relations (9) describe asymptotically the number of levels of the operator $B_{\alpha}(\pm)$ in the gap Λ . This is in agreement with the concept that most of the levels emerge from a neighbourhood of the edge $\lambda_{\pm}$ and always stay close to it. On the contrary, if $v \in L_{d/2}(\mathbb{R}^d)$ the levels are generated by the high-energy region and their distribution at large $\alpha > 0$ correspond to large negative energies.
(c) Theorems $2^{(\pm)}$, $3^{(\pm)}$ simplify the problem but give no explicit answers. The cases with the asymptotics $\Delta_q^{(\pm)}(M_{\pm},V_{\pm}^{0},0) = \delta_q^{(\pm)}(M_{\pm},V_{\pm}^{0},0)$ have to be studied.
(d) Little is known about the quantities (2),(3) for $\chi \in (\lambda_-,\lambda_+)$ when $2q > d$. A related result is given in [7].

References

1 M.Sh.Birman, M.Z.Solomyak, in *Schroedinger Operators, Standard and Non-Standard* (P.Exner, P.Šeba, eds.), World Scientific, Singapore 1989; pp.3-18.

2 M.Sh.Birman, M.Z.Solomyak, in *Proc. Conf. on Integral Equations and Inverse Problems* (Varna 1989), in print.

3 M.Sh.Birman, M.Z.Solomyak, in *Order, Disorder and*

Chaos in Quantum Systems (P.Exner, P.Seba, eds.), Birkhäuser, Basel 1990 ; pp.3-16.

4 M.Sh.Birman, in *Order, Disorder and Chaos in Quantum Systems* (P.Exner, P.Šeba, eds.), Birkhauser, Basel 1990 ; pp.17-25.

5 M.Sh.Birman, M.Z.Solomyak, in these proceedings.

6 R.Hempel, J.Reine Angew.Math. **399** (1989),38-59.

7 S.Alama, P.A.Deift, R.Hempel, Commun.Math.Phys. **121** (1989), 291-321.

8 A.V.Sobolev, Sov.J.Funct.Anal.Appl., to be published.

Discrete negative spectrum under non-regular perturbations (polyharmonic operators, Schroedinger operators with a magnetic field, periodic operators)

M.Sh.Birman and M.Z.Solomyak

Department of Mathematical Physics
Leningrad State University
198904 Leningrad, USSR

Let A be a self-adjoint semibounded operator on a Hilbert space H and $N(\lambda;A)$ be the number of points of its spectrum situated to the left of a given point $\lambda \in \mathbb{R}$. In our previous papers [1-3] we have proposed some approaches to estimation of the growth rate of $N(\lambda;A(\alpha V))$, $\lambda \le 0$, as $\alpha \longrightarrow \infty$, for the Schroedinger operator

$$A(\alpha V) = -\Delta - \alpha V \quad , \quad V = \overline{V} \in L_{1,loc}(\mathbb{R}^d) \tag{1}$$

on $L_2(\mathbb{R}^d)$. The rate appears to depend on the properties of V. Our methods apply essentially to the "non-regular" case ($d \ge 3$ and $V \notin L_{d/2}$) ; on one hand we proceed from some general Hardy- type inequalities, and on the other hand we use the well-known Rosenbljum-Lieb-Cwikel estimate

$$\mathcal{N}(0;A(\alpha V)) \le C(d)\ \alpha^{d/2} \int V_+^{d/2} dx \quad , \ d \ge 3 \ , \tag{2}$$
$$V_+ = (V+|V|)/2$$

(see the original papers [4-6] and the books [7,8]).

The aim of the present report is to extend the results of [1-3] to some other classes of operators. Let us describe briefly the contents of the paper. After some technical preliminaries collected in the following section, in the Sections 2-4 the classes of operators mentioned in the title are consecutively considered.

1. Let us consider the following operator,

$$A_s(\alpha V) = (-\Delta)^s - \alpha V \quad , \quad V = \overline{V} \in L_{1,loc}(\mathbb{R}^d) \quad , \quad s \in \mathbb{R}_+$$

on $L_2(\mathbb{R}^d)$. The associated quadratic form is

$$a_s(\alpha V)[u] = \int (|\nabla_s u|^2 - \alpha V|u|^2)\ dx \ .$$

To derive properties of $\mathcal{N}(\lambda, A_s(\alpha V))$, we need some preliminary considerations :

Hardy weights. A function $\psi \in L_{1,loc}(\mathbb{R}^d)$ is said to be a *Hardy weight of order* $s > 0$ (we shall write $\psi \in HW_s \equiv HW_s(\mathbb{R}^d)$) if the inequality

$$\int |\psi||u|^2dx \le M_0(\psi,s) \int |\nabla_s u|^2dx \quad , \quad u \in C_0^\infty(\mathbb{R}^d) \tag{3}$$

is valid. For $2s \ge d$ the class $HW_s(\mathbb{R}^d)$ contains the function $\psi \equiv 0$ only, while for $2s < d$ we have $|x|^{-2s} \in HW_s(\mathbb{R}^d)$.

A function $\psi \in L_{1,loc}$ is said to be a *weak Hardy weight of order* s (we write $\psi \in WHW_s \equiv WHW_s(\mathbb{R}^d)$) if the inequality

$$\int |\psi||u|^2dx \le M(\psi,s) \int (|\nabla_s u|^2+|u|^2)\, dx \quad , \quad u \in C_0^\infty(\mathbb{R}^d) \tag{4}$$

holds. It is clear that $L_\infty \subset WHW_s$ so the class WHW_s is non-trivial for any $s>0$.

There exists an analytic description of the classes HW_s and WHW_s in terms of capacity - see [9]. Here we are going to formulate some sufficient conditions for a function ψ to belong to these classes which would involve some integral norms instead of capacity ; if $2s > d$, then this condition for $WHW_s(\mathbb{R}^d)$ turns out to be necessary at the same time. First we describe the function classes we shall use :

(i) Let $\Omega \subset \mathbb{R}^d$ be a measurable set. The class $L_{p,w}(\Omega)$ *(the weak L_p-class)* consists of all measurable functions f on Ω for which one has

$$\|f\|_{p,\Omega} := \sup \left(t \left[mes\{ x\in\Omega : |f(x)|>t \}\right]^{1/p} \right) < \infty .$$

The functional $\|\cdot\|_{p,\Omega}$ defines a quasi-norm on $L_{p,w}(\Omega)$; we shall omit the index Ω if $\Omega = \mathbb{R}^d$. Notice that one can define in the same way the $L_{p,w}$-classes related to an arbitrary measure space.

(ii) Let Q be the unit cube in $\mathbb{R}^d$ and $Q_n = n+Q$, $n \in \mathbb{Z}^d$. The class $\ell_q(L_p)$, $0 < p,q \le \infty$, consists of all measurable functions f on $\mathbb{R}^d$ for which one has

$$\|f\|_{q(p)} := \left\| \left\{ \|f\|_{L_p(Q_n)} \right\}_{n\in\mathbb{Z}^d} \right\|_{\ell_q} < \infty . \tag{5}$$

Replacing $\|f\|_{L_p(Q_n)}$ by $|f|_{p,Q_n}$ in (5) we obtain the class $\ell_q(L_{p,w})$ with the quasi-norm $|f|_{q(p)}$.

The following results are not new (cf.[9]) ; we present them here in a form convenient for a further use. In what follows we shall denote $\varkappa := d/2s$ systematically.

Theorem 1. If $2s < d$, then $L_{\varkappa,w}(\mathbb{R}^d) \subset HW_s$ and

$$M_0(\psi,s) \le c_0(d,s)\ |\psi|_{\varkappa} .$$

Theorem 2. (a) If $2s < d$, then $\ell_\infty(L_{\varkappa,w}) \subset WHW_s$ and

$$M(\psi,s) \le c(d,s)\ |\psi|_{\infty(\varkappa)} .$$

(b) If $2s = d$, then $\ell_\infty(L_r) \subset WHW_s$ for each $r > 1$ and

$$M(\psi,s) \le c(d,r)\ \|\psi\|_{\infty(r)} .$$

(c) If $2s > d$, then $\ell_\infty(L_1) = WHW_s$ and

$$c_1(d,s)\ \|\psi\|_{\infty(1)} \le M(\psi,s) \le c_2(d,s)\ \|\psi\|_{\infty(1)}\ ,\quad c_1(d,s) > 0 .$$

Regular estimates. An estimate of $N(\lambda; A_s(\alpha V))$ is said to be *regular* if it guarantees the "quasi-classical" growth rate $O(\alpha^{\varkappa})$ as $\alpha \longrightarrow \infty$. The results we state below are well known. For $2s < d$ they generalize the estimate (2); they have been obtained in [4] for s integer and in [7] for s non-integer. Furthermore, for $s < d/2$ they follow from the results of [6] as well. Finally, for $2s \geq d$ these results have been obtained in [10].

Theorem 3. (a) If $2s < d$ and $V_+ \in L_{\varkappa}$, then

$$N(0; A_s(\alpha V)) \leq C(d,s)\, \alpha^{\varkappa} \int V_+^{\varkappa}\, dx .$$

(b) If $2s = d$ and $V_+ \in \ell_1(L_r)$ with some $r > 1$, then

$$N(-1; A_s(\alpha V)) \leq C(d,r)\, \alpha\, \|V_+\|_{1(r)} . \tag{6}$$

(c) If $2s > d$ and $V_+ \in \ell_{\varkappa}(L_1)$, then

$$N(-1; A_s(\alpha V)) \leq C(d,s)\, \alpha^{\varkappa}\, \|V_+\|_{\varkappa(1)} . \tag{7}$$

Notice that estimates of $N(0; A_s(\alpha V))$ homogeneous with respect to α cannot exist in the case $2s \geq d$ because of the operator $(-\Delta)^s$ has then a zero resonance. One can derive estimates, however, that contain an additional constant term on the *rhs*. Let us present an estimate of this type for $2s > d$:

Theorem 4. Let $2s > d$, $s-d/2 \notin \mathbb{N}$. Then the following inequality is valid,

$$\mathcal{N}(0;A_s(\alpha V)) \le m(d,s) + C(d,s)\,\alpha^{\varkappa}\sum_{n\in\mathbb{Z}}\Big(\int_{2^n<|x|<2^{n+1}} |x|^{2s-d}\, V_+(x)\, dx\Big)^{\varkappa}, \tag{8}$$

$m(d,s)$ being dimension of the space of polynomials on $\mathbb{R}^d$ of degree less than $(d/2)-s$.

Theorems 3 and 4 allow to justify the asymptotic formula

$$\mathcal{N}(\lambda;A_s(\alpha V)) \sim (2\pi)^{-d}\, v_d\, \alpha^{\varkappa} \int V_+^{\varkappa}\, dx\ , \quad \alpha \longrightarrow \infty\ , \tag{9}$$

where v_d denotes the volume of unit ball in $\mathbb{R}^d$. Namely, the following result is valid.

Theorem 5. (a) If $2s < d$ and $V_+ \in L_{\varkappa}$, or $2s > d$ and the sum on the *rhs* of (8) is finite, then the formula (9) holds for an arbitrary $\lambda \le 0$.
(b) If $2s = d$ and $V_+ \in \ell_1(L_r)$ with some $r > 1$, or $2s > d$ and $V_+ \in \ell_{\varkappa}(L_1)$, then the formula (9) is valid for an arbitrary $\lambda < 0$.

Notice that in the case $2s > d$, $\lambda < 0$ the second result is stronger than the first one.

2. Combining the results of Theorems 1-4 with the methods developed in [1,2], we get certain non-regular estimates for $\mathcal{N}(\lambda;A_s(\alpha V))$. First we are going to describe some results obtained by means of linear interpolation. Occasionally we include a functional parameter into our statements.

Theorem 6. Let $2s < d$, $\psi \in HW_s(\mathbb{R}^d)$, $\psi > 0$, $M_0(\psi,s) \le 1$. Then for $\varkappa < q < \infty$ the following implications are valid :

(a) If $V_+/\psi \in L_{q,w}(\psi^{\varkappa})$, then

$$N(0;A_s(\alpha V)) \le C(d,s,q)\ \alpha^q \sup_{t>0} \left[t^q \int_{V>t\psi} \psi^{\varkappa}\, dx \right] . \tag{10}$$

(b) If $\int_{V>t\psi} \psi^{\varkappa}\, dx = o(t^{-q})$ as $t \longrightarrow 0+$, $t \longrightarrow \infty$, then $N(0;A_s(\alpha V)) = o(\alpha^q)$ as $\alpha \longrightarrow \infty$.

Notice that finiteness of the supremum on the *rhs* of (10) is exactly the condition for V_+/ψ to belong to $L_{q,w}(\psi^{\varkappa})$.

Theorem 7. Let $2s < d$, $\psi \in WHW_s(\mathbb{R}^d)$, $\psi > 0$, $M(\psi,s) \le 1$. Then the both assertions of Theorem 6 remain valid for $N(-1;A_s(\alpha V))$.

On account of Theorem 1, one can take an arbitrary normalized positive function $\psi \in L_{\varkappa,w}$ as a functional parameter in Theorem 6. In this case the hypothesis of the first assertion of Theorem 6 is fulfilled, e.g., with any potential $V \in L_{1,loc}$ for which $V_+ \le c\psi(1+|\ln\ \psi|)^{-1/q}$. Notice that for the potential $V = |x|^{-2s}(1+|\ln|x||)^{-1/q}$ the estimate (10) has an exact order.

Let us single out the particular case of Theorem 7 corresponding to the choice $\psi \equiv 1$. This estimate can be easily extended to all $\lambda < 0$.

Corollary. Let $2s < d$, $\varkappa < q < \infty$. Then
(a) If $V_+ \in L_{q,w}$, then

$$N(-\gamma;A_s(\alpha V)) \le C(d,s)\ \alpha^q\ \gamma^{\varkappa-q} \sup_{t>0} \left[t^q\ \mathrm{mes}\{\ x\ :\ V(x)>t\ \} \right].$$

(b) If $\mathrm{mes}\{\ x\ :\ V(x)>t\ \} = o(t^{-q})$ as $t \longrightarrow 0+$, $t \longrightarrow \infty$, then $N(-\gamma;A_s(\alpha V)) = o(\alpha^q)$ as $\alpha \longrightarrow \infty$ for all $\gamma > 0$.

Theorem 8. Let $2s > d$, $\psi \in WHW_s(\mathbb{R}^d)$, $\psi > 0$, $M(\psi,s) \le 1$. Then the following assertions are valid for $1 < q < \infty$:

(a) If $V_+/\psi \in L_{q,w}(\psi)$, then

$$\mathcal{N}(-1;A_s(\alpha V)) \le C(d,s,q) \sup_{t>0} \left\{ t^q \int_{V>t\psi} \psi \, dx \right\} ;$$

(b) If $\int_{V>t\psi} \psi \, dx = o(t^{-q})$ as $t \longrightarrow 0+$, $t \longrightarrow \infty$, then

$\mathcal{N}(-1;A_s(\alpha V)) = o(\alpha^q)$ as $\alpha \longrightarrow \infty$.

The condition $q > 1$ in this theorem is rather unnatural. In fact, the natural lower bound for q would be the quasi- classical order $\varkappa = d/2s$ (which is < 1). The next theorem which includes also the case $2s = d$ allows us to take an arbitrary $q > \varkappa$. On the other hand, it contains no functional parameter.

Theorem 9. Let $2s \ge d$, $r > 1$ if $2s = d$, and $r = 1$ if $2s > d$. Then for $\varkappa < q < \infty$ one has : (a) it holds

$$\mathcal{N}(-1;A_s(\alpha V)) \le C(d,s,r) \, \alpha^q \sup_{t>0} \left\{ t^q \operatorname{card}\left\{ n\in\mathbb{Z}^d : \int_{Q_n} V_+^r \, dx > t^r \right\}\right\} .$$

(b) If $\operatorname{card}(n : \int_{Q_n} V_+^r \, dx > t^r) = o(t^q)$ as $t \longrightarrow 0$, then

$$\mathcal{N}(-1;A_s(\alpha V)) = o(\alpha^q) \quad \text{as} \quad \alpha \longrightarrow \infty .$$

In the case $2s > d$ one can deduce certain non-regular estimates for $\mathcal{N}(0;A_s(\alpha V))$ from Theorem 4, however, they look rather complicated and we shall not present them here. Some other non-regular estimates have been obtained by Egorov and Kondratiev [11].

Our proofs of Theorems 6-9 use linear interpolation. Another method to estimate $N(.)$ has been proposed in [2] and we shall present here the simplest result one can obtain by this method. Some further results for the case s=1 can be found in [2].

Theorem 10. Let $2s < d$, $\psi \in HW_s(\mathbb{R}^d)$, $\psi > 0$, $M_0(\psi,s) \leq 1$. Then the following estimate is valid :

$$N(0;A_s(\alpha V)) \leq C(d,s)\ \alpha^{\varkappa} \int_{V>\alpha\psi} V_+^{\varkappa}\, dx .$$

Notice that this estimate can give a non-powerlike order of growth for $N(.)$.

3. Let study now Schroedinger operator with a magnetic field,

$$A(\vec{a},\alpha V) = -(\nabla - i\vec{a})^2 - \alpha V , \tag{11}$$

where $\vec{a} \in L_{2,loc}(\mathbb{R}^d)$, $V \in L_{1,loc}(\mathbb{R}^d)$. Basic facts concerning the operator (11) are given, e.g., in [12]. In particular, the following implications are valid :

(i) if $d \geq 3$ and $\psi \in HW_1(\mathbb{R}^d)$, then

$$\int |\psi|\,|u|^2 dx \leq M_0(\psi)\ (A(\vec{a},0)u,u)\ ,\quad u \in C_0^{\infty}\ ; \tag{12}$$

(ii) if $d \geq 1$, $\psi \in WHW_1(\mathbb{R}^d)$, then

$$\int |\psi|\,|u|^2 dx \leq M(\psi)\ \left((A(\vec{a},0)u,u) + \|u\|^2\right)\ ,\quad u \in C_0^{\infty}\ . \tag{13}$$

The constants M_0, M in (12),(13) are the same as in (3),(4) for s=1. Furthermore the estimate

$$\mathcal{N}(0,A(\vec{a},\alpha V)) \le C(d)\ \alpha^{d/2} \int V_+^{d/2}\, dx\ ,\quad d \ge 3\ , \tag{14}$$

holds with the same constant as in (2).

The methods developed in [1,2] and used in section 2 are "abstract" in the sense that the results discussed above are, in fact, consequences of some basic estimates. The estimates (12)-(14) play basic role for the operator (11), and therefore all the results for the operator (1) based on the estimates (2)-(4) remain valid for the operator (11). The following result represents a particular case of this general statement.

Theorem 11. Let $d \ge 3$. Then the results of Theorem 6,7, and 10 (for s=1) remain valid for the operator (11), with the same constants in the estimates.

The authors are grateful to G.Raikov for a stimulating discussion on the material of this section.

4. In conclusion, let us consider the operator

$$Au \equiv A(a,f;\alpha V)u := -\mathrm{div}(a\nabla u) + fu - \alpha Vu\ , \tag{15}$$

$a = a(x)$ being a positive-definite matrix function on $\mathbb{R}^d$ of the class C^1 , and $f = f(x)$ being a continuous function. We assume a,f to be periodic with the common-period lattice $\mathbb{Z}^d$. Without loss of generality we may assume that the bottom of the spectrum for the operator $A(a,f;0)$ is $\lambda = 0$.

Notice that the positive periodic solution $\omega(x)$ of the equation $-\mathrm{div}(a\nabla u)+fu = 0$ is continuous. The estimates similar to (12)-(14) hold for the operator (15) (cf.[3],where the case $a\equiv 1$ has been treated). If $d=1,2$, then estimates similar to (6),(7) (for s=1) hold as well, however, in contrast to the operator (11), the constants in the estimates for the operator (15) depend on the function ω (more precisely, on its maximum and minimum values).

We conclude from the estimates of the type (6),(7),(12)-(14) that all the results for the operator (1) we have discussed above extend to the operator $A(a,f;\alpha V)$ as well (up to constants). For instance, the following assertion is valid :

Theorem 12. Let $d \geq 1$ and let the hypotheses of any of the Theorems 3,4,6-10 (for s=1) be satisfied for the potential V . Then the assertions of the respective theorem are valid, with $N(\lambda;A_1(\alpha V))$ replaced by $N(\lambda;A(a,f;\alpha V))$ and an additional factor in the estimates depending on $\min \omega$, $\max \omega$ only.

References

1 M.S.Birman, M.Z.Solomyak, in *Schroedinger Operators, Standard and Non-Standard* (Dubna 1988, ed. by P.Exner and P.Šeba), World Scientic, Singapore 1989 ; pp.3-18.

2 M.S.Birman, M.Z.Solomyak, in *Proc.Conf.on Integral Equations and Inverse Problems* (Varna 1989), in print.

3 M.S.Birman, M.Z.Solomyak, in *Order, Disorder and Chaos in Quantum Systems* (Dubna 1989, ed. by P.Exner and H.Neidhardt), Birkhauser,Basel 1990 ; pp.3-16.

4 G.V.Rosenbljum, Proc.Acad.Sci USSR **202** (1972), 1012-1015.

5 E.H.Lieb, Bull.Amer.Math.Soc.**82** (1976),751-753.

6 M.Cwikel, Ann.Math.**106** (1977),93-100.

7 M.S.Birman, M.Z.Solomyak, Amer.Math.Soc.Transl., Ser.2, **114** (1980).

8 M.Reed, B.Simon : *Methods of Modern Mathematical Physics IV*, Academic Press, New York 1978.

9 V.G.Mazja, T.O.Shaposhnikova : *Multiplicators in Spaces of Differentiable Functions*, Leningrad

University Press 1986 (in Russian). Engl.transl.: Pitman (Advanced Publ.Program), 1985.

10 M.S.Birman, V.V.Borzov, in *Problems in Mathematical Physics 5*, Leningrad University Press 1971, pp.24-38. Engl.transl.: Topics in Math.Phys., No.5, Plenum Press, New York 1972, pp.19-30.

11 Yu.V.Egorov, V.A.Kondratiev, Sov.Math.Sbornik **181** (1990), 147-166.

12 J.Avron, I.Herbst, B.Simon, Duke Math.J. **45** (1978), 847-883.

BOUNDS ON THE ORDER OF VANISHING OF EIGENFUNCTIONS OF SCHRÖDINGER OPERATORS

M. HOFFMANN-OSTENHOF
Institut für Mathematik, Universität Wien
A-1090 Wien, Austria

and

T. HOFFMANNN-OSTENHOF
Institut für Theoretische Chemie, Universität Wien
A-1090 Wien, Austria

ABSTRACT

Some recent results concerning zeros of quantum eigenfunctions are discussed.

We consider real valued continuous L^2-solutions ψ, $\psi \not\equiv 0$ of the Schrödinger equation

$$(-\Delta + V - E)\psi = 0 \qquad \text{in } R^n, n \geq 2 \tag{1}$$

with suitable assumptions on the potential $V : R^n \to R$. We define the order of vanishing of ψ in x_0 by

$$l_{x_0}(\psi) = \sup\{\alpha : \limsup_{R \to 0} R^{-2\alpha-n} \int_{|x-x_0|<R} \psi^2 dx < \infty\} \tag{2}$$

We have recently shown [HO^{2}1] that for a large class of potentials ($V \in K^{n,\delta}_{loc}(R^n)$, $\delta > 0$, see [AS] or [S] for a definition) which include e.g. Coulombic many particle potentials, $l_{x_0}(\psi)$ must be an integer (provided strong unique continuation holds i.e. $l_{x_0}(\psi) < \infty$) and that in a neighbourhood of x_0

$$\psi(x) = P_M(x - x_0) + o(|x - x_0|^M), \quad M := l_{x_0}(\psi) \tag{3}$$

where $P_M \not\equiv 0$ denotes a homogeneous harmonic polynomial of degree M. For earlier results in this spirit see e.g.[B, CF].

Estimates of $l_{x_0}(\psi)$ were recently obtained for Laplacians on compact Riemannian manifolds by Donelly and Fefferman [DF] in their seminal investigations on the length of nodal lines, and such bounds might be considered as quantitative versions of unique continuation. Specifically it was demonstrated in [DF] that if $-\Delta u_k = \lambda_k u_k$ on such a manifold $\mathcal{M}$ that

$$\sup_{x\in M} l_x(u_k) \le C(\mathcal{M})\sqrt{\lambda_k}$$

which is asymptotically correct for $k \to \infty$.

For solutions of (1) the situation can be more complicated. In particular, it is not even clear whether $l_{x_0}(\psi)$ is bounded in R^n, $n > 2$ for a fixed L^2-solution of (1).

To get some intuition consider the hydrogen atom so that (in suitable units) (1) becomes $(-\Delta - 2|x|^{-1} - E_k)\psi_k = 0$ in R^3 with $E_k = -k^{-2}$. A simple consideration shows that for $x_0 = 0$, $l_0(\psi_k) \le k-1 = |E_k|^{-1/2} - 1$ and that for maximal angular momentum this is an equality.

We shall give now a bound to $l_{x_0}(\psi)$ [HO^{2}2] under some unfortunately rather restrictive assumptions on V. Specifically we shall bound $l_{x_0}(\psi)$ by the ground state energy of an associated radially symmetric 2-dimensional eigenvalue problem.

We assume that $V : R^n \to R$, $n \ge 2$ has the following properties: (A) For some $x_0 \in R^n$, $V \in C^1(R^n\backslash\{x_0\})$ and $|x-x_0|^{\gamma-\delta}V$, $|x-x_0|^{\gamma-\delta}(x-x_0)\nabla V \in L^{\infty}_{loc}(R^n)$ for some $\delta > 0$, with $\gamma = 2$ for $n \ge 4$, and $\gamma = \frac{n}{2}$ for $n = 2, 3$.

Further we assume that either

(B1) $V \to 0$, $r|\nabla V| \to 0$ $\quad (r := |x|)$ for $r \to \infty$, or

(B2) $C_1 r^{\beta_0} \le V \le C_2 r^{\beta}$ and $C_1 r^{\beta_0} \le |(x-x_0).\nabla V| \le C_2 r^{\beta}$ for sufficiently large r with some C_1, $C_2 > 0$, and some $0 < \beta_0 \le \beta$.

Theorem 1 Let V satisfy (A) and either (B1) or (B2). Let $\psi \in L^2$ satisfy (1) with $E \ne 0$. Suppose the radially symmetric function $W_{x_0}(r)$ satisfies

(i) $r^{2-\delta}W_{x_0}(r) \to 0$ for $r \to 0$ for some $\delta > 0$, (ii) $\sup(-W_{x_0}, 0)$ is bounded for $r > 0$, and (iii) $W_{x_0}(|x-x_0|) \le V(x) + \frac{1}{2}(x-x_0).\nabla V(x) \quad \forall x \in R^n$.

Define on R^2 the selfadjoint operator $\mathcal{K}(k) = -k\Delta + W_{x_0}(|x|)$, W_{x_0} is now considered as a radial function $W_{x_0} : R^2 \to R$, then

(a) $\mathcal{K}(k)$ is selfadjoint for $k > n-2$ on $L^2(R^2)$ with C_0^∞ as form core and

(b) there exists a unique $k_0 > n-2$ such that inf spectrum $(\mathcal{K}(k_0)) = E$ and

$$l_{x_0}(\psi) \le \frac{1}{2}(k_0 - n + 2) \tag{4}$$

Remarks:(i) For every $l \in N$ and each $n \geq 2$ one can find a radial potential such that we have equality in (4).

(ii) For the 2-dimensional case one can generally via topological considerations obtain upper bounds to $l_{x_0}(\psi)$. (see e.g. [HO² 2] for a discussion).

(iii) If we consider a radial problem $(-\Delta + c_\alpha r^\alpha - E)\psi_E = 0$ in R^n with $\mathrm{sgn}\alpha = \mathrm{sgn}(c_\alpha), \alpha \neq 0$, $\alpha > -2$ this implies that $l_0(\psi_E) = \mathcal{O}(|E|^{1+2/\alpha})$ for $E \to 0$ if $\alpha < 0$ resp. $E \to \infty$ if $\alpha > 0$.

These asymptotics hold also if V satisfies $V + \frac{1}{2}x\nabla V \geq c_\alpha r^\alpha$. But as a simple ODE analysis for the radial case shows

$$l_0(\psi_E) = \mathcal{O}(|E|^{1/2+1/\alpha}).$$

The proof [HO² 2] follows ideas already developped in [F,H,HO²] where L^2-lower bounds to the asymptotics of solutions to Schrödinger equations were obtained.

Finally we want to give another relation between the order of vanishing of an eigenfunction and the spectral properties of the corresponding Hamiltonians. We suppose that V is chosen such that (3) holds. Further we note that in polar coordinates $P_M(r\omega) = r^M Y_M(w)$ with $w = x/r$, where $Y_M(w)$ is a surface harmonic and hence an eigenfunction of the Laplace Beltrami operator on the sphere S^{n-1}, so that $-\Delta_{S^{n-1}} Y_M = M(M+n-2)Y_M$. Moreover the eigenspace of this eigenvalue has dimension 1 for $M = 0$, n for $M = 1$, and

$$h(M) = \binom{n+M-1}{n-1} - \binom{n+M-3}{n-1}$$

for $M \geq 2$.

Theorem 2 Suppose N linear independent L^2-solutions $\psi_E^{(i)}$, $i = 1, \ldots N$ satisfy (1), then if

$$N \geq \sum_{m=0}^{M-1} h(m) + 1$$

there is for every $x_0 \in R^n$ a function $\psi \in \mathrm{span}[\psi_E^{(i)}, i = 1, \ldots N]$ such that $l_{x_0}(\psi) \geq M$.

This estimate can be turned around to give an upper bound to the multiplicity of N of an eigenvalue with orthogonal eigenfunction $\psi_E^{(1)}, i = 1, \ldots N$ provided a bound to $\sup l_{x_0}(\psi_E : \psi_E \in \mathrm{span}[\psi_E^{(i)}])$ is known, which is the case if Theorem 1 applies.

Theorem 2 follows by simple linear algebra and (3), and is in fact a local result.

REFERENCES

[AS] M.Aizenman, B.Simon, Commun.Pure.Appl.Math. 35, 209-273 (1982)

[B] L.Bers, Commun.Pure Appl.Math. 8, 473-496 (1955)

[CF] L.A.Cafarelli, A.Friedmann, J.Diff.Equ. 60, 420-433 (1985)

[DF] H.Donelly, Ch.Fefferman, Invent.Math. 93, 161-183 (1988)

[F,H,HO^2] R.Froese, I.Herbst, M.Hoffmann-Ostenhof, T.Hoffmann-Ostenhof, Proc.Roy.Soc.Edinburgh 95A, 25-38 (1983)

[HO^{2}1] M.Hoffmann-Ostenhof, T.Hoffmann-Ostenhof, Local properties of solutions to the Schrödinger equation, to be submitted

[HO^{2}2] M.Hoffmann-Ostenhof, T.Hoffmann-Ostenhof, Bounds on the order of vanishing of eigenfunctions of the Schrödinger equation, submitted

[S] B.Simon, Bull.Am.Math.Soc. 7, 447-526 (1982)

DYNAMICAL BEHAVIOUR FOR SOME MULTIPLE WELL SCHRÖDINGER OPERATORS

J.M. Combes (*)
Centre de Physique Théorique - CNRS - Luminy, Case 907
13288 Marseille Cedex 9 - FRANCE

and

I.M. Sigal
Department of Mathematics, University of Toronto

Abstract : Schrödinger operators $H = -\hbar^2\Delta + V$ *on* $L^2(\mathbb{R}^n)$ with "rare" resonating potential wells have exponentially decaying Green's functions.
This in turn implies finiteness of the diffusion constant. Unusual dynamical behaviour of logarithmic type occurs in particular if the resonating wells obey some separation laws.

1. Introduction

There has been considerable effort recently to describe quantum dynamics in deterministic or random disordered media. Here "disordered" has to be understood in the sense that the potential has strongly fluctuating local properties which can make propagation of a particle to infinity either impossible or diffusive in a sense to be specified below. Well-known examples are provided by random distributions of wells and barriers, quasi-periodic potentials or hierarchical models as considered by [JMS]. In the first case one expects Anderson localisation at high disorder (see e.g. [HM]), in the second one recurrent behaviour as analyzed by Avron and Simon [A.S.1,2] whereas for the last one some unusual time logarithmic behaviour has been proved to hold by [JMS].
A quantity of interest for such disordered media is the diffusion constant D ; whenever it exists it is defined by $< x^2(t) > \sim Dt(t \to \infty)$ where $< x^2(t) > = < \psi_t, x^2\psi_t >$ and $\psi_t = e^{-itH}\psi_0$, ψ_0 being an initial state with a given energy support ; if ψ_0 is a scattering state D is expected to be ∞ whereas it is zero if ψ_0 is a finite linear combination of eigenstates of H. These 2 extreme cases don't give much intuition about mechanisms leading to a finite D. Of course one expects that spectrum must be singular in some sense e.g. recurrent absolutely continuous, singular continuous or dense pure point ; but even in this last case the only general result which is known is a recent one by B. Simon [S]

(*) and "PHYMAT", Université de Toulon et du Var

stating that $< x^2(t) > /t^2 \rightarrow 0$ as $t \rightarrow \pm\infty$ if ψ_0 is a linear combination of eigenstates. Fortunately there exist an expression for D in terms of the Green's function of H, which reads as follows :

$$D = \lim_{\varepsilon \downarrow 0} \varepsilon^2 \int \| \ xR(E + i\varepsilon)\psi_0 \ \|^2 \, dE \tag{1}$$

where $R(z) = (H - z)^{-1}$ is the resolvent of H.
This can be derived easily under the assumption that D exists by noticing that one has then

$$D = \lim_{\varepsilon \downarrow 0} \varepsilon^2 \int_0^\infty e^{-\frac{t}{\varepsilon}} \ \| \ x\psi_t \ \|^2 \, dt$$

and (1) follows from Parseval equality. It is natural to define a diffusion constant at energy E as

$$D(E) = \lim_{\varepsilon \downarrow 0} \varepsilon^2 \ \| \ xR(E + i\varepsilon)\psi_0 \ \|^2 \tag{2}$$

Notice however that $D(E)$ for fixed E has no direct dynamical interpretation ; for example if E is an eigenvalue of H and ψ_0 an eigenvector then $D(E)$ is finite but non zero although $D = 0$ in this case. Such an interpretation can be given only if $D(E)$ is averaged over E as in (1) or over the potential V in case V is random thus leading the well-known Kubo's formula (see e.g. [MKS]). However it is of obvious interest to describe conditions on the potential V leading to a finite $D(E)$ for a given E ; this holds in particular if the kernel of $\varepsilon R(E + i\varepsilon)$ has exponential decay. To prove this when E is in the essential spectrum of H is a difficult problem. Results of this genre has been obtained in particular by Helffer and Sjostrand [HS] in the semi-classical limit or by Fröhlich and Spencer [FS] for discrete random Hamiltonians. Our contribution to this problem is somewhat related to these works.

In fact two essential sets of ideas are used to establish the decay properties. First the concept of resonant well or more generally of resonant domain; for a given energy E a domain Λ is non resonant if the spectrum of H suitably restricted to Λ, e.g. by imposing Dirichlet boundary conditions on $\partial\Lambda$, does not contain E. A classification of domains can be obtained by introducing a decreasing sequence of scales d_k; then Λ is said to be resonant to order k at energy E if $d_k \leq dist(\sigma(H_\Lambda), \ E) \leq d_{k+1}$ where H_Λ is the local Hamiltonian for Λ. Second a property which can be called "Wegner's principle" stating that resonant regions are unlikely ; the formulation of this in the random case is a statement of the genre ([W]):

$$Prob(dist(\sigma(H_\Lambda), E < \varepsilon) \leq C \mid \Lambda \mid \varepsilon^\gamma$$

for some $\gamma \in (0, 1]$. This can be proved in many discrete models with suitable assumptions on the random potential ; however a proof for genuine continuous models, like e.g. a Poisson distribution of wells in $\mathbb{R}^n$ is to our knowledge still laking. Wegner's principle makes plausible the following type of assumption which is formulated in terms of an increasing sequence of length scales $(l_k)_k$: resonant regions to order k are spaced by distances at least l_k. Of course any precise statement will involve a connection between the energy and length scales ; usualy the d_k's are exponentialy small with respect to the l_k's. Here "distance" is taken in a more general sense than the Euclidean one ; it could be e.g. Agmon's distance ([Ag] and (4) below) or a suitable modification of it. It is this property which for suitable scales d_k and l_k makes propagation over large distances difficult, a fact reflected by decay properties. Assumptions of this genre also play an essential role in Simon-Spencer proof [SS] that there is no absolutely continuous spectrum in some models with rare resonant domains. They have appeared first in [FS] in their analysis of random potentials and were further studied and modified in [JMS], [FMSS] and [KVD]. To obtain global decay one has to combine local decay estimates originating from the non resonant property with the help of the well-known geometric resolvent equation (see (10) below) ; local decay estimates for non resonant regions are obtained easily e.g. from standard Combes-Thomas [RS] or Agmon's arguments. The use of suitable decompositon of configuration space is also the essence of Frölich-Spencer deterministic decay estimates.

To conclude we will describe anomalous dynamical behaviour occuring when $D = 0$ and the potential V has wells with uncreasing distances at infinity. Some time ago Jona-Lasinio, Martinelli and Scoppola [JMS] showed the remarkable result that for some "hierarchical" distribution of wells one has a logarithmic behaviour $< x^2(t) >\sim C(Log\ t)^\beta$ for some $\beta > 0$. It turns out that such anomalous behaviour is a general feature of models where propagation is due essentialy to tunneling over larger and larger distances.

2. Decay properties of Green's functions

There are two standard ways to prove exponential decay for the kernel of $(H - z)^{-1}$. One is in terms of Euclidean metric and uses the complex boost method [RS] ; it works for energies outside the spectrum of H. The other one is due to Agmon [Ag]; for a reference energy E^+ it provides estimates in the classically forbidden region $G(E^+) = \{x,\ V(x) > E^+\}$. Here the decay is in terms of the geodesic distance ρ (so called "Agmon's distance") associated to the metric

$$ds^2 = \hbar^{-2}(V - E^+)_+ dx^2 \quad ,\text{where } (V - E^+)_+ = sup(V - E^+, 0) \tag{4}$$

To give the flavour of the type of result one gets, it is convenient to introduce "cut-off operators" C_Ω associated to an open region Ω through :

$$\| C_\Omega u \|^2 = \hbar^2 \int_\Omega | \nabla u |^2 + (V - E^+)_+ | u |^2 \quad , u \in \mathcal{H}^1(\Omega) \tag{5}$$

Then elementary energy inequalities imply :

$$\| C_{\Omega_1} R_0(z) C_{\Omega_2} \| \leq e^{-\rho(\Omega_1,\Omega_2)} \tag{6}$$

where : $Re z < E^+$ and $H_0 = -\hbar^2\Delta + V_0$ has "all the wells filled up" at energy E^+ i.e $V_0 = sup(V, E^+)$.

Notice that at the price of introducing prefactors on the r.h.s. of (6) one can obtain also estimates on $\chi_{\Omega_1} R_0(z) \chi_{\Omega_2}$ with χ_Ω the characteristic function of Ω.

The strategy followed now is, starting from (6) to add inductively more and more resonant wells to V_0. Using a KAM like method this can be done in a geometric perturbative way provided small denominators arising from resonant wells or domains are compensated by the geometric perturbation terms which are exponentially small due to tunneling in the distance between wells. Wells at energy E^+ are defined by the decomposition into connected components :

$$(V - E^+)_- = \bigcup_a V_a \tag{7}$$

We assume that for a given energy $E < E^+$ wells can be classified into groups C_k, $k = 1, 2, ..,$ satisfying the following properties : there exists a non decreasing sequence of length scales (l_k) and a non increasing sequence of energy scales (d_k) with $d_0 = 1$ such that :

(8i) $\sum_k \frac{l_k}{l_{k+1}} < \infty$ and $\sum_k \frac{|\log d_k|}{l_k} < \infty$

(8ii) $\rho(a, a') > l_j$ if $a \in C_l, a' \in C_k$ and $k, l \geq j$

(8iii) Let H_j be the Hamiltonian obtained from H by filling up all wells $a \in C_k$, $k > j$ (i.e. replace V_a by E^+ if $a \in C_k$). Let $H_j^{(a)}$ be $H_{j-1} + V_a$ on the ball B_a with center a and radius $\frac{1}{4} l_j$ (and arbitrary outside this ball) ; then :

$$dist\left(E, \sigma\left(H_j^{(a)}\right)\right) > d_j.$$

So the j-well a is an impurity which makes the ball B_a which contains only k wells, $k < j$, resonant to order j. Then one has :

Theorem 1

Assume (8i, ii, iii) ; then there exist constants $C_n, n = 1, \ldots, 4$, and α, $0 < \alpha < 1$ such that if $C_1\ l_j < l_{j+1}$ and $-C_2 \log\ d_j < l_j$ for all j one has $\forall z = E + i\varepsilon, \varepsilon > 0$:

(9i) $dist(\sigma(H_j), E) > C_3 d_j$

(9ii) For all $j \geq 0$, and all open sets Ω_0, Ω_1 with $\rho(\Omega_0, \Omega_1) \geq \frac{1}{2} l_j$:

$$\| C_{\Omega_1} R_j(z) C_{\Omega_0} \| \leq e^{-\alpha_j\, \rho(\Omega_0, \Omega_1)}$$

$$\text{with } \alpha_j = \left(1 - C_1 \frac{l_{j-1}}{l_j} + C_2 \frac{\log\ d_j}{l_j}\right) \alpha_{j-1}$$

(9iii) $\| C_{\Omega_1} R(z) C_{\Omega_0} \| \leq C_4 \varepsilon^{-1} e^{-\alpha\rho(\Omega_0, \Omega_1)}$
if Ω_1 and Ω_0 are open sets with $diam\ \Omega_0 \leq l_0$.

The proof of this theorem is inductive starting from $j = 0$ for which estimate (6) holds. Assuming it holds for $j - 1$ one can use a perturbative removal of j wells between Ω_0 and Ω_1 ; since $-C_2\ \log\ d_j < l_j$ the small denominators of order d_j^{-1} are compensated

by propagators of a j-well free medium between Ω_0 and Ω_1. A detailled proof of (9i, ii) is lengthy and will appear elsewhere [CS].
Essential in the inductive proof of (9i, ii) is the geometric resolvent equation (GRE) (see e.g. [BCD], [Sig]) :

$$JR(z) = R_0(z)J + R_0(z)W(J)R(z) \tag{10}$$

where J is a C^1 function, $W(J) = [\Delta, J]$ and R, R_0 are the resolvents of any pair of Hamiltonians H, H_0 such that $Hu = H_0 u$ for all u vanishing outside the support of J ; H_0 is called a local Hamiltonian for the support of J.
As an intermediate result let us consider operators $C_{\Omega_1} R_{j_0}(z) C_{\Omega_0}$ when $\frac{1}{2}l_j \leq \rho(\Omega_0, \Omega_1) < \frac{1}{2}l_{j+1}$ with $j < j_0$ and $diam\Omega_0 \leq l_0$. Then the ball of center Ω_0 and radius $\rho(\Omega_0, \Omega_1)$ contains at the most one k-well with $j+1 \leq k \leq j_0$. Removing this well by a straight forward pertubative argument using (10) gives:

$$\| C_{\Omega_1} R_{j_0}(z) C_{\Omega_0} \| \leq C d_{j_0}^{-1} e^{-\alpha\rho(\Omega_0,\Omega_1)} \tag{11}$$

with $\alpha = \lim_{j\to\infty} \alpha_j$ and C some uniform constant ; notice $\alpha > 0$ by assumption. Then one can obtain (9iii) from (11) using (10) as follows : for a given $\varepsilon > 0$ consider the geodesic ball B_ε of center Ω_0 and radius ρ_ε such that $exp(-\alpha\rho_\varepsilon) \leq \varepsilon\ exp(-\alpha\rho(\Omega_0, \Omega_1))$. Let J_ε be a suitable C^1 function equal to one on B_ε ; then it is easy to show using (10) and a suitable factorization of $W(J)$ (see e.g.[BCD]) that :

$$\| C_{\Omega_1} R(z) C_{\Omega_0} \| \leq \| C_{\Omega_1} R_\varepsilon(z) C_{\Omega_0} \| + \| C_{\Omega_1} R(z) C_{\Omega_\varepsilon} \| \ \| C_{\Omega_\varepsilon} R_\varepsilon(z) C_{\Omega_0} \| \tag{12}$$

Here Ω_ε is the support of J'_ε and R_ε is the resolvent of a local Hamiltonian for the support of J_ε ; so there exists an integer j_ε such that this Hamiltonian has only k-wells with $k \leq j_\varepsilon$, hence one can choose $R_\varepsilon = R_{j_\varepsilon}$. Then it follows from (11) and (12) that :

$$\| C_{\Omega_1} R(E + i\varepsilon) C_{\Omega_0} \| \leq C d_{j_\varepsilon}^{-1} e^{-\alpha\rho(\Omega_0,\Omega_1)} \tag{13}$$

Since $d_{j_\varepsilon} \geq \varepsilon$ one gets (9iii).

Under a mild equivalence assumption at infinity between Euclidean and Agmon's distance one immediatly obtains from (iii).

Theorem 2

Under the assumptions of Theorem 1 the diffusion constant $D(E)$ is finite for any ψ_0 with compact support.

Let us briefly coment on this result when V is random. Then according to Kubo's formula one averages $D(E)$ over all potential configurations. If a property like (13) holds with probability one and recalling that d_{j_ε} is $dist(\sigma(H_\varepsilon), E)$ where H_ε is a local Hamiltonian for B_ε, then provided a Wegner's principle (3) holds it is very elementary to check that the averaged $D(E)$ is zero. This is essentially Fröhlich-Spencer strategy to prove localisation ; of course an essential difficulty is to show that (13) holds with probability one and presently we have no result in this direction.

3. Subdiffusive behaviour

We now describe a class of models for which the diffusion constant as given by (1) is zero for initial states ψ_0 well localized both in energy and position. Furthermore upper bounds can be given for the long time behaviour of $< x^2(t) >$ which are of unusual type; in particular Sinai's diffusion $< x^2(t) >\sim C log^2 t$ as observed in [JMS] for hierarchical models appears as a natural upper bound for a general class of models in which tunneling is less ans less effective at infinity (see also [CK] for a related model).

Let $B(0,r) = \{x \in \mathbb{R}^\nu, \rho(0,x) < r\}$ and H_r^{ext} denote any Hamiltonian such that $H_r^{ext} u = Hu$ for all $u \in \mathcal{D}(H)$, $supp\ u \cap B(0,r) = \emptyset$. For example H_r^{ext} can have all the wells inside $B(0,r)$ filled up at energy E^+.
Our assumption reads :

For all $E \in I = [E^-, E^+]$, all $\varepsilon > 0$, there exists $\alpha > 0$ and $r(\varepsilon)$ such that :

$$\| C_{\Omega_1} R_{r(\varepsilon)}^{ext}(E + i\varepsilon) C_{\Omega_0} \| \leq C\varepsilon^{-1} e^{-\alpha\rho(\Omega_1,\Omega_0)} \tag{14}$$

for some constant C depending only on Ω_0.
We also assume that $\rho(0,x)$ and $| x |$ are equivalent as $| x |\to \infty$.

Then one has

Theorem 3
Let $\psi_0 = f(H)\phi_0$ where f is a smooth function with support in a set J strictly contained in I and ϕ_0 has compact support. Then one has $\| x e^{iHt}\psi_0 \|^2 \leq C \left(sup\left(r\left(\frac{1}{t}\right), \frac{1}{\alpha} log\ t\right)\right)^2$ as $t \to \infty$.
In particular $D = 0$ if $r^2\left(\frac{1}{t}\right)/t \to 0$ as $t \to \infty$.

Before giving some sketch of the proof let us describe some simple examples of models, outside those satisfying assumptions of theorem 1 for all $E \in I$, for which our assumptions hold. The simplest one is a distribution of wells, identical or not, whose separation obeys a law

$$\rho(a,a') \geq inf(f(\rho(a,0)), f(\rho(a',0)) \tag{15}$$

for some increasing function f with $f(\rho) \geq \rho$. Then given $\varepsilon > 0$ one has $\rho(a,a') > C \mid log\ \varepsilon \mid$ for all wells a, a' lying outside $B(\varepsilon) = B(0, C \mid log\ \varepsilon \mid)$. Thus we can apply Theorem 1 for the Hamiltonian exterior to $B(\varepsilon)$ (notice that here one has only one type of well) ; in particular (9iii) holds where α can be taken larger than 1/2 by choosing the constant C above large enough ; thus (14) holds.
Let us mention that if in this model all the wells are identical then it is possible to show using RAGE theorem [RS] that H has pure point spectrum in $(-\infty, E^+)$. A related but more general class of models is obtained by assuming that the law (15) is satisfied not by wells but by clusters of wells ; these clusters can be choosen rather arbitrarily but one

needs to modify the distance in such a way that they have zero diameter and one wants the new distance to also be equivalent to euclidean distance at infinity. This new class includes in particular the hierarchical models considered in [JMS]. In the situations so described one has $r\left(\frac{1}{t}\right) \sim C \mid log\ t \mid$ hence $< x^2(t) > \leq C(log\ t)^2$ as $t \to \infty$.

We conclude with an outline of the proof of Theorem 3. Let $n(t)$ be the smallest integer with $n(t) \geq 2sup\left(r\left(\frac{1}{t}\right), \frac{2log\ t}{\alpha}\right)$; then :

$$< \psi_t, \rho^2(0,x)\psi_t > \leq n(t)^2 + \sum_{n>n(t)} (n+1)^2 \parallel J_n\psi_t \parallel^2 \tag{16}$$

where J_n is a C^1 function equal to zero in $B(0,n)$ and one outside $B(0,n+1)$. By the fundamental theorem of calculus

$$\begin{aligned} \parallel J_n\psi_t \parallel^2 &= \parallel J_n\psi_0 \parallel^2 + \int_0^t < \psi_\tau, W(J_n)\psi_\tau > d\tau \\ &\leq \parallel J_n\psi_0 \parallel^2 + 2\int_0^t \parallel C_n\psi_\tau \parallel^2 d\tau \end{aligned}$$

where C_n is given by (5) with $\Omega = Supp\ J_n'$. Thus

$$\parallel J_n\psi_t \parallel^2 \leq \parallel J_n\psi_0 \parallel^2 + \frac{e}{2\pi}\int_{-\infty}^{+\infty} \parallel C_n R\left(E + \frac{i}{2t}\right)\psi_0 \parallel^2 dE \tag{17}$$

Since ψ_0 has been choosen in such a way that is has exponential decay one has $\sum_n (n+1)^2 \parallel J_n\psi_0 \parallel^2 < \infty$. Furthermore one has by the GRE (10) :

$$C_n\ R(z) = C_n\ R^{ext}(z)J_{n/2} + C_n\ R^{ext}(z)W(J_{n/2})R(z)$$

where R^{ext} denotes the resolvent for the exterior of $B\left(0, \frac{n(t)}{2}\right)$.
Thus for $z = E + \frac{i}{2t}$, $E \in I$, one has from (17) :

$$\parallel C_n\ R(z)\psi_0 \parallel^2 \leq Ct^2 \left(\parallel J_{n/2}\psi_0 \parallel^2 + e^{-2n}t^2 \parallel \psi_0 \parallel^2\right)$$

for some constant C independent of $E \in I$; accordingly

$$\sum_{n \geq n(t)} (n+1)^2 \int_I \parallel C_n\ R\left(E + \frac{i}{2t}\right)\psi_0 \parallel^2 dE < C < \infty$$

by our choice of $n(t)$ where C is now independent of t.
Now if $E \notin I$ it is enough to write $R\left(E + \frac{i}{2t}\right)\psi_0$ as a Cauchy integral $\oint \frac{R(z)}{z-E-\frac{i}{2t}}\psi_0 dz$ along a suitable contour enclosing J to obtain in the same way boundedness of $\sum_{n \geq n(t)} (n+1)^2 \int_{\mathbb{R}\setminus I} \parallel C_n\ R\left(E + \frac{i}{2t}\right)\psi_0 \parallel^2 dE$ uniformly in t. This together with (16) implies $< \psi_t, \rho^2(0,x)\psi_t > \leq 2n(t)^2$.

References

[Ag] S. Agmon : Lectures on exponential decay of solutions of second-order elliptic equations. Mathematical notes 29, Princeton University Press.

[AS 1] J. Avron, B. Simon : Transient and recurrent spectrum. J. Func. Anal **43**, 1 (1981).

[AS 2] J. Avron, B. Simon : Almost periodic Schrödinger operators I : Limit periodic potentials. Comm. Math. Phys. **82**, 101, (1981).

[BCD] P. Briet, J.M. Combes, P. Duclos : Spectral stability under tunneling. Comm. Math. Phys. **126**, 133 (1989).

[CK] H.A. Ceccatto, W.P. Keirstead : Critical eigenfunctions in a quantum hierarchical system. J. Phys. A. Math. Gene **21**, 75 (1988).

[CS] J.M. Combes, I.M. Sigal : Diffusive and subdiffusive behaviour in quantum mechanics. To appear.

[FS] J. Fröhlich, T. Spencer : Absence of diffusion in the Anderson tight binding Model for large disorder or low energy. Comm. Math. Phys. **88**, 151 (1983).

[FMSS] J. Fröhlich, F. Martinelli, E. Scoppola, T. Spencer : Constructive proof of loca lization in the Anderson tight binding model. Comm. Math. Phys. **101**, 21 (1985).

[HM] H. Holden, F. Martinelli : Absence of diffusion near the botton of the spectrum for a Schrödinger operator on $L^2(\mathbb{R}^\nu)$. Comm. Math. Phys. **93**, 197 (1984).

[HS] B. Helffer, J. Sjostrand : Multiple wells in the semi-classical limit : Interaction through non-resonant wells. Mathematisch Nachrichten **124**, 263(1985).

[JMS] G. Jona-Lasinio, F. Martinelli, E. Scoppola : Multiple tunnelings in d-dimensions: a quantum particle in a hierarchical potential. Ann. Inst. H. Poincaré **42**, 73 (1985).

[KVD] A. Klein, H. Von Dreyfus : A new proof of localization in the Anderson tight binding model. Comm. Math. Phys. **124**, 283 (1989).

[MKS] A. Mac Kane, M. Stone : Localization as an alternative to Goldstone's theorem. Ann. Phys. **131**, 36 (1981).

[RS] M. Reed, B. Simon : Methods of Modern Mathematical Physics IV. Analysis of operators. Academic Press New-York (1978).

[Sig] I.M. Sigal : Geometric parametrices and the many body Birman-Schwinger principle. Duke math. **50**, 517 (1983).

[S] B. Simon : Absence of ballistic motion. Comm. Math. Phys. **134**, 209 (1990).

[SS] B. Simon, T. Spencer : Trace class perturbations and the absence of absolutely continuous spectrum. Comm. Math. Phys.**125**, 113 (1989).

[W] F. Wegner : Bounds on the density of states in disordered systems. Z. Phys. **B44**, 9 (1981).

PROPERTIES OF SCHRÖDINGER OPERATORS WITH SINGULAR CONTINUOUS SPECTRUM

D J GILBERT
Department of Applied Mathematics
*University of Hull, Hull HU6 7RX, UK**

ABSTRACT

Various issues relating to one-dimensional Schrödinger operators with singular continuous spectrum are explored, including the correlation between the behaviour of solutions of the Schrödinger equation and spectral properties, the stability of the singular continuous spectrum under perturbation and appropriate formulations of the eigenfunction expansion when singular continuous spectrum is present.

1. Introduction

The importance of including the singular continuous spectrum in any comprehensive study of Schrödinger operators is now widely appreciated. Recent work has shown that apparently quite innocuous potentials can give rise to singular continuous spectrum [1], and interest in many types of spectrum formerly regarded as abnormal has been stimulated by the study of operators with almost periodic and random potentials [2].

This paper explores certain aspects of one-dimensional Schrödinger operators with singular continuous spectrum, with a view to exposing some of the underlying relationships between the spectral properties of the Hamiltonian, the asymptotic behaviour of solutions of the corresponding Schrödinger equations and the nature of the integral kernels occuring in the eigenfunction expansions. In addition, the stability of the singular continuous spectrum under certain types of perturbation is considered, with examples. The treatment of these topics here is necessarily brief; some further details may be found using the references at the end of the text. Those results of the author which are mentioned for the first time will be presented more fully in future publications.

2. Mathematical Preliminaries

Consider the one-dimensional time-independent Schrödinger equation

$$Lu(r,z) = -\frac{d^2u(r,z)}{dr^2} + V(r)u(r,z) = z\,u(r,z) \qquad (1)$$

where $r \in (a,b)$, $-\infty \leq a < r < b \leq \infty$, $z \in \mathbb{C}$, and the potential $V(r)$ is locally integrable. The differential expression L may be classified in terms of Weyl's limit point, limit circle theory as follows [3]:

(I) L is in the limit circle case at both a and b,

(II) L is in the limit circle case at just one endpoint and in the limit circle case at the remaining endpoint,

(III) L is in the limit point case at both a and b.

* now at Division of Applicable Mathematics, School of Engineering IT, Sheffield City Polytechnic, Sheffield, S1 1WB, UK

Cases (I), (II) and (III) correspond to deficiency indices (2,2), (1,1) and (0,0) respectively of the associated minimal operator L_0; thus in cases (I) and (II) a range of self-adjoint extensions of L_0 can be obtained, while in case (III), L_0 is essentially self-adjoint.

Let H denote a self-adjoint extension of L_0. If L is in case (I), H has isolated point spectrum with multiplicity 1 or 2 and no finite accumulation points: since there is no singular continuous component, this case will not be further considered here. If L is in case (II) or (III), the spectrum can contain absolutely continuous, singular continuous and pure point parts; the point spectrum need not be isolated but can be dense, embedded in the continuous spectrum, and so on. The spectrum is always simple in case (II), but may be doubly degenerate in case (III) if absolutely continuous spectrum is present [4]. Distinct self-adjoint extensions of the same minimal operator differ in respect of the supports of their singular spectra [5], but their absolutely continuous and essential spectra are the same [6,7].

For the purposes of investigating the spectrum, it is convenient to use a spectral measure generated by a spectral function or matrix, rather than the resolvent operator. Spectral functions and positive semi-definite spectral matrices have the forms:

$$\rho(\lambda) = \langle E_\lambda f, f\rangle \ , \quad (\rho(\lambda)) = \begin{bmatrix} \langle E_\lambda g, g\rangle & \langle E_\lambda g, h\rangle \\ \langle E_\lambda h, g\rangle & \langle E_\lambda h, h\rangle \end{bmatrix}$$

respectively, where E_λ denotes the spectral family, f is a generating vector and g and h are a generating basis for H. It is customary to employ a spectral function when L is in case (II) and one endpoint is regular, and a spectral matrix when L is in case (II) and both endpoints are singular, or when L is in case (III); however, a spectral function may be used whenever the spectrum is simple. The spectral measures generated by the spectral functions of a given Hamiltonian are all equivalent in the measure theoretic sense, so that the particular choice of generating vector is unimportant [8]; similarly, all measures generated by the sum of the diagonal entries of the spectral matrix are equivalent. One may therefore refer without ambiguity to the spectral measure of H, while bearing in mind that strictly speaking this is an equivalence class. Note that, if L is in case (III), such a measure gives no information about the multiplicity of H, which may be 1 or 2.

The spectrum of H is the set of non-constancy points of $\rho(\lambda)$ or $(\rho(\lambda))$, as appropriate; however, a more precise characterisation of where the spectrum is concentrated may be obtained using minimal supports of the spectral measure, which are defined as follows:

Definition 1 : A subset S of $\mathbb{R}$ is said to be a *minimal support* of a positive measure ι if the following conditions are satisfied:

(i) $\iota(\mathbb{R} \setminus S) = 0$

(ii) If S_0 is a subset of S with $\iota(S_0) = 0$ then $\kappa(S_0) = 0$, where κ denotes Lebesgue measure.

Minimal supports of the absolutely continuous, singular continuous and pure point parts ι_{ac}, ι_{sc} and ι_p respectively of ι may be defined in a similar way. Note that minimal supports of ι are unique up to sets of Lebesgue and ι-measure zero.

A minimal support of a singular spectral measure has Lebesgue measure zero, and this reflects the singular nature of the spectrum; on the other hand, the singular spectrum, being a closed set, may have positive, even infinite, Lebesgue measure [1,9]. A minimal support of a spectral measure is not necessarily a subset of the spectrum; however, there always exists a minimal support of a given spectral measure whose closure is the spectrum [10].

3. Subordinacy and Spectral Properties

A method of spectral analysis which is particularly illuminating for the singular continuous spectrum and uses minimal supports arises from the theory of subordinate solutions [10,11]. A solution of Lu = zu which is subordinate at a limit point endpoint is essentially a solution which is smaller than all other linearly independent solutions of the same equation in the sense of limiting ratios of Hilbert space norms.

Definition 2 : Let L be a differential expression defined on (a,b) as in Eq 1. A solution $u_s(r,z)$ of Lu = zu, $z \in \mathbb{C}$, is said to be *subordinate* at a if for every solution u(r,z) of Lu = zu which is not a scalar multiple of $u_s(r,z)$

$$\lim_{r \downarrow a} \frac{\int_r^c |u_s(r,z)|^2 \, dr}{\int_r^c |u(r,z)|^2 \, dr} = 0$$

where c is some arbitrary interior point of (a,b).

In a similar way, we may define solutions of Lu = zu which are subordinate at b. Evidently, for each fixed $z \in \mathbb{C}$, a subordinate solution is unique up to a multiplicative constant; moreover, if a (respectively b) is a limit point endpoint and if $z \in \mathbb{C} \setminus \mathbb{R}$, then there always exists a solution of Lu = zu which is square integrable at a (respectively b) and hence is also subordinate at a (respectively b). The principal significance, however, of subordinate solutions lies in the fact that solutions of Lu = zu which are subordinate but not square integrable at one or both endpoints can exist when $z = x \in \mathbb{R}$ is a point of the spectrum of H; indeed such solutions may be used to characterise the minimal supports of the singular continuous spectrum.

The precise relationship between the existence of subordinate solutions and the spectral properties of the Hamiltonian are given by the following theorems:

Theorem 1 : Let L be in case (II) and let H be a self-adjoint operator generated by L together with a boundary condition at the limit circle endpoint. If μ is a spectral measure for H, then minimal supports of μ_{ac}, μ_{sc} and μ_p respectively are as follows:

$M_{ac}(H)$ = {$x \in \mathbb{R}$: no solution of Lu = xu exists which is subordinate at the limit point endpoint}

$M_{sc}(H)$ = {$x \in \mathbb{R}$: a non-$L_2(a,b)$ solution of Lu = xu exists which is subordinate at the limit point endpoint and satisfies the boundary condition at the limit circle endpoint}

$M_p(H)$ = {$x \in \mathbb{R}$: there exists an $L_2(a,b)$ solution of Lu = xu which satisfies the boundary condition at the limit circle endpoint}.

Theorem 2 : Let L be in case (III) and let H be the self-adjoint extension of L_0. If μ is a spectral measure for H, then minimal supports of μ_{ac}, μ_{sc} and μ_p respectively are as follows:

$M_{ac}(H) = \{x \in \mathbb{R}$: no solution of Lu = xu exists which is subordinate at a$\}$
$\cup \{x \in \mathbb{R}$: no solution of Lu = xu exists which is subordinate at b$\}$
$M_{sc}(H) = \{x \in \mathbb{R}$: there exists a non-$L_2(a,b)$ solution of Lu = xu which is subordinate at both a and b$\}$
$M_p(H) = \{x \in \mathbb{R}$: there exists an $L_2(a,b)$ solution of Lu = xu$\}$.

Let $M(H) = M_{ac}(H) \cup M_{sc}(H) \cup M_p(H)$. If $x \in \mathbb{R} \setminus M(H)$ in case (II), then there exists a solution of Lu = xu which is subordinate at the limit point endpoint, but does not satisfy the boundary condition at the limit circle endpoint. If $x \in \mathbb{R} \setminus M(H)$ in case (III), then there exist two linearly independent solutions of Lu = xu, $x \in \mathbb{R}$, one of which is subordinate at a and one of which is subordinate at b. Solutions of Lu = xu, $x \in \mathbb{R}$, which are subordinate at a (respectively b) are the pointwise limits as $z \to x$ normally of suitably normalised solutions of Lu = zu, $z \in \mathbb{C} \setminus \mathbb{R}$, which are square integrable at a (respectively b). Thus the relative smallness of the solutions of Lu = zu, $z \in \mathbb{C} \setminus \mathbb{R}$, which are square integrable at a limit point endpoint, is reflected in the relative smallness of the corresponding subordinate solutions of Lu = xu, whenever such subordinate solutions exist.

It should be noted that it is perfectly possible for the absolutely continuous or the singular continuous spectrum to be empty, even if the minimal supports $M_{ac}(H)$ or $M_{sc}(H)$ respectively are not. A simple example which illustrates this point is obtained by taking $a = -\infty$, $b = \infty$, $V(r) = 0$; the spectrum is purely absolutely continuous, but $M_{sc}(H) = \{0\} \neq \emptyset$. A necessary, but not sufficient, condition for there to be singular continuous spectrum is that $M_{sc}(H)$ should be an uncountable set; a necessary and sufficient condition for there to be absolutely continuous spectrum is that $M_{ac}(H)$ should have positive Lebesgue measure [10].

A natural consequence of Theorem 1 is that whenever L is in case (II), the real line may be decomposed into 3 disjoint sets, each reflecting distinct subordinacy properties of the corresponding Schrödinger equations. Note that here and subsequently the dependence of H on particular boundary conditions α, β etc at a limit circle endpoint is indicated explicitly by such notations as H_α, $M_{ac}(H_\beta)$, as required.

Corollary to Theorem 1 : If L is in case (II), then

$$\mathbb{R} = S_1 \cup S_2 \cup S_3$$

where $S_i \cap S_j = \emptyset$ if $i \neq j$, and

$S_1 = \{x \in \mathbb{R}$: no solution of Lu = xu exists which is subordinate at the limit point endpoint$\}$
$S_2 = \{x \in \mathbb{R}$: there exists a non-$L_2(a,b)$ solution of Lu = xu which is subordinate at the limit point endpoint$\}$
$S_3 = \{x \in \mathbb{R}$: there exists an $L_2(a,b)$ solution of Lu = xu$\}$.

Moreover, for every boundary condition α at the limit circle endpoint

$$M_{ac}(H_\alpha) = S_1 \quad , \quad M_{sc}(H_\alpha) \subseteq S_2 \quad , \quad M_p(H_\alpha) \subseteq S_3$$

and, whenever α and β are distinct boundary conditions at the limit circle endpoint

$$M_{sc}(H_\alpha) \cap M_{sc}(H_\beta) = \emptyset \quad , \quad M_p(H_\alpha) \cap M_p(H_\beta) = \emptyset.$$

Theorem 2, together with a theorem due to Kac [4], enables the multiplicity of the spectrum in case (III) to be related to the subordinacy properties of solutions of the Schrödinger equation.

Corollary to Theorem 2 : If L is in case (III), then the spectral multiplicity of H is 2 if and only if the set $M_a \cap M_b$ has positive Lebesgue measure, where for k = a,b,

$M_k = \{x \in \mathbb{R}$: no solution of $Lu = xu$ exists which is subordinate at $k\}$.

An important consequence of Kac's result, which is also evident from Theorem 2 and the above corollary, is that the singular continuous spectrum is always simple for all one-dimensional Schrödinger operators. This contrasts with the point spectrum which may be doubly degenerate if L is in case (I) (consider $L = -d^2/dr^2$ on $[0,1]$ with coupled boundary conditions), and the absolutely continuous spectrum, which may be doubly degenerate if L is in case (III) (consider $L = -d^2/dr^2$ on $(-\infty,\infty)$).

An immediate consequence of the above corollary is that if L is in case (III), the spectrum of H is simple if and only if for Lebesgue almost all x in $\mathbb{R}$, there exists a solution of $Lu = xu$ which is subordinate at one or both of the endpoints a and b. An alternative necessary and sufficient condition for H to have simple spectrum can be obtained by introducing the concept of a distinguished solution of $Lu = xu$ on (a,b).

Definition 3 : A solution $u_d(r,z)$ of $Lu = zu$, $z \in \mathbb{C}$, is said to be *distinguished* when L is in case (III) if either

(i) $u_d(r,z)$ is subordinate at both a and b, or
(ii) $u_d(r,z)$ is subordinate at one endpoint and no subordinate solution exists at the remaining endpoint.

Thus a distinguished solution may be regarded in some generalised sense as a unique smaller solution of $Lu = zu$ on (a,b); note that distinguished solutions can only exist if $z = x \in \mathbb{R}$. The following result may now be deduced from Definition 1, Theorem 2 and the Corollary to Theorem 2.

Corollary 3 : If L is in case (III), then the spectrum of H is simple if and only if there exists a minimal support $M(H)$ of the spectral measure satisfying $M(H) = \{x \in \mathbb{R}$: a distinguished solution of $Lu = xu$ exists$\}$.

4. Perturbation of the Singular Continuous Spectrum

The singular continuous spectrum is notoriously unstable under perturbation. This was first noted by Aronszajn, who considered the stability of the various parts of the spectrum under a change of boundary condition at a regular endpoint; we briefly summarise two of his examples, which are well known [5].

Example 1 : $S \subseteq \mathbb{R}$ is a closed, nowhere dense set with no isolated points and Lebesgue measure zero. A continuous non-decreasing function $\rho(\lambda)$ is constructed on $\mathbb{R}$ in such a way that its upper derivative $\overline{\rho'(\lambda)} = \infty$ for all λ in S and $\rho'(\lambda) = 0$ for all λ in $\mathbb{R} \setminus S$. It then follows from the inverse method of

Gel'fand and Levitan [12] that a continuous function V(r) on $(0,\infty)$ and a boundary condition α at $r = 0$ exist such that $\rho(\lambda)$ is the spectral function of the self-adjoint operator H_α generated by $L = -d^2/dr^2 + V(r)$ on $(0,\infty)$ with boundary condition α. The spectrum $\Sigma(H_\alpha)$ of H_α is therefore purely singular continuous with $M_{sc}(H_\alpha) \subseteq S$ and $\Sigma(H_\alpha) = \bar{S}$; however, for every distinct boundary condition β at $r = 0$, $\Sigma(H_\beta)$ is purely discrete with $M_p(H_\beta) \cap S = \emptyset$ and $\Sigma(H_\beta) = M_p(H_\beta) \cup S$.

Example 2 : S is an arbitrary countable set which is dense in $\mathbb{R}$, and $\rho(\lambda)$ is a non-decreasing function on $\mathbb{R}$ which is discontinuous at every point of S but otherwise constant. Provided the discontinuities of $\rho(\lambda)$ have been suitably chosen,

(i) it may be deduced from the inverse method of Gel'fand and Levitan [12] that a continuous function V(r) and a boundary condition α at $r = 0$ exist such that $\rho(\lambda)$ is the spectral function of the self-adjoint operator H_α generated by $L = -d^2/dr^2$ on $(0,\infty)$ with boundary condition α; $\Sigma(H_\alpha)$ is thus purely singular continuous with $M_{sc}(H_\beta) \cap S = \emptyset$ and $\Sigma(H_\beta) = \mathbb{R}$
(ii) if β is any distinct boundary condition at $r = 0$, then $\Sigma(H_\beta)$ is purely singular continuous with $M_{sc}(H_\beta) \cap S = \emptyset$ and $\Sigma(H_\beta) = \mathbb{R}$.

The following examples, with L also in case (II), show that it is perfectly possible for the singular continuous spectrum to be completely stable under a change of boundary condition at a regular endpoint. To establish these results, the conditions in Pearson's method of inductive construction of potentials with singular continous spectrum [1] have been adapted and strengthened [13].

Example 3 : Let V(r) be a sequence of identical potential bumps of the form

$$V(r) = k\,, \qquad r \in I_n$$

$$V(r) = 0\,, \qquad r \in (0,\infty) \setminus \bigcup_{n=1}^{\infty} I_n$$

where $I_n = [a_n, a_n+d]$ for some $d > 0$ and all $n \in \mathbb{N}$, $a_{n+1} > a_n$ for all $n \in \mathbb{N}$, and $I_m \cap I_n = \emptyset$ if $m \neq n$. Then provided the separation between the bumps increases sufficiently rapidly with n, the spectrum of H_α is purely singular continuous on $(0,\infty)$ for every boundary condition α at $r = 0$, with $(0,\infty) \subseteq \Sigma(H_\alpha)$. This result generalises a well-known example due to Pearson [1], which was established for a Dirichlet boundary condition at 0 only.

Example 4 : Let V(r) be a sequence of equal barriers and wells of the form

$$V(r) = k\,, \qquad r \in I_n$$

$$V(r) = -k\,, \qquad r \in I'_n$$

where the intervals $\{I_n\}$ and $\{I'_n\}$ satisfy $|I_n| = |I'_n|$ for all $n \in \mathbb{N}$, $\bigcup_{n=1}^{\infty} (I_n \cup I'_n) = (0,\infty)$, and the right hand endpoints of I_n and I'_n are the left hand endpoints of I'_n and I_{n+1} respectively for all $n \in \mathbb{N}$. Then provided the lengths $|I_n|$, $|I'_n|$ of the intervals I_n, I'_n increase sufficiently rapidly with n, the spectrum of H_α is purely singular continuous on $(-k,k)$ for every boundary condition α at $r = 0$, with $(-k,k) \subseteq \Sigma(H_\alpha)$.

Note that, although the singular continuous spectra on $(0,\infty)$ and $(-k,k)$ respectively in Examples 3 and 4 are invariant under a change of boundary condition at 0, the same is not true of the corresponding minimal supports (see Corollary to Theorem 2).

Now suppose that L is in case (III) and that self-adjoint operators H_a and H_b are generated from L on (a,c) and (c,b) respectively, using a boundary condition at c (a<c<b). If the spectrum of H_a is singular and H_b has purely singular continuous spectrum on an interval I, then the fact that the essential spectrum of H is the union of the essential spectra of H_a and H_b [14] suggests that the singular continuous spectrum of H on I will be non-empty.

The examples below show that this conjecture is false in general, but that if the spectrum of H_b on I is purely singular continuous for all boundary conditions at c, then the conjecture is true.

Example 5 : Let $a = -\infty$, $c = 0$, $b = \infty$, and let H_b be as H_α in Example 1. We may choose H_a with the same boundary condition α at 0, such that H_a has isolated point spectrum with no finite accumulation points. Then it is not hard to show (cf Gilbert [11]) that the spectrum of H is isolated pure point with $M_p(H) \cap S = \emptyset$ and $\Sigma(H) = M_p(H) \cup S$.

Example 6 : Let $a = -\infty$, $c = 0$, $b = \infty$, let H_b be as H_α in Example 3 for some boundary condition α and suppose that the corresponding operator H_a has purely singular spectrum on $(0,\infty)$. Then, since H is a finite rank perturbation of $H_a \oplus H_b$, H has no absolutely continuous spectrum on $(0,\infty)$ [6]; moreover, $(0,\infty)$ is contained in the essential spectrum of H_b, and hence also in the essential spectrum of H [14]. Since H_α has singular continuous spectrum on $(0,\infty)$ for every boundary condition α at $r = 0$, no solutions of $Lu = xu$ are in $L_2(0,\infty)$ for any $x > 0$; hence H has no eigenvalues in $(0,\infty)$. It follows that $\Sigma(H)$ is purely singular continuous on $(0,\infty)$, with $(0,\infty) \subseteq \Sigma(H)$.

It is clear that the arguments in Example 6 can be applied to any analogous situation; for example, if H_b is as H_α in Example 4, and if the corresponding operator H_a has purely singular spectrum on $(-k,k)$, then $\Sigma(H)$ is purely singular continuous on $(-k,k)$, with $(-k,k) \subseteq \Sigma(H)$.

5. Eigenfunction Expansions

If the spectrum of a self-adjoint operator H consists solely of pure point and absolutely continuous parts, then the general form of the Weyl-Kodaira eigenfunction expansion [3] can be conveniently expressed in terms of a finite or countable sum corresponding to the discrete eigenvalues, together with an integral part, where integration is with respect to Lebesgue measure on $M_{ac}(H)$, corresponding to the absolutely continuous part. This reformulation is particularly useful when there is purely absolutely continuous spectrum on the positive half line, with $M_{ac}(H) = (0,\infty)$, together with a finite or countable set of eigenvalues on $(-\infty,0]$, possibly accumulating at 0, but otherwise isolated [15]. However, where the spectrum of H is purely singular continuous, or contains a singular continuous part, appropriate forms of the eigenfunction expansion are less obvious. Nonetheless, alternative forms can be derived, which may be more convenient when singular continuous spectrum is present.

As noted earlier, singular continuous spectrum is always simple; hence the eigenfunction expansion associated with a self-adjoint operator with singular continuous spectrum can be expressed in terms of a spectral function rather than a spectral matrix. The reduction of the Weyl–Kodaira expansion to a simpler form in the case of simple spectrum was first noted by Kac [4]. The following result shows that, if L is in case (III), the kernels in such a simplified expansion are solutions of $Lu = \lambda u$, $\lambda \in \mathbb{R}$, which are subordinate at one or both endpoints.

Theorem 3 : Let L be in case (III) and suppose the spectrum of H is simple. If $f(r) \in L_2(a,b)$, then

$$f(r) = \underset{\omega\to\infty}{\mathrm{l.i.m}} \int_{-\omega}^{\omega} u_s(r,\lambda)G(\lambda)d\rho(\lambda) \tag{2}$$

where

$$G(\lambda) = \underset{\substack{\tau\uparrow b\\ \sigma\downarrow a}}{\mathrm{l.i.m}} \int_{\sigma}^{\tau} u_s(r,\lambda)f(r)dr \tag{3}$$

$\rho(\lambda)$ is a spectral function for H and $u_s(r,\lambda)$ is a suitably normalised solution of $Lu = \lambda u$ which is subordinate at one or both of a and b. Convergence in Eqs 2 and 3 is in $L_2(a,b)$ and $L_2^{\rho}(-\infty,\infty)$ respectively.

It follows from Corollary 3 that only those values of λ for which there exists a distinguished solution of $Lu = \lambda u$ actively contribute to the expansion in Eq 2; for the remaining λ, there is no spectral measure. It therefore seems plausible to regard the distinguished solutions of the Schrödinger equation as the natural eigenfunctions of H in the case of simple spectrum.

If L is in case (II), then the spectrum of H is always simple and a similar expansion holds; however, in this case $u_s(r,\lambda)$ satisfies the boundary condition at the limit circle endpoint, and only those values of λ for which either $u_s(r,\lambda)$ is subordinate or no subordinate solution exists at the limit point endpoint, actively contribute to the expansion. The existence of such an expansion when L is regular at the limit circle endpoint is well-known [16].

If H has non-empty singular continuous spectrum, then the spectral function $\rho(\lambda)$ in Eq 2 or the spectral matrix in the Weyl–Kodaira expansion has a singular continuous component; these forms of the expansion may therefore be of limited usefulness when singular continuous spectrum is present. The following limiting form of the expansion, stated here for L in case (III), avoids this problem and has quite general application.

Theorem 4 [13] : Let L be in case (III). If $f(r) \in L_2(a,b)$ then

$$f(r) = \underset{\omega\to\infty}{\mathrm{l.i.m}} \lim_{y\downarrow 0} \frac{1}{2\pi} \int_{-\omega}^{\omega} \sum_{i=1}^{2} \overline{v_i(r,z)}G(z)d\lambda \tag{4}$$

where $z = \lambda + iy$, $y > 0$ and $\{G_1(z),G_2(z)\}$ satisfies

$$\lim_{\substack{\tau\uparrow b\\ \sigma\downarrow a}} \lim_{\omega\to\infty} \lim_{y\downarrow 0} \int_{-\omega}^{\omega} \sum_{i=1}^{2} \left|G_i(z) - \int_{\sigma}^{\tau} v_i(r,z)f(r)dr\right|^2 d\lambda = 0$$

$v_1(r,z)$ and $v_2(r,z)$ are suitably normalised solutions of $Lu = zu$, $\mathrm{Im}\, z > 0$, which are in $L_2(a,c)$ and $L_2(c,b)$ respectively, and convergence in Eq 4 is in $L_2(a,b)$.

Note that the kernels $v_1(r,z)$ and $v_2(r,z)$ in this limiting form of the expansion are solutions of $Lu = zu$, $z \in \mathbb{C} \setminus \mathbb{R}$, which are square integrable at just one of the endpoints a and b. It may be shown that solutions of $Lu = xu$, $x \in \mathbb{R}$, which are subordinate at a and/or b are the pointwise limits as $z = x + iy \to x$ normally of $k_1(z)v_1(r,z)$ and/or $k_2(z)v_2(r,z)$ respectively, where $k_1(z)$ and $k_2(z)$ are complex valued functions which are analytic on $\mathbb{C} \setminus \mathbb{R}$.

Acknowledgements

The author wishes to thank Professor D B Pearson for useful discussions concerning this work, which has been supported by the Science and Engineering Research Council.

References

1. D B Pearson, *Commun Math Phys* **60** (1978) 13–36.
2. H L Cycon, R G Froese, W Kirsch and B Simon, *Schrödinger Operators* (Springer–Verlag, Berlin, 1987).
3. K Kodaira, *Amer J Math* **71** (1949) 921–945.
4. I S Kac, *Soviet Math* **3** (1962) 1035–1039.
5. N A Aronszajn, *Amer J Math* **79** (1957) 597–610.
6. T Kato, *J Math Soc Japan* **9** (1957) 239–249.
7. J Weidmann, *Linear Operators in Hilbert Space* (Springer–Verlag, New York 1980) 246.
8. A I Plesner and V A Rohlin, *Amer Math Soc Transl* **62** (1967) 116.
9. S N Naboko, *Theor Math Phys* (Jan 1987) 646–653.
10. D J Gilbert and D B Pearson, *J Math Anal Appl* **128** (1987) 30–56.
11. D J Gilbert, *Proc R Soc Edinb* **112A** (1989) 213–229.
12. I M Gel'fand and B M Levitan, *Amer Math Soc Transl* **1** (1955) 257.
13. D J Gilbert, to be published.
14. K G Wolfson, *Amer J Math* **72** (1950) 713–719.
15. T A Green and O E Lanford, *J Math Phys* **1** (1960) 139–148.
16. E A Coddington and N Levinson, *Theory of Ordinary Differential Equations* (McGraw–Hill, New York, 1955) 232.

ABOUT REGULARIZATION AND CONVERGENCE FOR SINGULAR PERTURBATIONS

Hagen Neidhardt
Karl-Weierstrass-Institut für Mathematik
Mohrenstr. 39, Berlin 1086, Germany

Valentin Zagrebnov
Laboratory of Theoretical Physics
Joint Institute for Nuclear Research
141 980 Dubna, U.S.S.R.

ABSTRACT

We present an abstract result on removing regularization for singular perturbations in the operator theory. Our main result concerns singular perturbations which are not (formally) semibounded from below.

1. Introduction

For very singular and nonpositive potentials $-V$, $V \geq 0$ in quantum mechanics it may happen that the Schrödinger operator $H = -\Delta - \frac{1}{\varkappa} V$, $\varkappa > 0$, makes no sense. But restricting the Schrödinger operator to a certain subset $\mathcal{D}$ it is quite possible that the Friedrichs extension can be well-understood in the form sense and, therefore, that the so defined operator can be regarded as related to the considered singular interaction. However, in order to be sure we have to show that so defined operator can be obtained by a cut-off approximation of the singular potential. For example, let

$$V(x) = \frac{1}{4}\frac{1}{|x|^{\alpha}}, \quad 1 \le \alpha \le 2, \quad x \in \mathbb{R}^1. \tag{1}$$

It is well-known that the expression $-\Delta - \frac{1}{\varkappa} V$, cannot well-defined on a core of the Laplace operator $-\Delta$. However, restricting $-\Delta - \frac{1}{\varkappa} V$ to the set $\mathcal{D} = C_0^\infty(\mathbb{R}^1 \setminus \{0\})$ the Friedrichs extension $\hat{H}_\varkappa$ of $(-\Delta - \frac{1}{\varkappa} V)|\mathcal{D}$ can be identified with the form sum of $-\Delta \dot{+} (-\frac{1}{\varkappa} V)$ provided $\varkappa > 0$ for $1 \le \alpha < 2$ and $\varkappa > 1$ for $\alpha = 2$. To show that the form extension is related to the potential (1) we have to find a sequence of bounded potentials $\{V_n\}_{n=1}^\infty$ converging pointwise to V such that the corresponding sequence of well-defined Schrödinger operators $H_{\varkappa,n} = -\Delta - \frac{1}{\varkappa} V_n$ converges in the strong resolvent sense to $\hat{H}_\varkappa$ as $n \to \infty$. This program was realized by Gesztesy [1] for $\alpha = 1$ and by Klaus [2] for $1 \le \alpha < 2$. The case $\alpha = 2$ was excluded. The affirmative solution of this problem surprises since the sequence of lower bounds $\lambda_{\varkappa,n} = \inf \sigma(H_{\varkappa,n})$ tends to $-\infty$ as $n \to \infty$, whereas $\hat{H}_\varkappa$ is semibounded.

Nenciu [3] has made an attempt to clarify this phenomenon on an abstract operator-theoretical level. However, it seems to us that his assumptions are to strong and that the paper does not make transparent what is really going on. In the following we make a further attempt to explain this phenomenon and find out that the effect is mainly connected with the so-called *admissibility* of the approximation. By the way we succeeded to include the case $\alpha = 2$.

2. Definitions and Results

Let $\nu \ge 0$ be a closed symmetric form defined on a complex separable Hilbert space $\mathfrak{h}$ and let A be its associated self-adjoint operator, i.e. $\nu(f,g) = (\sqrt{A}f, \sqrt{A}g)$, $f,g \in \mathrm{dom}(\sqrt{A})$. Let $\mathcal{D}$ be a dense subset of $\mathrm{dom}(\nu)$. Since ν is a closed symmetric form the restricted form $\nu_0 = \nu|\mathrm{dom}(\nu)$ is closable, too. Denoting by $\hat{\nu}$ the closure of ν_0

we have $\hat{\nu} \geq 0$, $\mathrm{dom}(\hat{\nu}) \subseteq \mathrm{dom}(\nu)$, and $\nu|\mathrm{dom}(\hat{\nu}) = \hat{\nu}$. The self-adjoint operator associated with $\hat{\nu}$ is denoted by $\hat{A}$.

Further, let $\gamma \geq 0$ be a symmetric form such that $\mathcal{D} \subseteq \mathrm{dom}(\gamma)$ and

$$\gamma(f,f) \leq a\nu(f,f) + b\|f\|^2, \quad f \in \mathcal{D}, \quad 0 \leq a,b. \tag{2}$$

Notice that ν can be replaced by ν_o and $\hat{\nu}$ in the estimate (2). But this immediately yields that γ can be continuously extended to a symmetric form $\hat{\gamma}$ which is defined on $\mathrm{dom}(\hat{\nu})$ and obeys $\hat{\gamma}(f,f) \leq a\nu(f,f) + b\|f\|^2$, $f \in \mathrm{dom}(\hat{\nu})$. The form γ is called relatively compact with respect to $\hat{\nu}$ if for some $\lambda < 0$ the bounded symmetric form $\beta(f,g) = \hat{\gamma}((\hat{A}-\lambda)^{-1/2}f,(\hat{A}-\lambda)^{-1/2}g)$, $f,g \in \mathfrak{h}$, defines a compact operator on $\mathfrak{h}$.

The KLMN Theorem yields that the perturbed symmetric form $\hat{\mu}_\varkappa$

$$\hat{\mu}_\varkappa(f,g) = \hat{\nu}(f,g) - \frac{1}{\varkappa}\hat{\gamma}(f,g), \quad f,g \in \mathrm{dom}(\hat{\mu}) = \mathrm{dom}(\hat{\nu}),$$

is closed and bounded from below for $\varkappa > a$. We denote the self-adjoint operator associated with $\hat{\mu}_\varkappa$ and the lower bound of $\hat{\mu}_\varkappa$ by $\hat{H}_\varkappa$ and $\hat{\lambda}_\varkappa$, respectively.

Definition 1 [3]. A sequence $\{\gamma_n\}_{n=1}^{\infty}$ of bounded nonnegative symmetric forms on $\mathfrak{h}$ is called a *regularizing sequence* of γ with respect to $\mathcal{D} \subseteq \mathrm{dom}(\gamma)$ if

(i) $0 \leq \gamma_1 \leq \gamma_2 \leq \ldots \leq \gamma_n \leq \ldots \leq \gamma$ and

(ii) $\lim_{n\to\infty} \gamma_n(f,f) = \gamma(f,f)$, $f \in \mathcal{D}$.

If $\{\gamma_n\}_{n=1}^{\infty}$ is a *regularizing sequence* of γ with respect to $\mathcal{D}$, then it is possible to introduce a family $\{\mu_{\varkappa,n}\}_{n=1}^{\infty}$ of closed symmetric forms defined by

$$\mu_{\varkappa,n}(f,g) = \nu(f,g) - \frac{1}{\varkappa}\gamma_n(f,g), \quad f,g \in \mathrm{dom}(\mu_{\varkappa,n}) = \mathrm{dom}(\nu),$$

$\varkappa > a$. On account of (i) $\{\mu_{\varkappa,n}\}_{n=1}^{\infty}$ is nonincreasing. Denoting by $\{H_{\varkappa,n}\}_{n=1}^{\infty}$ the corresponding family of self-adjoint operators, in general, the regularity of $\{\gamma_n\}_{n=1}^{\infty}$ is not sufficient for the strong resolvent convergence of $\{H_{\varkappa,n}\}_{n=1}^{\infty}$ to $\hat{H}_{\varkappa}$ as $n \to \infty$.

Definition 2. Let $\{\gamma_n\}_{n=1}^{\infty}$ be a *regularizing sequence* of γ with respect to $\mathcal{D} \subseteq \mathrm{dom}(\nu)$. The sequence $\{\gamma_n\}_{n=1}^{\infty}$ is called $\hat{\nu}$*-admissible* if for every nontrivial $f \in \mathcal{M} = \{h \in \mathrm{dom}(\nu)$: $\nu(h,k) + (h,k) = 0,\ \forall k \in \mathrm{dom}(\hat{\nu})\ \}$ we have

$$\sup_n \gamma_n(f,f) = +\infty.$$

However, having regularity and admissibility the desired conclusion can be drawn.

Theorem 3. *Let γ be a symmetric form obeying (2) and let $\{\gamma_n\}_{n=1}^{\infty}$ be a regularizing sequence of γ with respect to $\mathcal{D}$. If $\{\gamma_n\}_{n=1}^{\infty}$ is $\hat{\nu}$-admissible and if there is an open set $\mathcal{O} \subset (a,\infty)$ and a $\lambda < 0$ such that $\lambda < \hat{\lambda}_{\varkappa}$ and $\lambda \in \rho(H_{\varkappa,n})$ for every $\varkappa \in \mathcal{O}$ and $n = 1,2,\ldots$, then*

$$\operatorname*{s-lim}_{n\to\infty} (H_{\varkappa,n} - z)^{-1} = (\hat{H}_{\varkappa} - z)^{-1},\ z \in \{\lambda\} \cup \mathbb{C}^1\setminus\mathbb{R}^1,\ \varkappa \in \mathcal{O},$$

The strong convergence can be replaced by the operator norm convergence if in addition the symmetric form γ is relatively compact with respect to the symmetric form $\hat{\nu}$ and $\dim(\mathcal{M}) < +\infty$.

Let us note that the admissibility is closed to the necessity for the desired conclusion. The previous theorem can be improved under additional assumptions which are often satisfied in applications.

Theorem 4. *Let γ be a symmetric form obeying (2) and let $m = \dim(\mathrm{dom}(\nu)/\mathrm{dom}(\hat{\nu})) < +\infty$. If $\{\gamma_n\}_{n=1}^{\infty}$ is a $\hat{\nu}$-admissible regularizing sequence of γ, then for every $\varkappa \in (a,\infty)$ except a finite set of at most m points there is a $\lambda < \hat{\lambda}_{\varkappa}$ such that $\lambda \in \rho(H_{\varkappa,n})$ for sufficiently large n and*

$$\operatorname*{s-lim}_{n\to\infty} (H_{\varkappa,n}-z)^{-1} = (\hat{H}_{\varkappa} - z)^{-1} \quad \textit{for } z \in \{\lambda\} \cup \mathbb{C}^1 \setminus \mathbb{R}^1.$$

The strong convergence can be replaced by the operator norm convergence if in addition the symmetric form γ is relatively compact with respect to the symmetric form $\hat{\nu}$.

Applying Theorem 4 to our potential (1) we get that the approximation $H_{\varkappa,n} = -\Delta - \frac{1}{\varkappa}\frac{1}{4}\frac{1}{|x|^{\alpha}}\chi_{\mathbb{R}^1\setminus[-\frac{1}{n},\frac{1}{n}]}(x)$, $x \in \mathbb{R}^1$, converges in the norm resolvent sense to $\hat{H}_{\varkappa} = -\Delta \dot{+} (-\frac{1}{\varkappa}\frac{1}{4}\frac{1}{|x|^{\alpha}})$ as $n \to \infty$ for $\varkappa > 0$ and $1 \le \alpha < 2$ and in the strong resolvent sense for $\varkappa > 1$ and $\alpha = 2$.

References

1. F. Gesztesy, J. Phys. A: Math. Gen. **13** (1980) 867-875.
2. M. Klaus, J. Phys. A: Math. Gen. **13** (1980) L295-L298.
3. G. Nenciu, Lett. Math. Phys. **7** (1983) 301-306.

ON THE STABILITY OF POSITIVE DIATOMIC MOLECULAR IONS

P. Duclos*
Centre de Physique Théorique - CNRS - Luminy, Case 907
13288 Marseille Cedex 9 - France

and

H. Hogreve**
Hahn Meitner Institut - Postfach 390128 - D-1000 Berlin 39

ABSTRACT

The stability of positive diatomic molecular ions, with equally charged nuclei, is investigated in the limit of infinite nuclear mass. It is proven that all electronic curves are monotonically decaying if the charge of the nuclei is bigger than 4 times the number of electrons. In addition, in the one-electron case, it is shown that for any charge smaller than 4 there exists an electronic curve which is not monotonically decaying. This value 4 equals precisely the critical charge for binding in a corresponding classical electrostatic system.

1. Introduction

If we consider a diatomic molecule with N electrons and equal charge Z of the nuclei, we expect for increasing Z that the molecule will become unstable and an effect experimentally observed as "Coulomb explosion"[11] will occur. Mathematically this can be rephrased as

$$\forall N \geq 0 \ , \ \exists Z_c \ , \ Z > Z_c \ \Rightarrow \ \sigma_d(\tilde{H}) = \phi \tag{1.1}$$

where $\sigma_d(\tilde{H})$ denotes the discrete spectrum of the Hamiltonian $\tilde{H}$ describing our (non relativistic) molecule.

A folk belief among quantum chemists is that whenever all the electronic curves of a molecule are monotonically decaying the molecule should be instable.

---

* and PHYMAT Université de Toulon et du Var, F83957 La Garde Cedex

** supported by the Sonderforschungsbereich 337 of the Deutsche Forschungsgemeinschaft

The electronic curves $E(R,Z)$ are just the discrete eigenvalues of the molecular hamiltonian H with fixed nuclei separated by a distance R, where H reads as

$$H = H(R,Z) = \sum_{k=1}^{N} h_k + \sum_{1\leq k<l\leq N} \omega_{kl} + \frac{Z^2}{R} \text{ on } L^2(\mathbb{R}^{3N}). \tag{1.2}$$

Here h_k is the one-electron hamiltonian for the kth electron

$$h_k = h_k(R,Z) = -\frac{1}{2}\Delta_k + v(x^{(k)}) , \tag{1.3}$$

v is the potential energy of one electron due to the nuclei

$$v = v_+ + v_- \ , \ v_\pm(x) = -Z\left[\left(x_1 \pm \frac{R}{2}\right)^2 + x_2^2 + x_3^2\right]^{-1/2} \tag{1.4}$$

and

$$\omega_{kl} = |x^{(k)} - x^{(l)}|^{-1} \tag{1.5}$$

is the potential energy due to the repulsion between the kth and l^{th} electrons.

Therefore we can formulate our question on stability as

$$\exists ? Z_c \geq 0 \ , \ Z > Z_c \text{ and } E(R,Z) \epsilon \sigma_d(H(R,Z)) \Rightarrow \frac{\partial E}{\partial R}(R,Z) < 0 \ . \tag{1.6}$$

Our answer to this question is the following :

Theorem 1 i) $\forall N \geq 0 \ , \ Z_c \leq 4N$

ii) If $N = 1$ then $Z_c = 4$.

We shall now make some comments on this result.

Comments

1) The electrons we have considered are in fact bosons since we do not include the Pauli principle in the definition of the Hamiltonian H. By standard arguments we know that if Eq. 1.6 has a solution for bosons, this solution is also valid for fermions.

2) The stability of positive molecular ions has already been studied by M.B. Ruskai[1] and J.Ph. Solovej[2]. They consider the diatomic molecule with dynamic and not necessarily equally charged nuclei. For equally charged nuclei and bosonic electrons their results are roughly $Z_c \leq 2N$ in the large N limit (Ruskai) and $Z_c \leq \frac{3}{1+a}N, a \sim \frac{Z}{2M}$, M being the mass of the nuclei (Solovej). G. Zhislin has also made a report on this subject in this conference.

3) Of course, our global upper bound $Z_c \leq 4N$ is far from being optimal when considering dynamic molecules. Even for infinite mass of the nuclei it is

not good, when compared to the one of Solovej $Z_c \leq 3N$. However concerning the global monotonicity of the electronic curves our result is optimal in the case $N = 1$; the statement ii) of the theorem must be understood as

$$\forall Z < 4 \ , \ \exists R > 0 \text{ and } E(R,Z) \epsilon \sigma_d(H(R,Z)) \ , \ \frac{\partial E}{\partial R}(R,Z) > 0 \qquad (1.7)$$

in addition to Eq. 1.6 with $Z_c = 4$.

4) Concerning the folk belief we mentioned above it is reasonable to make the following conjectures.

Conjecture 1 If all electronic curves of a diatomic molecule are monotonically decaying then the dynamic molecule has no bound state.

Conjecture 2 If for a fixed symmetry the ground state electronic curve of a diatomic molecule is monotonically decaying then the dynamic molecule has no bound state in this sector of symmetry.

We expect to soon publish a proof of conjecture 1. Conjecture 2, if true, would be a very efficient tool to study the stability of molecules.

5) The only lower bound on Z_c that we know is provided by the study of stability of negative molecular ions[3,4]. From the result of Lieb[3] one deduces $Z_c \geq \frac{N-2}{4}$.

6) There exists already some results by E.H. Lieb[5,6] and B. Simon[6] on the monotonicity of the electronic curves ; unfortunately since the nuclear repulsion is not included they are not directly applicable to the stability problem.

2. Upper bound on the critical charge of the nuclei

We shall only discuss the one-electron case and indicate briefly how to treat the many-electron case. Since we use the derivative of the electronic curves to study their monotonicity we need the

lemma 1[7,8]. There exists a labelling of the discrete spectrum of $H(R,Z)$,

$$\sigma_d(H(R,Z)) = \{E_1(R,Z) \leq E_2(R,Z) \leq \ldots \ \leq E_n(R,Z) \leq \ldots\}$$

such that each $E_n(.,Z)$ is real analytic on all of $\mathbb{R}_+^* := \{R \epsilon \mathbb{R}, R > 0\}$.

We shall study all the E_n's on the same footing so in the sequel $E = E(R,Z)$ will denote any $E_n(R,Z)$.

We shall use two different methods to analyse the sign of $\frac{\partial E}{\partial R}$: one for the small R values, the other for the large one.

lemma 2. Lower bound on the size of the molecule.

One has : $\forall Z > 0\ ,\ R \le \frac{15}{12N}\ \Rightarrow\ \frac{\partial E}{\partial R}(R,Z) < 0$.

The proof of this lemma, based on simple lower bounds on the spectrum of $H(R,Z)$, will not be included here due to the lack of space. In the rest of this section we analyse the sign of $\dfrac{\partial E}{\partial R}(R,Z)$ for $R > \dfrac{15}{12N}$.

Proof of Theorem 1 i). The one-electron case.

By the Feynman-Hellman theorem one has

$$\frac{\partial E}{\partial R}(R,Z) = -\frac{Z^2}{R^2} + < \frac{\partial v}{\partial R} > \tag{2.1}$$

where <> denotes expectation value with respect to the normalised eigenvector associated to $E(R,Z)$. The first idea to try is

$$\frac{\partial E}{\partial R}(R,Z) \le -\frac{Z^2}{R^2} + ||\frac{\partial v}{\partial R}(R,Z)||_\infty \tag{2.2}$$

Unfortunately $\dfrac{\partial v}{\partial R}$ being unbounded Eq. 2.2 gives no information. So the next attempt consists in eliminating the singularity of $\dfrac{\partial v}{\partial R}$; using virial theorem one also has

$$\frac{\partial E}{\partial R}(R,Z) = -\frac{Z^2}{R^2} + \left\langle \frac{\partial v}{\partial R} - i[A,H] \right\rangle \tag{2.3}$$

for suitably well behaved selfadjoint operators A. We shall look for an operator A such that $\dfrac{\partial v}{\partial R} - i[A,H]$ will be bounded and as small as possible. In the construction of A we are guided by the following identities

$$\frac{\partial v}{\partial R} = \frac{1}{2}\frac{\partial v_+}{\partial x_1} - \frac{1}{2}\frac{\partial v_-}{\partial x_1} = \frac{i}{2}[p_1, v_+ - v_-]$$

$$i\left[\frac{1}{2}p_1, H\right] = \frac{i}{2}[p_1, v_+ + v_-] \tag{2.4}$$

where $p_j := \frac{1}{i}\frac{\partial}{\partial x_j}$. The natural idea to try is to construct A out of $\frac{1}{2}p_1$ as follows

$$A := \frac{1}{2}\{fp_1f - gp_1g\}\,; \tag{2.5}$$

here f and g are two real functions of x_1 satisfying

$$f+g=1\ ,\ g(x_1)=f(-x_1)\ ,\ f=1 \text{ if } x_1 \le -\frac{R}{2}\ ,\ f=0 \text{ if } x_1 \ge \frac{R}{2}. \tag{2.6}$$

With such an A we have at least formally

$$\frac{\partial v}{\partial R} - i[A,H] = gv'_+ - fv'_- + p_1f'p_1 - \frac{1}{4}f''' \tag{2.7}$$

where the prime denotes the derivative with respect to x_1. Choosing f with $f' \leq 0$ leads to the estimate

$$\frac{\partial v}{\partial R} - i[A,H] \leq g v'_+ - f v'_- - \frac{1}{4} f''' \tag{2.8}$$

which r.h.s. is now bounded provided g and f vanish quickly enough at the singularity of v'_+ and v'_-. The precise choice of f we make is

$$f(x_1) = \frac{1}{16}\left(1 - \left(\frac{2x_1}{R}\right)\right)^3 \left(3\left(\frac{2x_1}{R}\right)^2 + 9\left(\frac{2x_1}{R}\right) + 8\right) \text{ if } |x_1| < \frac{R}{2}. \tag{2.9}$$

Then one has

$$\frac{\partial v}{\partial R} - i[A,H] \leq \begin{cases} \frac{Z}{R^2} \text{ if } |x_1| \geq \frac{R}{2} \\ \frac{Z}{R^2}\left(4 - 12\frac{x_1^2}{R^2}\right) - \frac{15}{2R^3}\left(1 - \frac{12x_1^2}{R^2}\right), \text{ if } |x_1| < \frac{R}{2} \end{cases} \tag{2.10}$$

Finally $\frac{\partial E}{\partial R} < 0$ is a consequence of $-\frac{Z^2}{R^2} +$ r.h.s. of Eq. 2.10 < 0 and solving this last inequality one finds $Z > 4$ provided $R > \frac{15}{12}$.

To make these formal manipulations rigorous notice that $C_0^\infty(I\!R^3)$ being a core of H and $D(A) \supset D(H)$, Eqs. 2.3, 2.4, 2.7, 2.8 are valid in the form sense on $D(H) \times D(H)$ which is sufficient for our purpose since the expectation value is taken in an eigenvector of H.

We end up this section by indicating how to handle the many-electron case.

Proof of Theorem 1 i). Sketch for the N-electron case.

If we take

$$A := \sum_{i=1}^N A_i \quad \text{with} \quad A_i := \frac{1}{2}\left\{f(x_1^{(i)})p_1^{(i)} f(x_1^{(i)}) - g(x_1^{(i)})p_1^{(i)} g(x_1^{(i)})\right\}$$

it is easy to check that

$$i\left[A, \sum_{k<l} \omega_{kl}\right] \leq 0. \tag{2.11}$$

Hence it will be sufficient to show that

$$-\frac{Z^2}{R^2} + \sum_{k=1}^N < \frac{\partial v}{\partial R}(x^{(k)}) - i[A_k, h_k] > < 0 \tag{2.12}$$

which shows that we have reduced the analysis to the one-electron case.

3. Lower bound on the critical charge of the nuclei for the one-electron case

Here we want to prove that the upper bound on the critical charge for $N = 1$ cannot be improved (see Eq. 1.7). The idea of the proof relies on the following observations : i) $Z = 4$ is the classical (i.e. not quantum) value corresponding to electrostatic equilibrium. In that case the electron sits exactly at the midpoint of the straight line joining the two nuclei. Quantum mechanically, if we would be able to construct an eigenvector well localised on that point, we expect the same effect. Indeed with $\psi^2(x) = \delta(x-0)$, the Dirac distribution at zero, we obtain

$$-\frac{Z^2}{R^2} + < \frac{\partial v}{\partial R} >= -\frac{Z^2}{R^2} + \frac{\partial v}{\partial R}(0) = -\frac{Z^2}{R^2} + \frac{4Z}{R^2} > 0 \Leftrightarrow Z < 4. \qquad (3.1)$$

Of course, it is not possible to have an eigenvector localised on one point, but if we could do it on a very small neighbourhood of 0 we could hope to approach the result of Eq. 3.1.
ii) The second observation is that the large R limit is equivalent to a semiclassical limit : by a scaling argument one has

$$h(R,Z) = h_1 \simeq \frac{Z}{R}\widehat{h} \ , \ \widehat{h} = \widehat{h}(R,Z) = -(RZ)^{-1}\frac{\Delta}{2} + \widehat{v}$$

$$\widehat{v} = \widehat{v}_+ + \widehat{v}_- \ , \ \widehat{v}_\pm(x) = -\left[(x_1 \pm 1)^2 + x_2^2 + x_3^2\right]^{-1/2} \qquad (3.2)$$

where $\simeq$ stands for unitary equivalence; we see that $(RZ)^{-1/2}$ plays the role of $\hbar$. At 0 the potential $\widehat{v}$ has a saddle point at energy -2; hence the point $(0,0)$ in the phase space $\mathbb{R}^3 \times \mathbb{R}^3$ corresponds to a recurent classical trajectory of the associated classical system. It is known[9] that the quantum probability concentrates in the limit $(RZ)^{-1/2} \downarrow 0$ to the classical probability which is here just the Dirac measure at zero. To proceed need information on the rate of concentration of this quantum probability :

lemma 3. For any $Z > 0$, there exists a sequence $\{R_n\}_{n \epsilon \mathbb{N}}$ in $\mathbb{R}_+^*$ such that

i) $\lim_{n\to\infty} R_n = +\infty$

ii) $-\frac{4Z}{R_n} \epsilon \sigma_d(h(Z,R_n))$

iii) The corresponding normalised eigenvectors ψ obey

$$\forall \varepsilon > 0 \quad ||\psi(R_n,Z)||_{|x|<\epsilon R_n} = 1 - O_\epsilon\left((lnR_n)^{-1}\right) (*).$$

To prove such a result we make profit of the separability of the hamiltonian h in elliptic coordinates (ξ,η,ϕ) and apply the analoguous one dimensional result[10] to

(*) $<>_B$ and $|| \ ||_B$ denote respectively the restriction to the set B of the expectation value and the norm.

the η-equation which has an asymptotic second order turning point with negative curvature.

Now coming back to our operator H with A defined in Sect. 2 we estimate

$$< \frac{\partial v}{\partial R} - i[A,H] > = < gv'_+ - fv'_- > + < Re\, p_1^2 f' + \frac{1}{4} f''' >$$

$$\geq < gv'_+ - fv'_- > + < f'(e-v) + \frac{1}{4} f''' > \qquad (3.3)$$

where $e = v(0) = -\frac{4Z}{R}$ is the value of the potential at the saddle point. It is straight forward to check that with our choice (see Eq. 2.9) of f one has

$$< f'(e-v) + \frac{1}{4} f''' > = O(R^{-2}Z) + O(R^{-3}) ; \qquad (3.4)$$

here it is crucial that f' vanishes fast enough at $\pm\frac{R}{2}$ to cancel the singularity of v.

Now let $B_\varepsilon := \{x \epsilon \mathbb{R}^3,\ |x| < \varepsilon \frac{R}{2}\}$; one has for $R = R_n$ given by lemma 3 and any $Z < 4$:

$$< gv'_+ - fv'_- > = < gv'_+ - fv'_- >_{B_\varepsilon} + < gv'_+ - fv'_- >_{\mathbb{R}^3 \setminus B_\varepsilon}$$

$$\geq \tfrac{4Z}{R_n^2}(1+\varepsilon^2)^{-3/2} ||\psi||^2_{B_\varepsilon} + O(R_n^{-2})(1 - ||\psi||^2_{B_\varepsilon})$$

$$\geq \tfrac{4Z}{R_n^2}(1+\varepsilon^2)^{-3/2} + O(R_n^{-2}(lnR_n)^{-1}).$$

Thus finally we have obtained

$$\forall 0 < Z < 4,\ \frac{\partial E}{\partial R}(R_n, Z) \geq -\frac{Z^2}{R_n^2} + \frac{4Z}{R_n^2}(1+\varepsilon^2)^{-3/2} + O(R_n^{-2}\,(lnR_n)^{-1})$$

from which we deduce easily Eq.1.7.

References

1. M.B. Ruskai, *Limits on Stability of Positive Molecular Ions*, *Lett. Math. Phys.* **18** (1989) 121-132 ; and *Abscence of Bound States in Extremely Asymmetric Positive Diatomic Molecules*, Preprint (1990).
2. J.Ph. Solovej, *Asymptotic Neutrality of Diatomic Molecules*, *Comm. Math. Phys.* **130** (1990) 185-204.
3. E.H. Lieb, *Bound on the Maximum Negative Ionization of Atoms and Molecules*, *Phys. Rev.* **A29** (1984) 3018-3028.
4. M.B. Ruskai, *Limit on the Excess Negative Charge of a Dynamic Diatomic Molecule*, *Ann. Inst. H. Poincaré A : Physique Théorique* **52** (1990) 397-414.

5. E.H. Lieb, *Monotonicity of the Electronic Contribution to the Born-Oppenheimer Energy*, *J. Phys. B : Atom. Molec. Phys.* **11** (18) (1978) L537-542.
6. E.H. Lieb and B. Simon, *Monotonicity of the Molecular Electronic Energy in the Nuclear Coordinates*, *J. Phys. B : Atom. Molec. Phys.* **15** (1982) L63-66.
7. P. Aventini and R. Seiler, *On the Electronic Spectrum of the Diatomic Molecular Ion*, *Comm. Math. Phys.* **41** (1975) 119-134.
8. J.M. Combes and R. Seiler, *Regularity and Asymptotic Properties of the Discrete Spectrum of Electronic Hamiltonians*, *Int. Jour. Quant. Chem.* **14** (1978) 213.
9. A. Knauf, *Coulombic periodic potentials : The quantum case*, *Annals of Physics* **191** (1989) 205-240.
10. P. Duclos and H. Hogreve, *On the Semiclassical Localization of Quantum Probability*, In preparation, and A Semiclassical Analysis of One Dimensional Quantum Localisation, In preparation.
11. O. Echt, *Multiply charged clusters In Physics and Chemistry of small clusters*, Edts P. Gena, B.K. Rao and F.N. Khanna, Plenum press New York (1987) 623-638.

ON THE STABILITY OF POSITIVE MOLECULAR IONS

S.A.Vugalter

Institute of Applied Physics,
Nizhni Novgorod, USSR

ABSTRACT

It is shown that a positive multiatomic molecular ion consisting of N electrons and k nuclei is unstable for a fixed large N if the nuclear charges are sufficiently large.

1. INTRODUCTION

Recently a considerable progress has been achieved in the study of stability of multiparticle systems with Coulomb interaction.

For atomic ions it was proved that extremely negative ions are unstable. The most precise estimates of maximum number of electrons that a nucleon can attach were obtained by Lieb, Sigal, Simon and Thirring [1]. They proved that a negative atomic ion with N electrons and charge of the nucleus Z is unstable if $N > N_0(\varepsilon)$ and $N > (1+\varepsilon)Z$ where ε is an arbitrary positive number.

For molecular first result was obtained in a recent paper by Ruskai [2] ; it was demonstrated that a positive diatomic molecular ion with N electrons and charges of the nuclei Z_1 and Z_2 is unstable if $Z_1 > 2N$, $Z_2 > 2N$ and N is large.

We are going to announce here a generalization of the Ruskai's result to the case of positive ions of multiatomic molecules.

2. DEFINITIONS

Let us consider a quantum system A of $N+k$ particles. Let $r_i \in \mathbb{R}^3$, $i = 1,\ldots,N+k$ be the position vector of the i-th particle. We assume $1,2,\ldots,k$ to be the nucleon numbers and $k+1,\ldots,N+k$ to be the numbers of electrons.

Let $M_i > 1$ and $Z_i > 1$ be the mass and charge of the i-th nucleus, and introduce

$$Z_0 := \min_{i=1,\ldots,k} Z_i ,$$

$$a := \min_{i=1,\ldots,k} Z_i M_i^{-1} , \quad b := \max_{i=1,\ldots,k} Z_i M_i^{-1} .$$

We assume that the masses and charges of the electrons equal 1 and -1 , respectively.

Let H_0 be the energy operator of the system A after invariant separation of the center-of-mass motion, with

$$R_0 = \left\{ r : r=(r_1,\ldots,r_{N+k}) , \sum_{i=1}^{k} M_i r_i + \sum_{i=k+1}^{N+k} r_i = 0 \right\}$$

being the relative-motion space of the system equipped with the inner product

$$(r,\tilde{r})_1 := \sum_{i=1}^{k} M_i(r_i,\tilde{r}_i) + \sum_{i=k+1}^{N+k} (r_i,\tilde{r}_i) .$$

It is easy to see that

$$H_0 = -\frac{1}{2}\Delta_0 + \sum_{\substack{i,j=1\\ i<j}}^{k} Z_iZ_j|r_{ij}|^{-1} - \sum_{i=1}^{k}\sum_{j=k+1}^{N+k} Z_i|r_{ij}|^{-1}$$

$$+ \sum_{\substack{i,j=k+1\\ i<j}}^{N+k} |r_{ij}|^{-1}$$

where $r_{ij} := r_i - r_j$ and Δ_0 is the Laplacian on R_0.

Let α be the type of irreducible representation of the group S of identical-particle permutations associated with our system, and denote by P^{α} the projection onto the subspace in $L_2(R_0)$ consisting of the functions which transform according to this representation under the group operation $T_g\Psi(r) = \Psi(g^{-1}r)$, $g\in S$, $r\in R_0$. We assume that the type α satisfies the Pauli principle. Finally, denote by H_0^{α} the corresponding part of the energy operator of the system, $H_0^{\alpha} := P^{\alpha}H_0$.

3. RESULTS

Our main conclusion is the following

THEOREM. To arbitrary $k \geq 2$, $a > 0$ and $b > 0$ one can find numbers $N_0 = N_0(k,a,b)$ and $d_0 = d_0(k,a,b)$ such that the discrete spectrum of the operator H_0^α is empty provided $Z_0 \geq d_0 N$ and $N \geq N_0$.

The proof which will be published elsewhere follows the same line as that by Ruskai. It is based on the variational principle combined with decomposition of the configuration space into regions corresponding to some decompositions of the system.

The main feature distinguishing our method is that the decomposition of R_0 is performed in two steps. The first of them concerns relative motion of the nuclei, and after that we decompose according to electron positions relative to fixed nuclei.

Our method allows us to estimate the value of the number $d_0(k,a,b)$ appearing in the theorem, but an optimal estimate has yet to be found.

REFERENCES

1. Lieb, E.H., Sigal, I.M., Simon, B. and Thirring, W., Commun.Math.Phys. 116,635 (1988).

2. Ruskai, M.B., Lett.Math.Phys. 18,121 (1989).

Limiting absorption principle and resonances for the Dirac operator

by

E.Balslev
University of Aarhus
(Denmark)
and
University of Virginia
(Charlottesville, USA)

and

B.Helffer *
University of Nantes
(France)
and
Ecole Normale Supérieure
(Paris,France)

* The present address of the second author is the Wissenschaftskolleg zu Berlin.

§0 Introduction

The stationary scattering theory for the Dirac operator

$$H_D = \left(\frac{\hbar c}{i}\right)\left(\sum_{j=1}^{3} \alpha_j \frac{\partial}{\partial x_j}\right) + \beta mc^2 + \sum_{j=1}^{3} \alpha_j A_j + V$$

is developed, using the trace formalism of Kuroda [Ku]. Here $\{\alpha_j\}_{j=1}^{3}$ and β are the Dirac matrices, and the magnetic and electric potentials A_j and V are real, short-range and with at most $|x|^{-1+\varepsilon}$-type singularities.

H_D is a self-adjoint operator in $L^2(\mathbb{R}^3,\mathbb{C}^4)$ with continuous spectrum $(-\infty,-mc^2]\cup[mc^2,\infty)$, empty singular continuous spectrum, and eigenvalues are finite-dimensional and accumulate at most at $\pm mc^2$. The trace operators $T_{D\pm}(\lambda)$, defined in terms of boundary values $R_D(\lambda\pm io)$ of the resolvent, yield spectral representations of the continuous part of H_D. The scattering matrix $S_D(\lambda)$ is a unitary operator on $L^2(S^2,\mathbb{C}^4)$ satisfying $T_{D+}(\lambda)=S_D(\lambda)T_{D-}(\lambda)$. The method of construction of $T_{D\pm}(\lambda)$ is via the Foldy-Wouthuysen transform, diagonalizing the free Dirac operator H_{OD}, and the connection between H_{OD} and the free Schrödinger operator $H_0=-\frac{h^2}{2m}\Delta$, given by $H_{OD}^2=2c^2H_0+m^2c^4$.

Under the further assumption that each of the functions A_j and V is a sum of a short-range potential, dilation-analytic in an angle $S_\alpha=\{z\,|\,|\mathrm{Arg}\,z|<\alpha\}$, and a potential which is $o(e^{-2b|x|})$, the $B(L^2(S^2,\mathbb{C}^4))$-valued function $S_D(\lambda)$ has a meromorphic extension from $[mc^2,\infty)$ to a region $\mathbb{O}^+_{\alpha,b}\subset\mathbb{C}^-$ and from $(-\infty,-mc^2]$ to $(-\mathbb{O}^+_{\alpha,b})\subset\mathbb{C}^+$. Also the resolvent $(H_D-\zeta)^{-1}$, acting between certain exponentially weighted spaces, has a meromorphic continuation from $\mathbb{C}^+$ across (mc^2,∞) to $\mathbb{O}^+_{\alpha,b}$ and from $\mathbb{C}^-$ across $(-\infty,-mc^2)$ to $(-\mathbb{O}^+_{\alpha,b})$. Moreover, the poles of the analytically continued scattering matrix and resolvent coincide and are of the same order. This extends the theory of [Ba-Sk] from the Schrödinger operator to the Dirac operator.

Resonance functions ϕ at a resonance λ are defined (cf. [Ba-Sk]) as solutions of the Dirac equation $(H_D-\lambda)\phi=0$, which lie in the range of the residue at λ of the analytically contiuned resolvent. The space F of resonance functions at λ is isomorphic to the null space of the inverse S-matrix at λ. Using this, we prove that F is even-dimensional, assuming $A_j=0$ and V diagonal.

Finally, we study the non-relativistic limit $c\to\infty$ of $H_D = H_D(c)$, assuming $A_j = 0$ and V of the form $V = \begin{pmatrix} V^+ & 0 \\ 0 & V^- \end{pmatrix}$. As $c\to\infty$, the spectrum of $H_D' = H_D - mc^2$ "tends" to $[0,\infty)$, and we expect to obtain in the limit the Pauli operator $H_+ = -\frac{\Delta}{2m} + V_+$ in $L^2(\mathbb{R}^3, \mathbb{C}^2)$. This is made precise as follows. For any $\zeta = z^2$ with $-\alpha<\mathrm{Arg}\, z<0$, $-b<\mathrm{Im}\, z<0$, ζ not a resonance of (H_0,H_+), the resolvent $R_D'(c,\zeta)$ converges as $c\to\infty$ to the operator $(H_+-\zeta)^{-1}P_+$, where $P_+ = \begin{pmatrix} 1 & 0 \\ 0 & 0 \end{pmatrix}$. The analytically continued scattering matrix $\mathfrak{S}_D'(c,\zeta)$ converges as $c\to\infty$ to $\mathfrak{S}(\zeta)p_+$, where $\mathfrak{S}(\zeta)$ is the analytically continued S-matrix of (H_0,H_+) and p_+ is the projection $\begin{pmatrix} 1 & 0 \\ 0 & 0 \end{pmatrix}$ in $L^2(S^2,\mathbb{C}^4)$. Moreover, $\mathfrak{S}_D'(c,\zeta)$ extended for $c=\infty$ by $\mathfrak{S}(\zeta)p_+$ is an analytic function of $x = c^{-1}$ near $x=0$. Generically, resonances of (H_0,H_+) continue analytically (in fact as functions of x^2) into resonances of $(H_{0D}(x), H_0(x))$ for x near 0. This extends to resonances results on eigenvalues proved by [Ve],[Hu],[G-N-P]. Note that while eigenvalues are always analytic, resonances may exceptionally have a branch point at $x=0$.

§1 Definitions and notations :

We denote by $\mathbb{C}^+$ and $\mathbb{C}^-$ the upper and lower half-plane respectively and set $\tilde{\mathbb{C}}^+ = \overline{\mathbb{C}^+} \setminus \{0\}$.

$\mathbb{C}_1 = \mathbb{C} \setminus \{ (-\infty,-1] \cup [1,\infty)\}$

Let $(1,\infty)_\pm$ denote the interval $(1,\infty)$ considered as part of the boundary of $\mathbb{C}^\pm$ and $(-\infty,-1)_\pm$ the interval $(-\infty,-1)$ considered as part of the boundary of $\mathbb{C}^\pm$.

Let $\tilde{\mathbb{C}}_1 = \mathbb{C}_1 \cup (-\infty,-1)_+ \cup (-\infty,-1)_- \cup (1,\infty)_+ \cup (1,\infty)_-$.

$\tilde{\mathbb{C}}_1^\pm = \mathbb{C}_1 \cup (-\infty,-1)_\mp \cup (1,\infty)_\pm$.

We denote by $\mathcal{H}$ the Hilbert space $L^2(\mathbb{R}^3)$ with norm $\| \,.\, \|_{\mathcal{H}}$ and inner product $(\,.\,,\,.\,)_{\mathcal{H}}$. The space of spinors $\mathcal{H}_4$ is given by :

$$\mathcal{H}_4 = \Sigma_{i=1}^4 \oplus \mathcal{H}_{(i)} = \{ f : \mathbb{R}^3 \to \mathbb{C}^4 \mid \| f \|^2_{\mathcal{H}_4} = \int_{\mathbb{R}^3} |f(x)|^2 dx < \infty \}$$

with inner product and norm :

$$(f,g)_4 = \int_{\mathbb{R}^3} (f(x),g(x))dx \quad , \quad \|f\|_4 = (f,f)_4^{1/2}$$

where $(.,.)$ and $|\,.\,|$ denote the inner product and norm on $\mathbb{C}^4$. The weighted Sobolev spaces $H^{t,s} = H^{t,s}(\mathbb{R}^3)$ are defined for $s,t \in \mathbb{R}$ by :

$$H^{t,s} = \{ f \mid \|f\|_{H^{t,s}} = \| (1+r^2)^{s/2} f \|_{H^t} < \infty \}$$

where $r = (\Sigma_i x_i^2)^{1/2}$, $\|f\|_{H^t} = (\int_{\mathbb{R}^3} <\xi>^{2t} . |\hat{f}(\xi)|^2 d\xi)^{1/2}$, $<\xi> = (1+|\xi|^2)^{1/2}$.

Of course $H^{t,0}$ is the classical Sobolev space H^t and we define the corresponding spinor spaces by :

$H_4^{t,s} = \Sigma_{i=1}^4 \oplus H_{(i)}^{t,s}$, $H_4^t = H_4^{t,0}$ with norm :

$\| f \|_{H_4^{t,s}} = (\Sigma_{i=1}^4 \| f_i \|^2_{H^{t,s}})^{1/2}$.

The exponentially weighted spaces $\mathcal{H}^b$ are defined for $b \in \mathbb{R}$ by :

$\mathcal{H}^b = \{ f \mid \| f \|_{\mathcal{H}^b} = \| e^{br} f \|_{\mathcal{H}} < \infty \}$

and the corresponding weighted Sobolev spaces $\mathcal{H}^{t,b}$ by :

$\mathcal{H}^{t,b} = \{ f \mid \| f \|_{\mathcal{H}^{t,b}} = \| e^{br} f \|_{H^t} < \infty \}$.

The corresponding spinor spaces are defined by

$\mathcal{H}_4^{t,b} = \Sigma_{i=1}^{4} \oplus \mathcal{H}_{(i)}^{t,b} = \{ f : \mathbb{R}^3 \to \mathbb{C}^4 \mid \| f \|_{\mathcal{H}_4^{t,b}} = (\Sigma_{i=1}^{4} \| f_i \|^2_{\mathcal{H}^{t,b}})^{1/2} < \infty \}$.

We set $h = L^2(S^2)$ and $h^4 = \Sigma_{i=1}^{4} \oplus h_{(i)}$.

The free Schrodinger operator H_0 in $\mathcal{H}_4$ is given by

$H_0 = \Sigma_{i=1}^{4} \oplus (-(1/2)\Delta)_i$ with domain H_4^2 , where we have set $\hbar = m = 1$.

In momentum representation we write :

$\hat{H}_0 = (1/2) (\Sigma_{i=1}^{4} \xi_i^2)$ with domain $H_4^{0,2}$.

We denote by $R_0(\eta)$ the free resolvent of H_0,

$R_0(\eta) = (H_0 - \eta)^{-1}$ for $\eta \notin \sigma(H_0) = [0,\infty)$.

The free Dirac operator H_{0D} in $\mathcal{H}_4$ is defined by

$H_{0D}\psi = (\hbar c/i)(\Sigma_{j=1}^{3} \alpha_j \partial\psi/\partial x_j) + \beta m c^2 \psi$ on $\mathcal{D}(H_{0D}) = H_4^1$,

where

$$\alpha_j = \begin{pmatrix} 0 & \sigma_j \\ \sigma_j & 0 \end{pmatrix}, \; j=1,2,3 \; ; \; \beta = \alpha_4 = \begin{pmatrix} 1 & 0 \\ 0 & -1 \end{pmatrix}$$

and the σ_j are the Pauli spin matrices :

$\sigma_1 = \begin{pmatrix} 0 & 1 \\ 1 & 0 \end{pmatrix}, \sigma_2 = \begin{pmatrix} 0 & -i \\ i & 0 \end{pmatrix}, \sigma_3 = \begin{pmatrix} 1 & 0 \\ 0 & -1 \end{pmatrix}$.

The matrices α_j satisfy

$$\alpha_j\alpha_k + \alpha_k\alpha_j = 2\,\delta_j^{\,k}\ ,\ j,k = 1,2,3,4 \qquad (1.1)$$

For simplicity we choose units such that $m=c=\hbar=1$ (we are not considering here the limits $\hbar\to 0$ (see [Pa] for this question) or $c\to\infty$ (see §7)), so that :

$$H_{oD}\psi = i^{-1}\Sigma_{j=1}^{3}\alpha_j\partial\psi/\partial x_j + \beta\,\psi$$

and in momentum representation :

$$\hat{H}_{oD} = \Sigma_{j=1}^{3}\alpha_j\xi_j + \beta \text{ on } H_4^{0,1}.$$

The operator H_{oD} is selfadjoint. From (1.1) we obtain the identity :

$$\hat{H}_{oD}^2 = \xi^2 + 1\ ,\ H_{oD}^2 = 2\,H_o + 1 \qquad (1.2)$$

It follows from (1.2) that the spectrum $\sigma(H_{oD})$ and resolvent set $\rho(H_{oD})$ are given by :

$$\sigma(H_{oD}) = (-\infty,-1] \cup [1,\infty)\ ,\ \rho\,(H_{oD}) = \mathbb{C}_1$$

Again by (1.2), we have :

$$(\hat{H}_{oD}-\zeta)\,(\hat{H}_{oD}+\zeta) = \xi^2+1-\zeta^2;\ (H_{oD}-\zeta)\,(H_{oD}+\zeta) = 2H_o + 1 - \zeta^2 \qquad (1.3)$$

By (1.3) the free resolvent $R_{oD}(\zeta)$ of H_{oD} is given by

$$R_{oD}(\zeta) = (1/2).\,(H_{oD}+\zeta)R_o((\zeta^2-1)/2) \text{ for } \zeta \in \mathbb{C}_1. \qquad (1.4)$$

Let us now introduce a potential V_D which will be a symmetric operator in $\mathcal{H}_4$ satisfying the following conditions, analogous to the conditions in [Ku] for Schrodinger operators :

(A_1) There exists $s > 1/2$ such that $V_D \in \mathcal{C}(H_4^{1,-s}, H_4^{0,s})$ (the class of compact operators from $H_4^{1,-s}$ into $H_4^{0,s}$)

(A_2) $\langle V_D f, f \rangle_4 = \int_{\mathbb{R}^3} ((V_D f)(x), f(x))dx \in \mathbb{R}$ for all $f \in H_4^{1,-s}$

(A_3) $V_D \in \mathcal{B}(H_4^{1,t}, H_4^{0,t+2s})$ (the class of bounded operators from $H_4^{1,t}$ into $H_4^{0,t+2s}$) for $t \in [-s,0]$.

The Dirac operator H_D is defined in $\mathcal{H}_4$ by

$H_D = H_{OD} + V_D$, $\mathcal{D}(H_D) = H_4^1$.

By (A1) , V_D is H_{OD}-compact, hence (using also (A2)) H_D is self-adjoint and the essential spectrum $\sigma_e(H_D)$ is given by

$\sigma_e(H_D) = \sigma(H_{OD}) = (-\infty,-1] \cup [1,\infty)$.

Finally, the discrete spectrum $\sigma_d(H_D)$ is contained in $(-1,1)$.

Remark 1.1 :

A potential of the type $V \otimes I$ satisfies (A1)-(A3) if V is multiplication by a real-valued function v on $\mathbb{R}^3$ satisfying the following conditions for some $\varepsilon, R, C > 0$:

$$\sup_{x \in \mathbb{R}^3} \int_{|x-y| \leq 1} |v(y)|^2 |x-y|^{-1-\varepsilon} dy < \infty$$

and $\int_{|x-y| \leq 1} |v(y)|^2 dy \leq C\, |x|^{-1-\varepsilon}$ for $|x| \geq R$

(cf $[Ba]_2$). Thus, local singularities of the type $|x_0 - y|^{-1+\delta}$ are allowed. A potential satisfying (A1)-(A3) which is not of this type is for example the magnetic vector potential $\Sigma_{i=1}^{3} \alpha_i A_i$, where A_1, A_2, A_3 are real-valued functions on $\mathbb{R}^3$ satisfying the above conditions.

§2 The Foldy-Wouthuysen transform.

The Foldy-Wouthuysen transform, diagonalizing the free Dirac operator, is defined according to [Bj-Dr], as follows. For fixed $\xi \in \mathbb{R}^3$ let $\hat{G}(\xi)$ be the unitary 4×4 matrix defined by :

$$\hat{G}(\xi) = \exp\,(\beta\alpha.\xi\ \theta(|\xi|)) \tag{2.1}$$

$$\text{where } \theta(t) = (2t)^{-1} \text{ Arctan } t \text{ for } t > 0 \tag{2.2}$$

The matrix $\hat{G}(\xi)$ transforms $\hat{H}_{OD}(\xi)$ into

$$\tilde{H}_{OD} = \hat{G}(\xi)\,\hat{H}_{OD}(\xi)\,\hat{G}^{-1}(\xi) = \begin{pmatrix} (1+\xi^2)^{1/2} & 0 \\ & \\ 0 & -(1+\xi^2)^{1/2} \end{pmatrix}$$

where $(1+\xi^2)^{1/2}$ is a 2×2 diagonal matrix.

Let us remark that $\theta(|\xi|)$ is smooth (real analytic), also at $\xi = 0$, since θ is an even function. The function θ has also the following properties :

$\lim_{t\to 0} \theta(t) = 1/2$, $\lim_{t\to\infty} \theta(t) = 0$, and $\theta(t)$ is decreasing. (2.3)

The Foldy-Wouthuysen transform in the ξ-representation is the unitary operator $\hat{G}$ on $\mathcal{H}_4$ defined by

$(\hat{G}\, g)(\xi) = \hat{G}(\xi)\, g(\xi)$.

On $\mathcal{H}_4$ it is given by

$G = \mathcal{F}^{-1}\hat{G}\mathcal{F}$

where $\mathcal{F} = \Sigma_{i=1}^{4} \oplus \mathcal{F}_{(i)}$, and $\mathcal{F}_{(i)}$ is the Fourier-Plancherel transform on $L^2_{(i)}(\mathbb{R}^3)$.

G transforms H_{0D} into

$$\tilde{H}_{0D} = G\, H_{0D} G^{-1} = \begin{pmatrix} (1+2H_0)^{1/2} & 0 \\ 0 & -(1+2H_0)^{1/2} \end{pmatrix}$$

Definition 2.1 :

Let $\mathcal{U}(\rho)$ be the dilation group on $\mathcal{H}_4$:

$(\mathcal{U}(\rho)\, g)(x) = \rho^{3/2} g(\rho x)$ for $\rho \in \mathbb{R}^+$, $g \in \mathcal{H}_4$.

Setting $G(\rho) = \mathcal{U}(\rho)\, G\, \mathcal{U}(\rho^{-1})$ for $\rho \in \mathbb{R}^+$, we have the

Lemma 2.2 :

$G(\rho)$ has an analytic extension $G(z)$ to the right half-plane $\{z|\ |\mathrm{Arg}\ z| < \pi/2\}$ as a $\mathcal{B}(\mathcal{H}_4)$-valued function and also as a $\mathcal{B}(H_4^{0,s})$-valued and a $\mathcal{B}(H_4^{-1,s})$-valued function.

§3 Limiting absorption principle

By the limiting absorption principle (L.A.P) for the free Schrodinger operator H_0 (Agmon [Ag]$_1$), the following limits exist for $s>1/2$ and $\mu \in (0,\infty)$ in the uniform topology of $\mathcal{B}(H_4^{0,s}, H_4^{2,-s})$, uniformly on compact sets ,

$R_0(\mu \pm i0) = \lim_{\varepsilon \to +0} R_0(\mu \pm i\varepsilon)$.

From this we obtain :

<u>Lemma 3.1 .</u>

The following limits exist in the uniform operator topology of $\mathcal{B}(H_4^{0,s}, H_4^{1,-s})$, uniformly on compact sets ,

$R_{0D}(\lambda \pm i0) = \lim_{\varepsilon \to 0} R_{0D}(\lambda \pm i\varepsilon) =$

$$= \begin{cases} (1/2)(H_{0D}+\lambda)R_0(((\lambda^2-1)/2) \pm i0) \text{ for } \lambda \in (1,\infty) \\ \\ (1/2)(H_{0D}+\lambda)R_0(((\lambda^2-1)/2) \mp i0) \text{ for } \lambda \in (-\infty,-1) \end{cases}$$

Let us recall that the free trace operator for the Schrodinger operator $T_0(\lambda)$ is defined on $H_4^{1,s}$ for $s > 1/2$ by

$$T_0(\mu)f = \gamma_0(\mu^{1/2})\, \hat{f}$$

where $\gamma_0(\rho)\, g = g(\rho,.)$ for $g \in \hat{H}_4^{1,s}$ (using polar coordinates)

Note that

$T_0(\mu) \in \mathcal{B}(H_4^{1,s}, h_4)$, $T_0^*(\mu) \in \mathcal{B}(h_4, H_4^{1+2,-s})$

Let us also recall that the spectral density $E'_0(\mu)$ for the free Schrödinger operator H_0 is given for $\mu > 0$ by :

$$E'_0(\mu) = (1/2i\pi)\,(R_0(\mu+i0) - R_0(\mu-i0)) = \tfrac{1}{2}\mu^{1/2}\, T_0^*(\mu)T_0(\mu) \qquad (3.1)$$

Let us now compute the spectral density for the free Dirac in the Foldy-Wouthuysen representation. The spectral density $\tilde{E}'_{0D}(\lambda)$ of $\tilde{H}_{0D}$ is given for $\lambda > 1$, setting $\mu = (\lambda^2-1)/2$ by

Let

$$q(\lambda,\xi) = [\, (1+\xi^2)^{1/2} + |\lambda|\,]^{1/2},$$

then

$$\tilde{T}^{\pm}_{oD}(\lambda) = (1/2)\,((\lambda^2-1)/2)^{1/4}\,\gamma_0(((\lambda^2-1)/2)^{1/2})\; q(\lambda,\xi)\mathcal{F}P_{\pm}$$

Definition 3.2 : The trace operator $T_{oD}(\lambda)$ for the free Dirac operator is given by

$$T^{\pm}_{oD}(\lambda) = (1/2)\,((\lambda^2-1)/2)^{1/4}\,\gamma_0(((\lambda^2-1)/2)^{1/2})\; q(\lambda,\xi)\mathcal{F}P_{\pm}G \in \mathcal{B}(H_4^{0,s}, \hbar_4) \tag{3.5}$$

$$T_{oD}(\lambda) = \begin{cases} T^{+}_{oD}(\lambda) \text{ for } \lambda > 1 \\ \\ T^{-}_{oD}(\lambda) \text{ for } \lambda < -1 \end{cases}$$

The adjoint $T^{\pm}_{oD}{}^{*}(\lambda)$ is given by :

$$T^{\pm}_{oD}{}^{*}(\lambda) = G^{*}(1/2)P_{\pm}\mathcal{F}^{*}((\lambda^2-1)/2)^{1/4}q(\lambda,\xi)\gamma_0{}^{*}(((\lambda^2-1)/2)^{1/2}) \in \mathcal{B}(\hbar_4, H_4^{1,-s}).$$

We have the identity

$$E_{oD}{}'(\lambda) = G^{*} \,.\, \tilde{E}_{oD}{}'(\lambda) \,.\, G = T_{oD}{}^{*}(\lambda) \,.\, T_{oD}(\lambda)$$

so $\{ T_{oD}(\lambda)\}$ yields a spectral representation of H_{oD} on $L^2((-\infty,-1)\cup(1,\infty);\hbar_4)$.

In order to obtain the L.A.P. for H_D we need to identify singular points, defined as follows.

Definition 3.3

For $\lambda \in (-\infty,-1) \cup (1,\infty)$ we set

$N_{\pm}(\lambda) = \{ \phi \in H_4^{0,s} \mid \phi + V_D R_{oD}(\lambda \pm i(\lambda/|\lambda|)0)\phi = 0 \}$,

$\Sigma_{\pm} = \{ \lambda \in (-\infty,-1) \cup (1,\infty) \mid N_{\pm}(\lambda) \neq \{0\} \}$,

$M(\lambda) = \{ \Psi \in H_4^1 \mid (H_D - \lambda)\Psi = 0 \}$

Lemma 3.4 (Agmon $[Ag]_1$).

Let $f \in H_4^{t,s}$ and assume : $T_0(\mu)f = 0$ for some $\mu > 0$.

Then $R_0(\mu \pm i0)f \in H_4^{t+2,s-1}$.

Theorem 3.5

$N_+(\lambda) = N_-(\lambda)$ and $N_{\pm}(\lambda)$ are isomorphic to $M(\lambda)$ via the maps

$N_{\pm}(\lambda) \ni \phi \to R_{oD}(\lambda \pm i(\lambda/|\lambda|)\,0)\,\phi = \Psi \in M(\lambda)$

$M(\lambda) \ni \Psi \to \phi = -V_D\,\Psi \in N_{\pm}(\lambda)$

Corollary 3.6:

$\Sigma_{\pm} = \sigma_p(H_D) \cap \{ (-\infty,-1) \cup (1,\infty) \}$.

Using Theorem 3.5 we can prove, following Kuroda [Ku]

Theorem 3.7

$\sigma_p(H_D) \cap \{ (-\infty,-1) \cup (1,\infty) \}$ consists of eigenvalues with finite-dimensional eigenspace, accumulating at most at $+1$ and -1.

From (A.1) and Theorems 3.5-3.7 follows

Theorem 3.8

The following limits exist for $\lambda \in \{(-\infty,-1) \cup (1,\infty)\} \setminus \sigma_p(H_D)$ in the uniform operator topology of $\mathfrak{B}(H_4^{0,s}, H_4^{1,-s})$, uniformly on compact sets,

$R_D(\lambda \pm i0) = \lim_{\varepsilon \to 0} R_D(\lambda \pm i\varepsilon) = R_{oD}(\lambda \pm i0)\,(1 + V_D R_{oD}(\lambda \pm i0))^{-1}$.

§4 The scattering matrix.

As in the case of the Schrodinger operator asymptotic completeness of the wave operators follows from the L.A.P. of section 3. We formulate this in terms of the perturbed trace operator $T_{D\pm}(\lambda)$ defined as follows

Definition 4.1: Trace operator for Dirac

For $\lambda \in ((-\infty,-1) \cup (1,\infty)) \setminus \sigma_p(H_D)$

$$T_{D\pm}(\lambda) = T_{0D}(\lambda)(1 - V_D R_D(\lambda \pm i(\lambda/|\lambda|).0)) \in \mathcal{B}(H_4^{0,s}, \mathfrak{h}_4) \tag{4.1}$$

We have $T_{D\pm}^*(\lambda) \in \mathcal{B}(\mathfrak{h}_4, H_4^{1,-s})$, and

$$E_D'(\lambda) = (1/2i\pi)(R_D(\lambda+i0) - R_D(\lambda-i0)) = T_{D\pm}^*(\lambda)\, T_{D\pm}(\lambda) \tag{4.2}$$

Thus, $\{T_{D+}(\lambda)\}$ as well as $\{T_{D-}(\lambda)\}$, $\lambda \in ((-\infty,-1) \cup (1,\infty)) \setminus \sigma_p(H_D)$, yield spectral representations of the absolutely continuous part $H_{D,ac}$ of H_D on $L^2((-\infty,-1) \cup (1,\infty); \mathfrak{h}_4)$.

Theorem 4.2:

(i) The operators $F_{D\pm}$, defined for $f \in \mathcal{H}_{4,ac} \cap H_4^{0,s}$ by :

$(F_{D\pm}f)(\lambda,.) = T_{D\pm}(\lambda)f$, $\lambda \in ((-\infty,-1) \cup (1,\infty)) \setminus \sigma_p(H_D)$,

extend uniquely to bounded operators mapping $\mathcal{H}_{4,ac}$ isometrically onto $L^2((-\infty,-1) \cup (1,\infty); \mathfrak{h}_4)$.

(ii) The scattering matrix $S_D(\lambda)$ is the unitary operator on $\mathfrak{h}_4$ defined for $\lambda \in ((-\infty,-1) \cup (1,\infty)) \setminus \sigma_p(H)$ by :

$S_D(\lambda)\, T_{D-}(\lambda) = T_{D+}(\lambda)$.

(iii) $S_D(\lambda)$ is given explicitly by

$$S_D(\lambda) = 1 - 2\pi i\, T_{0D}(\lambda)(V_D - V_D R_D(\lambda + i(\lambda/|\lambda|).0)V_D)\, T_{0D}^*(\lambda) \qquad (4.3)$$

and

$$\text{(iv)}\ S_D^{-1}(\lambda) = 1 + 2\pi i\, T_{0D}(\lambda)(V_D - V_D R_D(\lambda - i(\lambda/|\lambda|).0)V_D)\, T_{0D}^*(\lambda) \quad (4.4)$$

We now assume that $V_D = V_{D1} + V_{D2}$, where V_{D1} and V_{D2} satisfy (A1)-(A3) and the further assumptions

(A4) $V_{D1}(\rho) = \mathcal{U}(\rho)V_{D1}\mathcal{U}(\rho^{-1})$ has a $C(H_4^{1,-s}, H_4^{0,s})$-valued analytic extension from $\mathbb{R}^+$ to $S_\alpha = \{z \mid |\text{Arg } z| < \alpha\}$, $0 < \alpha \leq \frac{\pi}{2}$.

(A5) $V_{D1}(k) = e^{ikr}V_{D1}e^{-ikr} \in C(H_4^{1,-s})$ for $k \in \mathbb{C}$ where $e^{ikr}\{f_i\}_{i=1}^4 = \{e^{ikr}f_i\}_{i=1}^4$.
(Note that (A5) follows from (A1) if V is multiplicative)

(A6) $V_{D2} \in C(H_4^{1,-b}, H_4^{0,b})$ for some $b > 0$.

Utilizing the methods of [Ba-Sk] and the analyticity properties of the Foldy-Wouthuysen transform given by Lemma 2.2, we obtain the following result.

<u>Theorem 4.3</u> Assume that $V_D = V_{D1} + V_{D2}$ satisfies (A1)-(A6). Then

(i) $R_D(\zeta)$ has a $\mathcal{B}(\mathcal{H}_4^{0,b}, \mathcal{H}_4^{1,-b})$-valued meromorphic continuation $\tilde{R}_D(\zeta)$ from $\mathbb{C}_1$ across $(-\infty,-1)_- \cup (1,\infty)_+$ into the region $\mathcal{O}_{\alpha,b}^- \cup \mathcal{O}_{\alpha,b}^+$, where $\mathcal{O}_{\alpha,b}^+$ is bounded by $(1,\infty)_+$ and the curves

$\mathcal{C}_1 = \{\zeta = (1+(t-ib)^2)^{1/2} \mid t \in \mathbb{R}^-\}$

and

$\mathcal{C}_2 = \{\zeta = (1+s^2 e^{-2i\alpha})^{1/2} \mid s > 0\}$

while $\mathcal{O}^-_{\alpha,b}$ is bounded by $(-\infty,-1)_-$ and the curves $(-\mathcal{C}_1)$ and $(-\mathcal{C}_2)$.

(ii) The S-matrix $S_D(\lambda)$ has a $\mathfrak{B}(\mathcal{h}_4)$-valued, meromorphic extension $\tilde{S}_D(\lambda)$ from $(-\infty,-1)_- \cup (1,\infty)_+$ to $\mathcal{O}^+_{\alpha,b} \cup \mathcal{O}^-_{\alpha,b}$.

(iii) The poles of $\tilde{R}(\zeta)$ and $\tilde{S}(\zeta)$ coincide and are of the same order. Similarly, $R_D(\zeta)$ has a $\mathfrak{B}(\mathcal{H}^{0,b}_4, \mathcal{H}^{1,-b}_4)$-valued meromorphic continuation $\tilde{\tilde{R}}_D(\zeta)$ from $\mathbb{C}_1$ across $(-\infty,-1)_+ \cup (1,\infty)_-$ into the region $\overline{\mathcal{O}^+_{\alpha,b}} \cup \overline{\mathcal{O}^-_{\alpha,b}}$. The inverse S-matrix $S_D^{-1}(\lambda)$ has a $\mathfrak{B}(\mathcal{h}_4)$-valued, meromorphic extension $\tilde{\tilde{S}}_D^{-1}(\lambda)$ from $(-\infty,-1)_+ \cup (1,\infty)_-$ into the region $\overline{\mathcal{O}^+_{\alpha,b}} \cup \overline{\mathcal{O}^-_{\alpha,b}}$, and the poles of $\tilde{\tilde{R}}_D(\lambda)$ and $\tilde{\tilde{S}}_D^{-1}(\lambda)$ coincide and are of the same order. They are conjugates of the poles of $\tilde{R}_D(\lambda)$ and $\tilde{S}_D(\lambda)$.

§5 Resonances and resonance functions.

The poles of $\tilde{S}_D(\lambda)$ are called resonances and those of $\tilde{\tilde{S}}_D^{-1}(\lambda)$ antiresonances. We define the resonance functions asociated with a resonance λ_0 as follows :

Definition 5.1 :

Let λ_0 be a resonance of H_D, and let Γ be a closed curve in $\mathcal{O}^+_{\alpha,b} \cup \mathcal{O}^-_{\alpha,b}$ separating λ_0 from other resonances. Set

$P_{\lambda_0} = -\frac{1}{2i\pi} \int_\Gamma \tilde{R}_D(\lambda)\, d\lambda \in \mathfrak{B}(\mathcal{H}^{0,b}_4, \mathcal{H}^{1,-b}_4)$. (We omit the reference to λ_0 if there is no ambiguity).

The space $\mathcal{F}$ of resonance functions is defined by :

$\mathcal{F} = \{ f \in \mathcal{R}(P) \mid (H_D - \lambda_0)f = 0\} \subset \mathcal{H}_4^{1,-b}$.

The space $\overline{\mathcal{F}}$ of anti-resonance functions is defined by :

$\overline{\mathcal{F}} = \{ g \in \mathcal{R}(P^*) \mid (H_D - \overline{\lambda}_0)g = 0\} \subset \mathcal{H}_4^{1,-b}$.

As in the case of the Schrödinger operator we have :

<u>Theorem 5.2:</u>

Let λ_0 be a resonance of H_D. Then

$\mathcal{N}(\tilde{S}_D^{-1}(\lambda_0)) \simeq \mathcal{F}$,

the isomorphism being given by

$\mathcal{N}(\tilde{S}_D^{-1}(\lambda_0)) \ni \sigma \mapsto T_D^*(\overline{\lambda}_0)\,\sigma = f \in \mathcal{F}$

and

$$\mathcal{N}(\tilde{\tilde{S}}_D(\overline{\lambda}_0)) \simeq \overline{\mathcal{F}},$$

the isomorphism being given by

$\mathcal{N}(\tilde{\tilde{S}}_D(\overline{\lambda}_0)) \ni \tau \to T_D^*(\lambda_0)\,\tau = g \in \overline{\mathcal{F}}$

Adapting a result of Parisse [Pa] in our context, we obtain

<u>Theorem 5.3</u>

Let λ_0 be a resonance of H_D. Let us assume that V_D is diagonal and real (i.e. commuting with the operator of complex conjugation Γ). Then the resonance space $\mathcal{F}$ defined above is of even dimension.

§6 The non relativistic limit $c \to \infty$

We introduce in this section again the parameter c (keeping $\hbar = 1$ and $m = 1$) and will study the behavior of the scattering matrix as $c \to \infty$ attached to the Dirac equation :

$$H_{oD} = (c/i)\left(\Sigma_{j=1}^{3} \alpha_j \partial_{x_j}\right) + \beta c^2 \tag{6.1}$$

To normalize correctly as $c \to \infty$, let us recall some definitions (following Hunziker [Hu] and references there):

Definition 6.1

$$H'_{oD} = H_{oD} - c^2 \; ; \; H'_D = H_D - c^2 \tag{6.2}$$

$$R'_{oD}(\zeta) = (H'_{oD} - \zeta)^{-1} \; ; \; R'_D(\zeta) = (H'_D - \zeta)^{-1} \tag{6.3}$$

(sometimes we shall write $R'_{oD}(\zeta,c)$ or $R'_D(\zeta,c)$ to recall the dependence on c)

$S'_D(\lambda,c)$ denotes the scattering matrix of the pair (H'_{oD}, H'_D) (6.4)

Of course, we have : $\sigma_e(H'_D) = \sigma_e(H'_{oD}) = (-\infty, -2c^2] \cup [0,\infty)$ and the negative part of the spectrum "tends" as $c \to \infty$ to $-\infty$.

Let us observe that we have for $\lambda > 0$

$$R'_{oD}(\lambda + i0,c) = (H_{oD} + \lambda + c^2)(2c^2)^{-1} R_0\left([(\lambda^2 + 2\lambda c^2)/2c^2] + i0\right) \tag{6.5}$$

$$R'_D(\lambda + i0,c) = R'_{oD}(\lambda + i0,c)\,(1 + V_D \cdot R'_{oD}(\lambda + i0,c))^{-1} \tag{6.6}$$

$$T'_{oD}(\lambda,c) = \gamma_0(\rho).\rho^{1/2}(2c)^{-1}\mathcal{F}\left((c^4 + 2c^2 H_0)^{1/2} + \lambda + c^2\right)^{1/2} P_+ G(c) \tag{6.7}$$

where

$$\rho = [(\lambda^2/(2c^2)) + \lambda]^{1/2} \tag{6.8}$$

$$T'_D(\lambda,c) = T'_{oD}(\lambda,c)\ (1 - V_D .\ R_D(\lambda + c^2 + i0,c)) \qquad (6.9)$$

$$S'_D(\lambda,c) = 1 - 2\pi i\ T'_{oD}(\lambda,c)\ (V_D - V_D .\ R_D(\lambda + c^2 + i0,c).V_D\ )\ T'_{oD}{}^{*}(\lambda,c) \qquad (6.10)$$

We assume that V_D is of the form :

$$V_D = \begin{pmatrix} V^+ & 0 \\ & \\ 0 & V^- \end{pmatrix} \qquad (6.11)$$

We set

$$\mathcal{H}_+ = P_+\mathcal{H}\ .\ H_0{}^+ = H_0P_+\ .\ H^+ = HP_+ = H_0{}^+ + V^+ \qquad (6.12)$$

where P_+ is defined after (3.3).

We have identified in a natural way the operators $H_0{}^+$ and H^+ with the operators $H_0{}'\,|\mathcal{H}_+$ and $H^+\,|\,\mathcal{H}_+$.

We let $x = c^{-1}$ and replace c by x^{-1} in the above formulas :

$$T'_{oD}(\lambda,x) =$$

$$= \gamma_0(\rho).(\rho/2)^{1/2}\mathcal{F}((1/4) + (1/2)x^2H_0)^{1/2} + (\lambda/2)x^2 + (1/2))^{1/2}P_+G(x^{-1}) \qquad (6.13)$$

where (with some obvious change of notations)

$$\rho = \rho(\lambda,x) = (\ (\lambda^2\ /2)x^2 + \lambda)^{1/2} \qquad (6.14)$$

$$R'_{oD}(\lambda + i0,x) =$$

$$= \left(\ (x/2i)(\ \Sigma_{j-1}^{3}\alpha_j\partial_{x_j}\ ) + P_+ + (\lambda\ /2)x^2\right)\ R_0(\lambda + (\lambda\ /2)x^2 + i0) \qquad (6.15)$$

$$T'_D(\lambda,x) = T'_{oD}(\lambda,x)\ (1 - V_D .\ R'_D(\lambda + i0,x)) \qquad (6.16)$$

$$S'_D(\lambda,x) = 1 - 2\pi i\ T'_{oD}(\lambda,x)\ (V_D - V_D .\ R'_D(\lambda + i0,x).V_D\ )\ T'_{oD}{}^{*}(\lambda,x) \qquad (6.17)$$

Let us now introduce some notations.

$\mathcal{O}_\alpha = \{z \in \mathbb{C} \mid -\alpha < \operatorname{Arg} z < 0\}$

$\mathcal{O}_b = \{\mathbb{C} \mid \operatorname{Re} z > 0, -b < \operatorname{Im} z < 0\}$

and

$$\mathcal{O}^{+\,'}_{\alpha,b} = \mathcal{O}^2_\alpha \cap \mathcal{O}^2_b \tag{6.18}$$

where $\mathcal{O}^2_\alpha = \mathcal{O}_{2\alpha}$

For $K > 0$, we introduce also : $\mathcal{O}^+_{\alpha,b,K} = \mathcal{O}^{+\,'}_{\alpha,b} \cap \{|z| < K\}$

Then we obtain the following:

<u>Theorem 6.2</u>

Assume that $V_D = V_{D1} + V_{D2}$ satisfies (A1) - (A6) for some pair (α, b) and (6.11).

Then, for each (β, c) s.t. $0 < \beta < \alpha$, $0 < c < b$ and each $K > 0$, there exists $r > 0$ s.t for each x s.t. $|x| < r$, the map $\lambda \to S'_D(\lambda, x)$ admits a meromorphic extension $\tilde{S}'_D(\zeta, x)$ from $0 < \lambda < K$ to $\mathcal{O}^+_{\beta,c,K} \subset \mathbb{C}^-$

For each $\zeta \in \mathbb{R}^+ \cup \mathcal{O}^{+\,'}_{\alpha,b}$ (ζ not a resonance of H^+), we have

$$\tilde{S}'_D(\zeta, x) \xrightarrow[x \to 0]{} \tilde{S}_0(\zeta) = \tilde{S}_+(\zeta) p_+ + p_- \tag{6.19}$$

where $\tilde{S}_+(\zeta)$ is the analytically extended scattering matrix of (H^+_0, H^+) and $p_\pm$ are the projections in $\hbar_4$ corresponding to $P_\pm$.

Moreover, r can be chosen such that the function $\tilde{S}(\zeta, x)$ defined by :

$$\tilde{S}(\zeta, x) = \begin{cases} \tilde{S}'(\zeta, x) & \text{for } x > 0 \\ \\ \tilde{S}_0(\zeta) & \text{for } x = 0 \end{cases} \tag{6.20}$$

is, for fixed ζ in $\mathcal{O}^{+\,'}_{\beta,c}$, an analytic function of x for $|x| < r$.

We finally consider the question of analyticity of resonances as functions of $x = c^{-1}$. By Theorem 4.3, ζ is a resonance of $H_D'(x)$ if and only if the kernel of $\tilde{S}_D^{-1}(\zeta, x)$ is not zero. Note that $\tilde{S}^{-1}(\zeta, x)$ equals the identity plus a compact operator-valued, analytic function of ζ and x for ζ near a resonance and x near 0. Using analytic perturbation theory, we obtain the following result, extending earlier results on eigenvalues proved by [Ve],[Hu],[G-N-P].

Theorem 6.3. Assume that $V_D = V_{D1} + V_{D2}$ satisfies (A1)-(A6) and (6.11). Then the resonances of $H_D'(x)$ are analytic in x near each point of $[0,\infty)$, except for at most a discrete set σ. If $0 \notin \sigma$, then the resonances are analytic in x^2.

Remark 6.4

Theorem 6.3 generalizes to the case of resonances earlier results on eigenvalues proved by [Ve], [Hu], [G-N-P]. Of course, in that case by Rellich's theorem they obtained the analyticity in x of the eigenvalues, but it is proved in the last reference that they are in fact analytic functions of x^2 in the neighborhood of 0. In the situation of the above theorem, it is also true if $0 \notin \sigma$. This follows immediately from the observation that the matrix β intertwines $H_D(c)$ and $H_D(-c)$. But we can also follow the proof of [G-N-P] to get that the matrix $A(\zeta,x)$ and therefore $d(\zeta,x)$ is in fact a function of $y = x^2$:

$\tilde{d}(\zeta,y) = d(\zeta,x)$.

REFERENCES

[Ag] S.Agmon: Spectral properties of Schrödinger operators and scattering theory
Ann.Scuola Norm.Sup. di Pisa (4) 2 (1975),151-218

[Ba]$_1$ E.Balslev: Analytic scattering theory for 2-body Schrödinger operator
J.Funct.Anal. 29,3 (1978), 375-396

[Ba]$_2$ E.Balslev: The singular spectrum of elliptic operators in $L^p(\mathbb{R}^n)$
Math.Scand.19 (1966), p.193-210

[Ba-He] E.Balslev-B.Helffer: Limiting absorption principle and resonances for the Dirac operator
Aarhus Universiteit Preprint, January 1990

[Ba-Sk] E.Balslev-E.Skibsted: Resonance theory of two-body Schrödinger operators
Ann.Inst.H.Poincare.Vol.51,2,1989,p.129-154

[Bj-Dr] J.D.Bjorken-S.D.Drell: Relativistic quantum mechanics
MacGraw-Hill book company 1964

[En-Th] V.Enss-B.Thaller: Asymptotic observables and Coulomb scattering for the Dirac operator
Ann.Inst.H.Poincaré Phys.Théor.45 (1986)

[G-N-P] D.R.Grigoire, G.Nenciu and R.Purice: On the nonrelativistic limit of the Dirac hamiltonian
Ann.Inst.Poincaré, section Physique théorique
Vol.51,no.3, 1989, p.231-263

[Ha] G.Hachem: Théorie spectrale le l'opérateur de Dirac dans un champ électromagnétique a croissance linéaire a l'infini
Thèse d'état à l'université Paris-Nord 1988

[He-Sj] B.Helffer-J.Sjöstrand: Résonances en limite semi-classique
Mémoire no.24/25 de la SMF (1986), Supplément du bulletin de la SMF Tome 114, Fasc.3

[Hu] W.Hunziker: [1] On the nonrelativistic limit of the Dirac theory
Commun.Math.Phys.vol.40,1975,pp 215-222
[2] Unpublished report mentioned in [1]

[Ku] S.Kuroda: An introduction to Scattering theory
Lecture notes 51 (1980), Matematisk Institut, Aarhus

[Pa] B.Parisse: Résonances paires pour l'opérateur de Dirac
C.R.Acad.Sci.Paris, t.310, Série I,p.265-268,1990
Preprint de l'école Normale supérieure 1990

[Se] P.Šeba: The complex scaling method for Dirac resonances
Lett.Math.Phys. 16(1988),51-59

[Ve] K.Veselic: Perturbation of pseudoresolvents and analyticity in 1/c in relativistic quantum mechanics Comm.in Math.Phys.vol.22,1971,pp 27-43

[We] R.A.Weder: Spectral properties of the Dirac Hamiltonian Ann.Soc.Sci.Bruxelles ser.1 87(1973), 341-355

[Ya] K.Yamada: Eigenfunction expansions and scattering theory for Dirac operators Pub.Res.Inst.Math.Sci.Kyoto Univ.11 (1976),651-689

Schrödinger Equation : Methods of Solution

SOLUTION OF OPERATOR EQUATIONS OF MOTION

Carl M. Bender
Department of Physics
Washington University
St. Louis, Missouri 63130

ABSTRACT

Given a quantum-mechanical Hamiltonian H(p,q) having one degree of freedom, we can construct the Heisenberg equations of motion $\dot{q} = -i[q,H]$, $\dot{p} = -i[p,H]$. It is a commonly held belief that such operator differential equations are intractable. However, a technique is presented in this talk that allows one to obtain exact, closed-form solutions for huge classes of Hamiltonians. This technique, which is a generalization of the classical action-angle variable methods, allows one to solve, albeit formally and implicitly, the operator-differential equations of two anharmonic oscillators whose Hamiltonians are $H = p^2/2 + q^4/4$ and $H = p^4/4 + q^4/4$.

INTRODUCTION

The time-evolution equations

$$\dot{q} = \frac{\partial H}{\partial p} \quad , \quad \dot{p} = -\frac{\partial H}{\partial q} \quad , \tag{1}$$

describe a classical dynamical system that evolves according to the Hamiltonian H(p,q). We assume that the solution to (1) satisfies the initial conditions $p(0) = p_0$, $q(0) = q_0$. Although (1) is equivalent to a single second-order differential equation, it is often possible to find a closed-form solution because the Hamiltonian H is a constant of the motion. Thus, we can in principle use the algebraic equation H(p,q) = E to solve for and eliminate one of the variables p(t) or q(t) in (1) and then to solve the resulting *first-order* differential equation for q(t) or p(t). For example, let us consider the Hamiltonian $H = p^2/2 + V(q)$, for which Eq. (1) takes the form

$$\dot{q} = p \quad , \quad \dot{p} = -V'(q) \quad . \tag{2}$$

Now, solving H(p,q) = E for p(t) and using the first equation of (2) gives $\dot{q} = \sqrt{2[E - V(q)]}$. This is a first-order separable equation whose implicit solution satisfying $q(0) = q_0$ is

$$F[q(t)] = t + F(q_0) \quad , \tag{3}$$

where

$$F(q) = \int^{q} \frac{dx}{\sqrt{2[E - V(x)]}} \quad . \tag{4}$$

The quantum equations of motion for the Hamiltonian H(p,q) are

$$\dot{q} = -i[q,H] \quad , \quad \dot{p} = -i[p,H] \quad , \tag{5}$$

where the operators q(t) and p(t) satisfy the equal-time commutation relation $[q(t),p(t)] = i$. It is the noncommutivity of q(t) and p(t) that makes (5) difficult to solve if it is a nonlinear system. Only the special case of the harmonic oscillator $H = p^2/2 + q^2/2$ gives rise to the easily solvable linear equations $\dot{q} = p$, $\dot{p} = -q$, whose exact solution is

$$q(t) = q_0 \cos t + p_0 \sin t \quad , \tag{6}$$

$$p(t) = p_0 \cos t - q_0 \sin t \quad . \tag{7}$$

We propose to solve operator equations that are nonlinear by constructing a quantum analogy to Eq. (3). Specifically, we will attempt to obtain a function F(p,q) of the operators p(t) and q(t) that satisfies

$$-i[F(p,q),H(p,q)] = \frac{d}{dt} F(p,q) = 1 \quad . \tag{8}$$

Note that F is not unique; $\bar{F} = F + \phi(H)$, where ϕ is an arbitrary function of the Hamiltonian, also satisfies (8). However, later on we will show how to find the simplest solution for F, which we will call the *minimal* solution. If such a function F can be found then of course the solution to (8),

$$F\big(p(t),q(t)\big) = t + F(p_0,q_0) \quad , \tag{9}$$

together with

$$H\big(p(t),q(t)\big) = H(p_0,q_0) \quad , \tag{10}$$

constitutes an exact *implicit* solution to the operator equations of motion (5). If we can then solve (9) and (10) simultaneously for p(t) and q(t) as functions of p_0, q_0, and t, we have an *explicit* solution to the equations of motion. Here are some simple examples:

Example 1: $H = (pe^q + e^q p)/2$. The operator equation of motion for q, $\dot{q} = e^q$, immediately suggests that F(p,q) in (8) is actually a function of q only: $F(p,q) = -e^{-q(t)}$. The *explicit* solution to the operator equations of motion is

$$q(t) = -\log(e^{-q_0} - t) \,, \quad p(t) = p_0 - t(p_0\, e^{q_0} + e^{q_0}\, p_0)/2 \quad .$$

Example 2: $H = pe^{aq}p$, where a is a constant. For this Hamiltonian, a function F satisfying (5) is

$$F = -\frac{1}{a} \frac{1}{\sqrt{H}}\, p(t)\, \frac{1}{\sqrt{H}} \quad .$$

The *explicit* solution to the operator equations of motion is

$$p(t) = -aH(p_0,q_0)t + p_0, \quad q(t) = -\frac{1}{a} \ln[e^{-aq_0} - 2p_0 t + a^2 H(p_0,q_0)t^2] \quad .$$

Example 3: Euler Hamiltonians. We define an Euler Hamiltonian to be one in which the operator p(t) is always accompanied by a multiple of q(t); that is, H = H(pq). In general, the operator equations of motion for Euler Hamiltonians can always be solved explicitly and in closed form.[1] This is true because the operator equations of motion for any Euler Hamiltonian can be written in the form

$$\dot{q} = q\sum_n \alpha_n (pq)^n \quad , \quad \dot{p} = -p\sum_n \alpha_n (qp)^n \quad .$$

The solution to these operator equations is

$$q = q_0 e^{\sum_n \alpha_n (p_0 q_0)^n t} \quad ,$$
$$p = p_0 e^{\sum_n \alpha_n (q_0 p_0)^n t} \quad .$$

For example, the operator equations for the Hamiltonian $H = apq^2p + bqp^2q + c(pqpq + qpqp) + d(q^2p^2 + p^2q^2)$ have the form

$$\dot{q} = \alpha qpq \ , \quad \dot{p} = -\alpha pqp \quad ,$$

where $\alpha = 2a + 2b + 4c + 4d$. The solution to these operator differential equations is

$$q(t) = q_0 e^{\alpha p_0 q_0 t} \ , \quad p(t) = p_0 e^{-\alpha q_0 p_0 t} \quad .$$

A particularly interesting example of an Euler Hamiltonian is

$$H = e^{c(pq + qp)} \ , \quad (|c|<\pi/2) \quad .$$

The solution to the operator differential equations for this Hamiltonian is

$$q(t) = q_0 e^{2t(\sin c)e^{2p_0q_0c}} \quad ,$$
$$p(t) = p_0 e^{-2t(\sin c)e^{2q_0p_0c}} \quad .$$

Example 4: $H = p^\alpha q^{2\beta} p^\alpha$. For this Hamiltonian it is easy to see that a function F satisfying (5) is

$$F(p,q) = \frac{1}{4\alpha - 4\beta} \frac{1}{\sqrt{H}} (pq + qp) \frac{1}{\sqrt{H}} \quad . \tag{11}$$

Note that (11) ceases to exist when $\alpha = \beta$. However, the special case $\alpha = \beta$ gives an *Euler* Hamiltonian whose solution is discussed in Example 3 above. Solving $F(p,q) = t + F(p_0,q_0)$ simultaneously with $H(p,q) = H(p_0,q_0)$ can be complicated. However, a relatively simple case arises when $\alpha = 1$ and $\beta = N/2$. Now the explicit solution is

$$q(t) = \{[p_0q_0 + (2-N)p_0q_0^N p_0 t]^{-1} p_0q_0^N p_0[q_0p_0 + (2-N)p_0q_0^N p_0 t]^{-1}\}^{1/(N-2)} \quad ,$$

$$p(t) = \{[q(t)]^{-N/2} p_0q_0^N p_0[q(t)]^{-N/2} - \frac{1}{4} N(N-2)[q(t)]^{-2}\}^{1/2} \quad .$$

Example 5: $H = \alpha p^\gamma + \beta q^{-\gamma}$. Here, the form of F(p,q) is similar to that in Example 4:

$$F(p,q) = \frac{1}{2\gamma}\,\frac{1}{\sqrt{H}}\,(pq + qp)\,\frac{1}{\sqrt{H}} \quad .$$

OPERATOR BASIS ELEMENTS AND THEIR ALGEBRA

Our objective is now to describe a general procedure for obtaining an operator F(p,q) that satisfies (8). To do so we introduce an operator basis. The basis elements $T_{m,n}$ are defined as the sum of all possible terms containing m factors of p and n factors of q multiplied by m!n!/(m+n)!. $T_{m,n}$ is thus a totally symmetric Hermitian object containing (m+n)!/(m!n!) individual terms. For example,

$$T_{0,0} = 1 \quad ,$$
$$T_{0,3} = q^3 \quad ,$$
$$T_{1,1} = (pq + qp)/2 \quad ,$$
$$T_{2,1} = (p^2q + pqp + qp^2)/3 \quad ,$$
$$T_{2,2} = (p^2q^2 + q^2p^2 + pqpq + qpqp + pq^2p + qp^2q)/6 \quad .$$

For two reasons, this seems to be a natural basis with which to express operators. First, $T_{m,n}$ contains positive powers of p and q, so it should be useful for constructing Taylor–like expansions of operators. Note that we can expand operators, regardless of whether they are symmetric. For example,

$$p^2q^3 = T_{2,3} - 3iT_{1,2} - 3T_{0,1}/2 \quad .$$

Second, $T_{m,n}$ satisfies an extremely useful set of commutation and anticommutation relations. Commuting with p or q has the effect of a lowering operator:

$$[T_{m,n},p] = in\,T_{m,n-1} \quad ,$$
$$[T_{m,n},q] = -im\,T_{m-1,n} \quad . \tag{12}$$

Anticommuting with p or q is analogous to applying a raising operator:

$$[T_{m,n},p]_+ = 2T_{m+1,n} \quad ,$$
$$[T_{m,n},q]_+ = 2T_{m,n+1} \quad . \tag{13}$$

The operator basis elements $T_{m,n}$ form a closed algebra under commutation and anticommutation; that is, the commutator or anticommutator of any two basis elements can be expressed as a sum of operator basis elements.

The basis elements have the crucial property that the totally symmetric operator $T_{m,n}$ can be completely reorganized using the commutation relation $[q,p] = i$ and recast in Weyl-ordered form

$$T_{m,n} = \frac{1}{2^m}\sum_{j=0}^{\infty}\binom{m}{j}\,p^jq^np^{m-j} = \frac{1}{2^n}\sum_{k=0}^{\infty}\binom{n}{k}\,q^kp^mq^{n-k} \quad . \tag{14}$$

The operators $T_{m,n}$ have many more remarkable properties that have allowed them to play a central role in previous studies involving finite-element lattice approximations,[2] operator ordering,[3] and Hahn polynomials.[4] To summarize briefly, the operator $T_{m,m}$ can be expressed as a polynomial P of degree m in terms of the operator $T_{1,1}$:

$$T_{m,m} = P_m(T_{1,1}) \quad , \tag{15}$$

where

$$P_0(x) = 1 \quad ,$$
$$P_1(x) = x \quad ,$$
$$P_2(x) = x^2 - 1 \quad ,$$
$$P_3(x) = x^3 - 5x \quad ,$$
$$P_4(x) = x^4 - 14x^2 + 9 \quad ,$$
$$P_5(x) = x^5 - 30x^3 + 89x \quad ,$$
$$P_6(x) = x^6 - 55x^4 + 439x^2 - 225 \quad .$$

Equation (15) is reminiscent of the defining equation for Chebyshev polynomials $C_n(x)$: $\cos(n\theta) = C_n(\cos\theta)$.

The Hahn polynomials exhibit the following properties: First, they satisfy a three-term recursion relation:

$$P_n(x) = xP_{n-1}(x) - (n-1)^2P_{n-2}(x) \quad .$$

Second, $P_n(x)$ satisfies a functional difference equation (rather than a differential equation):

$$(1-ix)P_n(x+2i) + (1+ix)P_n(x-2i) = (4n+2)P_n(x) \quad .$$

Third, the Hahn polynomials have a simple generating function G(x):

$$\frac{e^{x \tan^{-1} t}}{\sqrt{1+t^2}} = \sum_{n=0}^{\infty} \frac{t^n}{n!} P_n(x) \quad .$$

Fourth, the Hahn polynomials are orthogonal on the line $-\infty<x<\infty$ with respect to the weight function $w(x) = \mathrm{sech}(\pi x/2)/2$:

$$\int_{-\infty}^{\infty} dx\; w(x)P_m(x)P_n(x) = \delta_{m,n}(n!)^2 \quad .$$

It is easy to expand functions in terms of Hahn polynomials. For example,

$$e^{cx} = \sum_{n=0}^{\infty} \frac{(\tan c)^n}{n! \cos c} P_n(x) \qquad (|c|<\pi/2) \quad .$$

Finally, the Hahn polynomials can be expressed in terms of a generalized hypergeometric function:

$$P_n(x) = \sqrt{n!}\,i^n\ {}_3F_2(-n,n+1,(1-ix)/4;1/2,1;1) \quad .$$

MINIMAL SOLUTION OF THE HARMONIC OSCILLATOR

Now we return to the problem of obtaining a solution to (8). We represent F(p,q) as an arbitrary sum of operator basis elements

$$F(p,q) = \sum_{m,n} \alpha_{m,n}\, T_{m,n} \quad , \tag{16}$$

where $\alpha_{m,n}$ are constants to be determined from the requirement in (8) that $-i[F(p,q),H(p,q)] = 1$. To illustrate, we begin by finding F(p,q) for the harmonic oscillator Hamiltonian $H(p,q) = p^2/2 + q^2/2$. Equations (12) and (13) make the computation very easy:

$$\frac{1}{i}[T_{m,n},\frac{1}{2}p^2] = \frac{1}{2i}\big[p,[T_{m,n},p]\big]_+ = \frac{n}{2}[p,T_{m,n-1}]_+ = n\, T_{m+1,n-1} \quad . \tag{17}$$

Similarly,

$$\frac{1}{i}[T_{m,n},\frac{1}{2}q^2] = -m\, T_{m-1,n+1} \quad . \tag{18}$$

Combining (16)-(18), we see that the commutation relation in (8) takes the form

$$\sum_{m,n} \alpha_{m,n}(nT_{m+1,n-1} - mT_{m-1,n+1}) = T_{0,0} \quad . \tag{19}$$

Hence, assuming completeness, we determine that the coefficients $\alpha_{m,n}$ satisfy the linear partial difference equation

$$(n+1)\alpha_{m-1,n+1} - (m+1)\alpha_{m+1,n-1} = \delta_{m,0}\,\delta_{n,0} \quad . \tag{20}$$

This partial difference relates pairs of coefficients $\alpha_{m,n}$. If we represent the coefficients as dots on an integer planar lattice then it is clear that next-to-nearest neighboring points on a diagonal whose slope is −1 are related by (20) (see Fig. 1). As we pointed out earlier, F is *not* uniquely determined. Thus, we are free to take the simplest particular solution to (20); namely, the one having the smallest possible number of nonvanishing coefficients $\alpha_{m,n}$. We call this the minimal solution. The minimal solution to (20) is given by

$$\alpha_{-2m-1,2m+1} = (-1)^m/(2m+1) \quad , \quad m = 0,1,2,3,\ldots \quad ,$$

and $\alpha_{m,n} = 0$ for other values of m,n. Thus, the formula for F(p,q) in (16) becomes

$$F(p,q) = \sum_{m=0}^{\infty} \frac{(-1)^m}{2m+1}\, T_{-2m-1,2m+1} \quad . \tag{21}$$

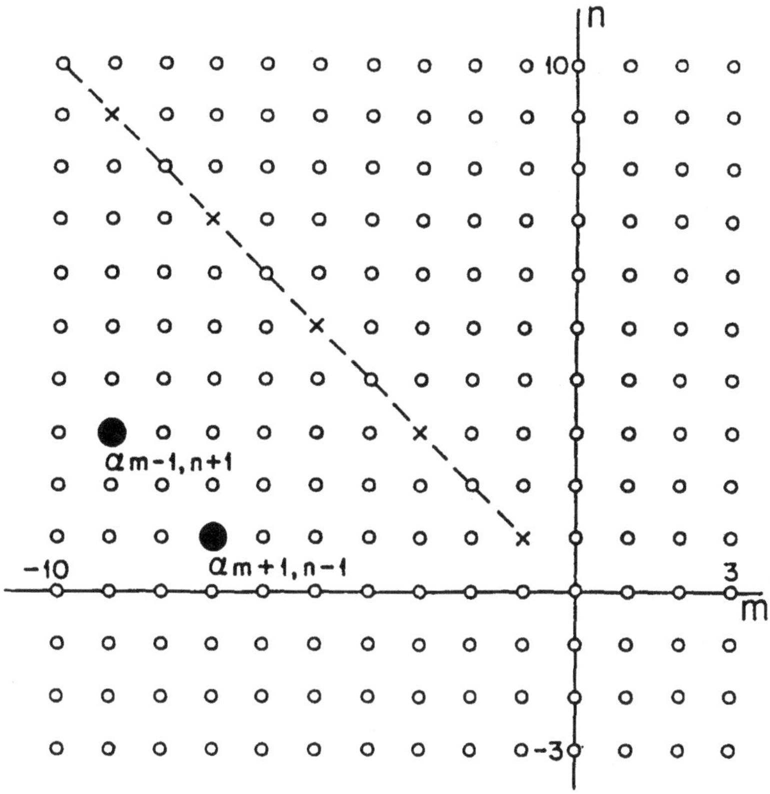

Figure 1: The partial difference equation (20) for the harmonic oscillator relates the coefficients $\alpha_{m-1,n+1}$ and $\alpha_{m+1,n-1}$. The minimal solution consists of all $\alpha_{m,n} = 0$ except $\alpha_{-2m-1,2m+1}$, m = 0,1,2,..., as indicated by crosses.

Observe that we are forced to generalize our initial assumption that, by analogy with Taylor series, our basis $T_{m,n}$ has $m,n \geq 0$. Apparently, a more accurate analogy is with Laurent series in which powers may be positive or negative. Fortunately, the formulas in (14) allow us to define $T_{m,n}$ when $m \geq 0$ and $n < 0$ and when $n \geq 0$ and $m < 0$. Moreover, the commutation and anticommutation relations in (12) and (13) continue to hold in this extended and singular basis.

In order to understand the formal series in (21) we return to the exact solution to the harmonic oscillator in (6) and (7). We divide (6) by (7) and let $Z(t) = q(t)[p(t)]^{-1}$:

$$\begin{aligned} Z(t) &= (q_0 \cos t + p_0 \sin t)(p_0 \cos t - q_0 \sin t)^{-1} \\ &= (q_0 \cos t + p_0 \sin t)p_0^{-1}\, p_0(p_0 \cos t - q_0 \sin t)^{-1} \\ &= (Z_0 \cos t + \sin t)[(p_0 \cos t - q_0 \sin t)p_0^{-1}]^{-1} \\ &= \frac{Z_0 + \tan t}{1 - Z_0 \tan t} \quad . \end{aligned} \tag{22}$$

Since Z(t) is a function of Z_0, Z(t) must commute with Z_0. Thus, we can treat (21) as a c–number algebraic equation and solve for tan t: $\tan t = [Z(t)-Z_0]/[1+Z(t)Z_0]$. Finally, taking the inverse tangent of this equation, we obtain

$$\arctan Z(t) = t + \arctan Z_0 \quad . \tag{23}$$

This equation is an instance of (9), so we identify

$$F(p,q) = \arctan\{q(t)[p(t)]^{-1}\} \quad . \tag{24}$$

Compare (21) and (24). Note that the coefficients α in (21) correspond exactly with the Taylor expansion of $\arctan x = \sum_{n=0}^{\infty} (-1)^n(2n+1)^{-1}\, x^{2n+1}$. However, these two functions F are not the same even though they satisfy the same commutation relation in (8). As expected, they differ by a function of the Hamiltonian H:

$$F_{Eq.(21)} - F_{Eq.(24)} = \frac{i}{2}\sum_{n=0}^{\infty} E_{2n}H^{-2n-1} = \frac{i}{2H}\int_0^{\infty} ds \, \frac{e^{-s}}{\cosh(\frac{s}{2H})} \, ,$$

where E_{2n} are Euler numbers. (The first few Euler numbers are $E_0 = 1$, $E_2 = -1$, $E_4 = 5$, $E_6 = -61$, $E_8 = 1385$, and so on.)

Even more important, the physical meaning of F(p,q) is now more evident: Classically, the harmonic oscillator describes an orbit in phase space which is a circle of radius $p_0^2 + q_0^2$. The singular quantity arc tan[q(t)/p(t)] is the *angle* θ of a point on this circle. Apparently, (21) is the generalization of the angle coordinate from a classical (c–number) theory to a quantum (operator) theory. This result clearly illustrates that the operator methods used here generalize the notion of classical action-angle variables to the realm of quantum mechanics.

MINIMAL SOLUTION FOR THE ANHARMONIC OSCILLATOR $H = p^2/2+q^4/4$

Next, we consider the anharmonic oscillator whose Hamiltonian is given by $H = p^2/2 + q^4/4$. To find F(p,q) we first evaluate the commutator

$$
\begin{aligned}
-i[T_{m,n},q^4] &= -\, m(q^3T_{m-1,n} + q^2T_{m-1,n}q + qT_{m-1,n}q^2 + T_{m-1,n}q^3) \\
&= -2m(q^2T_{m-1,n+1} + T_{m-1,n+1}q^2) \\
&= -m[q,[q,T_{m-1,n+1}]_+]_+ - m[q,[q,T_{m-1,n+1}]]_+ \\
&= -m(2[q,T_{m-1,n+2}]_+ + (m-1)i[q,T_{m-2,n+1}]) \\
&= -4mT_{m-1,n+3} + m(m-1)(m-2)T_{m-3,n+1} \quad .
\end{aligned}
\tag{25}
$$

Combining the result in (17) with that in (25), and assuming that the operators $T_{m,n}$ form a complete basis gives a recursion relation for $\alpha_{m,n}$ in (16):

$$
(n+1)\alpha_{m-1,n+1} - (m+1)\alpha_{m+1,n-3} + \frac{(m+3)(m+2)(m+1)}{4}\alpha_{m+3,n-1} = \delta_{m,0}\delta_{n,0} \,. \tag{26}
$$

This partial difference equation relates triples of points on the integer lattice in Fig. 2 whose points correspond to coefficients $\alpha_{m,n}$. A careful analysis of (26) shows that a minimal solution exists consisting of certain nonzero values of $\alpha_{m,n}$. These nonzero values of $\alpha_{m,n}$ correspond to points in Fig. 2 lying in the quadrant $m<0$, $n>0$ and forming a triangular network as indicated in the figure.

A first step in solving (26) involves mapping the triangular network of nonvanishing values of $\alpha_{m,n}$ onto a triangular domain. On this domain the partial difference equation (26) works in much the same fashion as the difference equation that generates the binomial coefficients in Pascal's triangle. The relevant transformation of the independent variables is

$$
M = -(n + 2m)/6 \,, \quad N=(n - m)/6 \quad .
$$

We define a new independent variable $A_{M,N}$ by

$$
A_{M,N} = \alpha_{m-1,n+1} \quad ,
$$

along with the constraint that $A_{M,N} = 0$ for $M < 0$, $N < 0$, and for $M > N$.

The advantage of the M,N variables over the m,n variables is that the partial difference equation is first order:

$$
\begin{aligned}
&(4N - 2M + 1)A_{M,N} + (2N + 2M - 1)A_{M,N-1} \\
&-\frac{1}{4}(2N + 2M - 1)(2N + 2M - 2)(2N + 2M - 3)A_{M-1,N-1} = \delta_{M,0}\delta_{N,0} \,.
\end{aligned}
\tag{27}
$$

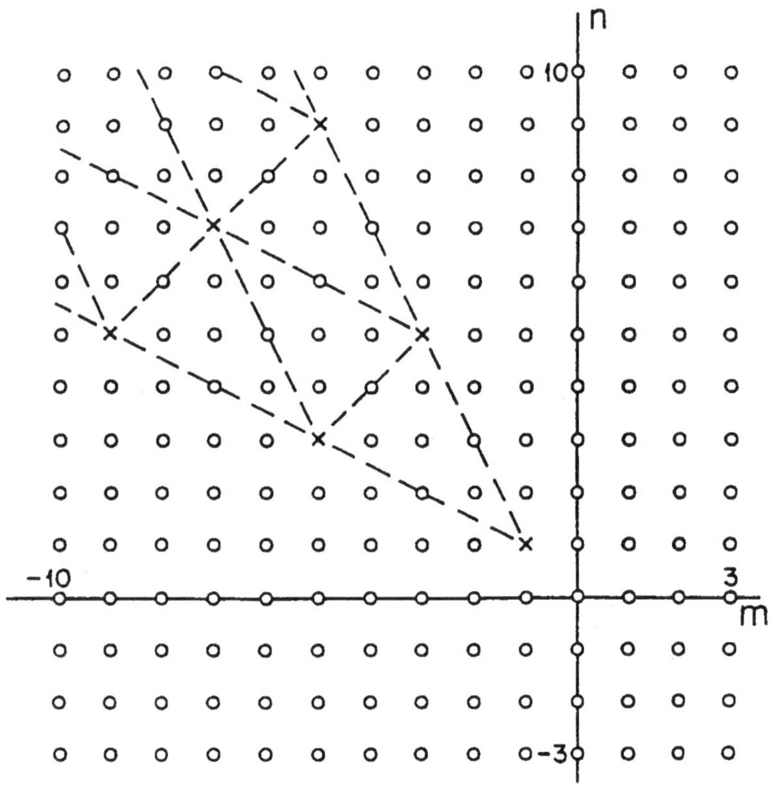

Figure 2: Triplets of points related by the partial difference equation (26) for the anharmonic oscillator with $H = \frac{1}{2} p^2 + \frac{1}{4} q^4$. The minimal solution consists of nonvanishing values of $\alpha_{m,n}$ indicated by crosses; all other $\alpha_{m,n}$'s vanish.

A further transformation, this time of the dependent variable, reduces the partial difference equation in (27) to one whose coefficients are linear functions of M and N. We define $B_{M,N} = 2^{-N}\Gamma(1/2)A_{M,N}/\Gamma(M + N + 1/2)$ and the difference equation becomes

$$(4N - 2M + 1)B_{M,N} + B_{M,N-1} - (N + M - 1) = \delta_{M,0}\delta_{N,0} \ . \tag{28}$$

The simplest way to express a general element $B_{M,N}$ is in terms of a generating function; to wit, we define the generating function

$$g(x,y) = \sum_{N=0}^{\infty}\sum_{M=0}^{N} B_{M,N}x^M y^N \ .$$

From (28) we can derive a linear first order partial differential equation satisfied by g(x,y):

$$-x(2 + xy)g_x + y(4 - xy)g_y + (1 + y - xy)g = 1 \ . \tag{29}$$

Using the method of characteristics we cna find the unique solution to this partial differential equation that satisfies the initial condition

$$g(0,y) = \sum_{N=0}^{\infty}(-y/4)^N\frac{\Gamma(5/4)}{\Gamma(N + 5/4)} = e^{-y/4}\int_0^1 du\ e^{u^4 y/4} \ .$$

The solution is

$$g(x,y) = \int_0^1 \frac{du}{2\sqrt{(u)}(1 - xyu)}\exp\left(\frac{2}{3x^2y} - \frac{1}{x} + \frac{(1 - xy)^{3/2}(xyu - 2/3)}{x^2y(1 - xyu)^{3/2}}\right) \ . \tag{30}$$

In terms of the generating function g(x,y),

$$B_{M,N} = \frac{1}{M!N!}\left(\frac{\partial}{\partial x}\right)^M\left(\frac{\partial}{\partial y}\right)^N g(x,y)\bigg|_{x=y=0} \ . \tag{31}$$

Using this expression for the coefficients $B_{M,N}$ in the formula (16) for F(p,q) gives a complete and exact minimal solution to the Heisenberg operator differential equations for the anharmonic oscillator with the Hamiltonian $H = p^2/2 + q^4/4$. The operator F is given by

$$F(p,q) = \sum_{N=0}^{\infty}\sum_{M=0}^{N} 2^N\frac{\Gamma(M + N + 1/2)}{\Gamma(1/2)}B_{M,N}T_{-2N-2M-1,4N-2M+1} \ . \tag{32}$$

MINIMAL SOLUTION FOR THE ANHARMONIC OSCILLATOR $H = p^4/4+q^4/4$

Next, we consider the anharmonic oscillator whose Hamiltonian is given by $H = p^4/4 + q^4/4$. This is the natural next step in our program of solving increasingly complex systems of quantum operator differential equations: $H = p^2/2 = q^2/2$ gives rise to a two-term partial difference equation, $H = p^2/2 = q^4/4$ gives rise to a three-term partial difference equation, and, as we will see, $H = p^4/4 = q^4/4$ gives rise to a four-term partial difference equation. Specifically, the partial difference equation for the coefficients $\alpha_{m,n}$ in (16) has the form

$$(n+1)\alpha_{m-3,n+1} + \frac{1}{4}(m+3)(m+2)(m+1)\alpha_{m+3,n-1} - (m+1)\alpha_{m+1,n-3} - \frac{1}{4}(n+3)(n+2)(n+1)\alpha_{m-1,n+3} = \delta_{m,0}\delta_{n,0}\ . \qquad \text{I}(33)$$

This partial difference equation relates quartets of points, which lie at vertices of rectangles, on the integer lattice in Fig. 3 whose points correspond to coefficients $\alpha_{m,n}$. Examining (33) we see that a minimal solution exists with nonzero values of $\alpha_{m,n}$ lying in the quadrant $m<0$, $n>0$, and forming a wedge-shaped network as indicated in Fig. 3.

To solve (33) we use the transformation

$$N = (n - m)/8\ , \quad M = -(n + m)/4\ ,$$

which maps points in the m,n plane for which the coefficient $\alpha_{m,n}$ is nonzero into a triangular lattice for which N runs from 0 to infinity and M runs from 0 to 2N. We then define a new dependent variable $C_{M,N}$ by

$$C_{M,N} = \alpha_{m-3,n+1}\ ,$$

along with the constraint that $C_{M,N} = 0$ for $M < 0$, $N < 0$, and $M > 2N$. $C_{M,N}$ satisfies a first order partial difference equation which can be simplified further by the transformation

$$D_{M,N} = 4^{-M}\frac{\Gamma(N - M/2 + 3/4)\Gamma(N - M/2 + 5/4)}{\Gamma(N + M/2 + 3/4)\Gamma(N + M/2 + 5/4)}C_{M,N}\ .$$

The equation satisfied by $D_{M,N}$ has linear coefficients:

$$(N + \frac{M}{2} + \frac{1}{4})D_{M,N} - (N - \frac{M}{2} + \frac{1}{2})D_{M-1,N} + (N - \frac{M}{2} - \frac{1}{4})D_{M,N-1} - (N + \frac{M}{2} - \frac{1}{2})D_{M-1,N-1} = \frac{1}{4}\delta_{M,0}\delta_{N,0}\ . \qquad (34)$$

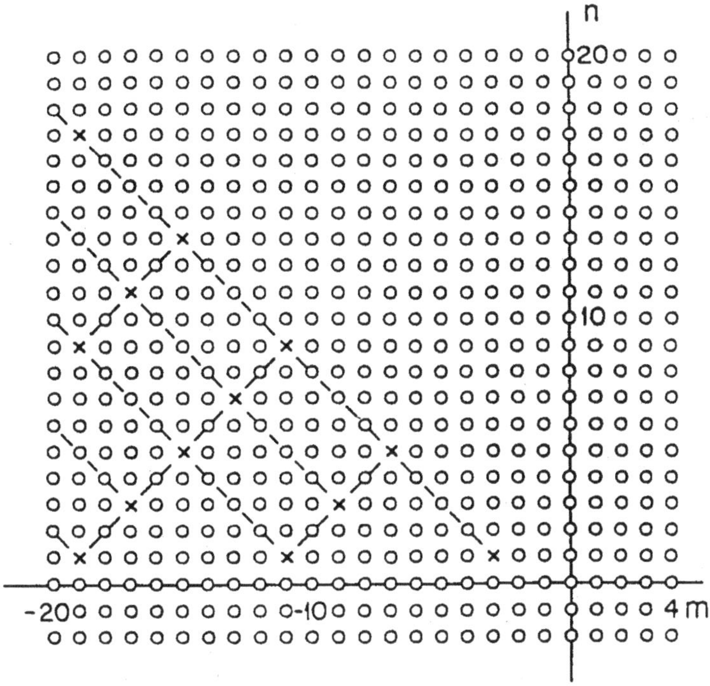

Figure 3: Quartets of points related by the partial difference equation (33) for the anharmonic oscillator with $H = \frac{1}{4} p^4 + \frac{1}{4} q^4$. The minimal solution consists of nonvanishing values of $\alpha_{m,n}$ indicated by crosses; all other $\alpha_{m,n}$'s vanish.

A general element $D_{M,N}$ in this array can be expressed in terms of a generating function h(x,y), where

$$h(x,y) = \sum_{N=0}^{\infty} \sum_{M=0}^{2N} D_{M,N} x^M y^N \quad .$$

From (34) we can derive a first order linear partial differential equation satisfied by h(x,y):

$$\frac{x}{2} \frac{1+x}{1-x} h_x + y \frac{1+y}{1-y} h_y + \frac{1 + 3y - 4xy}{4(1-x)(1-y)} h = \frac{1}{4(1-x)(1-y)} \quad . \tag{35}$$

The unique solution to this partial differential equation that satisfies the initial condition

$$h(0,y) = \sum_{N=0}^{\infty} (-y)^N \frac{\Gamma(5/4)\Gamma(N+3/4)}{\Gamma(3/4)\Gamma(N+5/4)} = \int_0^1 \frac{du}{\sqrt{(1+y)^2 - 4yu^4}}$$

is

$$h(x,y) = \int_0^1 \frac{du}{\sqrt{(1 + y)^2 - 4yu^4(1 + x - xu^2)^2}} . \tag{36}$$

In terms of this generating function,

$$D_{M,N} = \frac{1}{M!N!}\left(\frac{\partial}{\partial x}\right)^M\left(\frac{\partial}{\partial y}\right)^N h(x,y)\Big|_{x=y=0} .$$

Using this expression for the coefficients $D_{M,N}$ in the formula (16) for F(p,q) gives a complete and exact minimal solution to the Heisenberg operator differential equations for the anharmonic oscillator whose Hamiltonian is $H = p^4/4 + q^4/4$. The operator F(p,q) satisfying (8) is

$$F(p,q) = \sum_{N=0}^{\infty}\sum_{M=0}^{2N} 4^M \frac{\Gamma(N+M/2+3/4)\Gamma(N+M/2+5/4)}{\Gamma(N-M/2+3/4)\Gamma(N-M/2+5/4)} D_{M,N} T_{-2N-4M-3,4N-2M+1}. \tag{37}$$

CONCLUDING REMARKS

The progress we have reported here is formal because the operator F(p,q) can be extremely singular as it contains in many cases arbitrary powers of 1/p(t) and/or 1/q(t). If F(p,q) did exist as an operator in a Hilbert space, then by virtue of (8), $U \equiv \exp[i\lambda F(p,q)]$ would be an energy raising operator. That is, if $H|E\rangle = E|E\rangle$, then $H(U|E\rangle) = (E + \lambda)(U|E\rangle)$. Clearly, for Hamiltonians with discrete spectra, U must map states out of the Hilbert space except for special discrete values of λ. An obvious direction for future research is to find a way to identify these special values of λ and thereby to use the operator solution to a quantum theory to compute such quantities as eigenvalues and also unequal-time commutation relations. One should also consider applying the ideas presented here to systems having more than one degree of freedom and especially to models that exhibit quantum chaos. It was our sole intention here to point out that operator differential equations are not inaccessible to exact analytic methods and that in many cases they can actually be solved in closed form.

REFERENCES

1. C. M. Bender and G. V. Dunne, Phys. Letters **B200**, 520 (1988).
2. C. M. Bender, L. R. Mead, and S. S. Pinsky, Phys. Rev. Lett. **56**, 2445 (1986).
3. C. M. Bender and G. V. Dunne, J. Math. Phys. **29**, 1727 (1988).
4. C. M. Bender, L. R. Mead, and S. S. Pinsky, J. Math. Phys. **28**, 509 (1987). For a review of the Hahn polynomials see W. Hahn, Math. Nachr. **2**, 4 (1949), N. M. Atakishiyev and S. K. Suslov, J. Phys. A **18**, 1583 (1985), and R. Askey, J. Phys. A **18**, L1017 (1985).

HILL DETERMINANTS - A REVIEW

MILOSLAV ZNOJIL

Institute of Nuclear Physics
Czechoslovak Academy of Sciences
CS - 250 68 Řež, Czechoslovakia

Abstract

Bound states in the not exactly solvable simple potentials may be constructed by means of the various versions of the so called method of Hill determinants. We review this set of techniques, with an emphasis on the rigorous proofs that the physical boundary conditions are satisfied.

1 Introduction

The role of exactly solvable models in quantum theory is obvious: They minimize computational troubles and maximize our intuitive insight into the basic phenomena. In this sense, the situation in the one-body quantum mechanics is particularly fortunate: The corresponding differential Schroedinger equation

$$-\Delta\psi + V\psi = E\psi \tag{1}$$

happens to be exactly solvable for $V(r) = gr^2$ (i.e., from an "applied" point of view, for each potential well near its minimum), for $V(r) = e/r$ (i.e., just for the Coulomb force which governs interactions in atoms and molecules) and, moreover[1], for the interaction $V(r) = f/r^4$.

The "next-to-solvable" systems may be treated perturbatively. Here, an alternative aggregate of the non-perturbative approaches which carry a common name of "Hill-determinant" (HD) methods will be reviewed. Their importance may be seen not only in an extension of applications of exact models beyond their limitations, but also in a possibility of analysis of any method by providing an alternative representation of the corresponding results. In this sense, the attention of HD methods ranges from the tests of realistic computations and complicated or divergent perturbative analyses up to some purely mathematical and methodical questions.

A typical domain of applications of the various versions of the general HD approach may be characterized as lying somewhere in between the exact and purely numerical or variational methods. The merits as well as shortcomings of both these extremes are inherited by the HD techniques. In their description, an emphasis will be laid upon a necessary improvement of the mathematical foundations of some of the reviewed HD prescriptions which are often — undeservedly — being recommended via mere purely intuitive arguments.

2 HILL DETERMINANTS IN THE FLOQUET THEORY

The so called Hill differential equation[2] appeared in the classical celestial mechanics and, virtually, gave rise to the name of the HD group of methods. In the quantum-mechanical setting, repulsion $V(r) = f/r^4$ leads to the same type of mathematics. In accord with the detailed analyses available in the literature[1-3], scattering amplitudes may be constructed, for such a repulsive core, in a more or less non-numerical manner, i.e., in terms of Mathieu special functions. The general theory of these functions is a good illustration of what is meant by Hill determinants in this (and, in general, in the so called Floquet-theory) context.

For a short review of the "next-to-Mathieu" Floquet-Hill determinants, let us employ the following superposition of the two exactly solvable forces,

$$V(r) = ar^{-1} + br^{-4}, \qquad b > 0. \tag{2}$$

After the so called Fivel[4] change of variables $r \to r^2$ in our basic differential equation (1), this most elementary force acquires a "canonical" form

$$V(r) = ar^2 + br^{-6}, \qquad a,\ b > 0. \tag{3}$$

The reader is recommended to find more details in our recent paper[5]. Here, let us only summmarize the main aspects of the method.

2.1 *The Idea*

In a way which forms a background of all the HD constructions we intend to review, an initial idea which leads to the introduction of "Hill determinants", lies in a tentative generalization of harmonic oscillators,

$$\psi(r) = polynomial(r) \times an\ auxiliary\ exponential. \tag{4}$$

In the other words, an assumption of analyticity of ψ (implying, virtually, that ψ may be expanded in a suitable power series in the variable r) is more or less natural in the next-to-solvable systems. Hence, for interactions of the type (2) or (3), a discovery[6] of the (particular) exact solutions of the type (4),

$$\psi(r) = \psi(r, \kappa) = r^{\kappa} \sum_{n=-M}^{N} h_n r^n \tag{5}$$

inspired also a study of the related infinite-series (i.e., Laurent-series) limits of these polynomial Ansaetze with $N \to \infty$ and $M \to \infty$.

The insertion of Eq.(5) into our differential Schroedinger equation (1) leads to the recurrences for h's, and the general Floquet theory relates these recurrences to their special (and, a priori, by far not unique) truncations of the form

$$Q \times \begin{pmatrix} h_{-M} \\ \vdots \\ h_N \end{pmatrix} = 0, \qquad M \to \infty, \quad N \to \infty \tag{6}$$

with a general doubly-infinite matrix Q. Now, the name of "Hill determinant" comes naturally from the related necessary condition

$$\det Q(E) = 0 \tag{7}$$

of solvability of the system (6).

2.2 *The Basic Properties of* $\det Q$

In analogy with Mathieu functions, we may expect that physical bound states have the general form

$$\psi(r) = c_1 \psi(r, \kappa_1) + c_2 \psi(r, \kappa_2). \tag{8}$$

Thus, our sample interaction (3) (as well as all its Lennard-Jones generalizations[7]) has to provide two roots κ of the HD equation (7) at each energy, in a vicinity of the real bound state at least. Such an expectation has been confirmed by the numerical tests[5], and a full formal parallelism with the exactly solvable Mathieu-function cases (cf., e.g., ref.[3]) was confirmed.

The freedom in our choice of Ansaetze (and, in particular, the presence of the "Floquet exponent" parameter κ) is compensated by restriction (7). This equation with an origin in the truncation of recurrences for h's remains rather puzzling at the first sight. It has no variational or other background. In accord with our study[5], it may be interpreted very easily. Briefly, we have to start form an asymptotic leading-order analysis of the recurrences, and prove that the very Ansatz (5) makes sense (i.e., converges in the double limit $M, N \to \infty$) if and only if we postulate the truncation or "boundary conditions" as represented by the matrix formula (6).

In our particular example (3), the above-mentioned recurrences read

$$Q_{nn-1} h_{n-1} + Q_{nn} h_n + Q_{nn+1} h_{n+1} = 0 \tag{9}$$

and remain very short — their explicit coefficients Q may be found in ref.[5]. As a difference equation, this relation also admits an easy asymptotic analysis and the closed leading-order solution

$$h_n = \begin{cases} c_1/[(n+n_1)!] + c_2(n+n_2)!, & n \gg 1 \\ c_3(|n|+n_3)! + c_4/[(|n|+n_4)!], & n \ll -1. \end{cases} \tag{10}$$

We may immediately see the consequences — we get

$$|h_{|n+1|}/h_{|n|}| \gg 1, \qquad |n| \gg 1 \tag{11}$$

so that the series (5) will diverge whenever c_2 or c_3 remains different from zero. Vice versa, an addition of the trivial boundary conditions

$$h_{-M-1} = 0, \qquad h_{N+1} = 0 \tag{12}$$

to our difference equation (9) guarantees the disappearance of both the divergence-causing components from Eq.(10)[8].

As long as the prescription (12) is equivalent to the truncation used in Eqs.(9) and (6), we may conclude that our present HD condition (7) has precisely the same meaning as in the theory of Mathieu functions: It is simply the condition which fixes the free parameters (namely, Floquet exponents) in our generalized harmonic oscillator (5).

3 HILL DETERMINANTS AND FROBENIUS METHOD

In the light of the almost classical conjecture by Biswas *et al*[9], we may use the above-described procedure also after a limiting transition to interactions without an essential singularity in the origin. Our second next-to-solvable superposition of the solvable forces

$$V(r) = ar^{-1} + br^2, \qquad b > 0 \tag{13}$$

parallels (2) as an interaction with mere weakly singular behaviour near $r = 0$. In the general mathematical framework, our assumption (5) may be reduced to a simpler $M = 0$ Ansatz

$$\psi(r) = Taylor\ series \times an\ auxiliary\ exponential. \tag{14}$$

By Singh *et al*[10], an existence of the harmonic-oscillator-like finite-dimensional and exact predecessors (4) of such an expansion was noticed for the simple example (13) in its Fivel-equivalent form

$$V(r) = ar^2 + br^4 + cr^6, \qquad c > 0. \tag{15}$$

Magyari[11] has extended this observation to all polynomial V's.

3.1 *Hill Determinants in the WKB Singh-Magyari Interpretation*

In an analogy with the preceding Floquet HD formalism, a necessary condition for termination (14) → (4) of all the Frobenius-inspired Ansaetze for wave functions is the asymptotically correct (i.e., WKB) specification of the auxiliary exponentials. In such a case, we are only left with the single free parameter (namely, energy) and any HD condition of the type (7) may only fix these energies.

It is amusing to notice that in such a formulation, a general HD ideology is even able to reproduce the ordinary harmonic-oscillator (HO, $V(r) = ar^2$) exact solutions. It is well known that the HO recurrences merely have a two-term structure so that the determinant of Q may disappear if and only if one of its diagonal elements becomes equal to zero. The obligatory termination (4) of the HO series (14) then becomes just a consequence of the Hill-determinant formal (i.e., trivial) secular equation (7). After all, this is the way harmonic oscillator is presented in almost any textbook.

In accord with Singh *et al*[10], interaction (15) is the next one to be considered as a candidate for description by the Ansatz (14) or (5) with $M = 0$. Due to the regularity of $V(r)$ in the origin, the parameters κ have to be replaced by the quantum number l (or rather $l + 1$) of the angular momentum,

$$\psi(r) = \psi_{WKB}(r) \sum_{n=0}^{N} h_n r^{n+l+1}. \tag{16}$$

The insertion of this formula in the standard radial form of Eq.(1) gives again the recurrences of the three-term form (9). The universal HD ideology seems to be applicable without any problems[10]. In fact, it was a rather unpleasant surprise to find out that the things are not so easy and that the HD equation (7) generated by the WKB Ansatz (16) fails to work in certain domains of couplings[12–18]. This is a price we pay for the uniqueness and extreme simplicity of the WKB factor in the wave function.

3.2 *Hill Determinants in the Ginsburg-Hautot Re-Interpretation*

A deeper explanation of the restricted applicability of the WKB-HD algorithm may be found in the quoted papers. In general, in the spirit of the HD philosophy[9,19], an introduction of free parameters in the Ansatz (16) helps and extends the applicability of equation (7) to all the "next-to-solvable" (e.g., polynomial) interactions[16–18].

In fact, a posteriori, an imanently restricted character of the WKB, parameter-free HD method should not be surprising, once we fail in finding a universal (for example, variational) background for truncations of the recurrences. As a remark, we may notice that some authors (e.g., Hautot[16]) incorporate even the variationally generated equations (7) under the same HD heading. By our opinion, such a union is superfluous: The variational principles are extremely efficient numerically while, at the same time, they are more or less unable to provide us with a purely non-numerical insight into the structure of solutions.

One of the simplest choices of an overall factor in the Frobenius Ansatz (14) is represented by a Gaussian,

$$\psi(r) = \psi_{Gauss}(r) \sum_{n=0}^{N} h_n r^{n+l+1}, \qquad \psi_{Gauss}(r) = \exp(-sr^2) \tag{17}$$

with a free parameter $s > 0$. An alternative decision may just let some[15,17] or all[16] of the WKB parameters vary. Both the "leading-order-WKB" and "non-WKB" Ansaetze dispense with the nice asymptotics of the $N = 0$ approximants. The value of s becomes a parameter which may be optimalized in advance[16].

In accord with the Hautot's[16] review, there is an interesting and simple explanation of properties of the recurent algorithms in general and of the whole HD technique in particular: Once we succeed in solving our HD difference Schroedinger equations of the t-th order (e.g., we have $t = 2$ in Eq.(9) above), we may often arrive at a complete set of the independent and asymptotically ordered solutions,

$$\begin{aligned} & h_n = c_1 h_n^{(1)} + \ldots + c_t h_n^{(t)} \\ & |h_n^{(1)}| \gg |h_n^{(2)}| \gg \ldots \gg |h_n^{(t)}|, \qquad n \gg 1. \end{aligned} \tag{18}$$

Then, we may employ the existing computational mathematics[20] and claim that the different possible truncations of the HD recurrences may rigorously define the different separate particular solutions in (18). Thus, in the q-th arrangement

$$\begin{pmatrix} Q_{n,0} & Q_{n,1} & \ldots & Q_{n,q-1} & & \\ Q_{n+1,0} & Q_{n+1,1} & \ldots & & Q_{n+1,q} & \\ & & \ldots & & & \\ & & \ldots & & Q_{n+M,M-1} & Q_{n+M,M} \end{pmatrix} \begin{pmatrix} h_0^{(q)} \\ h_1^{(q)} \\ \vdots \\ h_M^{(q)} \end{pmatrix} = 0 \tag{19}$$

(where Q has precisely $q-1$ nonzero upper diagonals) we get the definition of the q-th solution in (18), i.e., a suppression of all the first $q-1$ asymptotically dominating components of h. In such a setting, a mathematical basis of the HD method in its Gaussian-like or free-parametric form may be seen in its interpretation as a generalized Miller-Oliver algorithm. More details may be found in refs[16–18].

4 THE GENERALIZED HD METHOD

An intuitive quantum-mechanical appeal of techniques of Sect. 3 [10,19] lies in a coincidence of zeros of Hill determinants det Q with zeros of the coefficient h_{N+1}. In the situations where the two quantities do not coincide, we might like to know which one of them "survives" the generalization. Recently, a nice test of this dichotomy has been proposed by Estrin *et al*[21].

4.1 *Hill Determinants of Estrin, Fernandez and Castro*

We may try to apply the overall HD strategy to one of the simplest possible asymmetric potential wells in one dimension, namely, to the ordinary anharmonic oscillator

$$-\ddot{\psi}(x) + V(x)\,\psi(x) = E\,\psi(x)$$
$$V(x) = ax + bx^2 + cx^3 + dx^4, \quad d > 0. \tag{20}$$

This example differs from the symmetric three-dimensional case by an absence of any angular-momentum-type symmetry. In principle, we may still use $l = -1$ or $l = 0$ as a quantum number of parity. Nevertheless, an absence of the $x \to -x$ symmetry in Eq.(20) indicates that our Ansatz will then contain l as (the second) free parameter again. Thus, we have to return to the bound states in the semi-numerical form (8). In such a setting, a success of the single HD Gaussian Ansatz

$$\psi(x) = \exp(-sx^2)(D_0 + D_1x + D_2x^2 + \ldots) \tag{21}$$

is not at all obvious in advance.

The HD numerical tests[21] confirmed the validity of the HD secular equation (7), originating now from the corresponding seven-term recurrences with $Q_{nn+1} = 0$,

$$Q_{nn-4}D_{n-4} + Q_{nn-3}D_{n-3} + \ldots + Q_{nn}D_n + Q_{nn+2}D_{n+2} = 0$$
$$n = 0, 1, \ldots, M, \qquad M \to \infty \tag{22}$$
$$D_{-1} = D_{-2} = \ldots = 0.$$

The coefficients Q are listed in the quoted paper and depend on the couplings and energy E.

Each choice of the second free parameter (say, D_1/D_0) enables us to define all the corresponding (unnormalized) D's step by step. A suitable superposition (8) of the two independent solutions (21) (presumably, with different parameters D_1/D_0 and at the same energy) has again to satisfy the standard physical boundary conditions,

$$\begin{aligned} \psi(x_L) &= 0, \qquad x_L \ll -1 \\ \psi(x_R) &= 0, \qquad x_R \gg 1. \end{aligned} \tag{23}$$

In the non-variational Hill-determinant spirit, we have to prove the alternative square-matrix truncation of our linear system (22),

$$D_{M+1} = 0, \quad D_{M+2} = 0, \qquad M \gg 1. \tag{24}$$

Estrin *et al*[21], unfortunately, failed to find the mathematical background of this "generalized HD" conjecture: Let us show that their reference to Hautot[16] ceases to be relevant in the case of pairs of boundary conditions of the type (23)[22].

4.2 *The Rigorous Proof*

In the first step, let us consider the dominant part of Eq.(22) in the asymptotic region,

$$-n(n+1)D_{n+1} + 2aD_{n-2} + 2dD_{n-5} = 0, \qquad n \gg 1 \tag{25}$$

and assume that the coordinates x are shifted (to $x + x_0$) in such a way that the coupling a becomes equal to zero. Then, we easily get the first-order $n \gg 1$ estimate

$$D_n = F_n^{[m]}/[(v^{[m]})^n G(n)^{1/3}], \qquad F_n^{[m]} = a\ constant\ +\ small\ corrections$$

$$v^{[m]} = \left(2de^{-2m\pi i}\right)^{-1/6}, \qquad m = 1, 2, \ldots, 6. \tag{26}$$

The improved $n \gg 1$ asymptotical equation

$$F_{n+1} - F_{n-5} = 2sv^2 n^{-1/3} F_{n-1} + 2cv^5 n^{-1/3} F_{n-4} + O(n^{-2/3}) \tag{27}$$

provides also the easy second-order solution

$$F_n^{[m]} = \exp[gn^{2/3} + O(n^{1/3})], \qquad g = g^{[m]} = sv^2 + cv^5. \tag{28}$$

Here, corrections partially remove the asymptotic degeneracy in the magnitude of the six different Taylor coefficients F and/or D.

For the positive couplings c, we get the desirable pattern

$$|D_n^{[6]}| \gg |D_n^{[3]}| \gg |D_n^{[1,5]}| \gg |D_n^{[2,4]}|, \qquad n \gg 1 \tag{29}$$

(cf. (18)) if and only if we choose a sufficiently large s,

$$s > c/|(2d)^{1/2}| > 0. \tag{30}$$

For the negative c's and for

$$s > -c/|(2d)^{1/2}| > 0 \tag{31}$$

the superscripts have only to be permuted in Eq.(29),

$$|D_n^{[3]}| \gg |D_n^{[6]}| \gg |D_n^{[2,4]}| \gg |D_n^{[1,5]}|, \qquad n \gg 1. \tag{32}$$

Moreover, we also discover another quadruplet of acceptable possibilities at the negative s's. All of them may produce the doublet of dominant solutions and may be characterized by the similar estimates, *viz.*,

$$|D_n^{[1,5]}| \gg \ldots \gg |D_n^{[3]}|, \qquad n \gg 1$$
$$s < -c/|(8d)^{1/2}| < 0 \tag{33}$$

and

$$|D_n^{[2,4]}| \gg \ldots \gg |D_n^{[6]}|, \qquad n \gg 1$$
$$s < c/|(8d)^{1/2}| < 0. \tag{34}$$

We may summarize: The rigorous analysis confirms the conjecture of Estrin *et al*[21] for a restricted range of parameters. This domain only partially overlaps with the one considered by the authors.

References

[1] R. G. Newton, *Scattering Theory of Waves and Particles* (Springer, New York, 1982), p. 398.

[2] H. Bateman and A. Erdelyi, *Higher Transcendental Functions III* (McGraw-Hill, New York, 1955).

[3] E. Vogt and G. H. Wanier, *Phys. Rev.* **95** (1954) 1190.

[4] D. C. Fivel, *Phys. Rev.* **142** (1966) 1219; see also ref. 1.

[5] M. Znojil, *J. Math. Phys.* **31** (1990) 108.

[6] M. Znojil, *J. Phys.* **A 15** (1982) 2111.

[7] M. Znojil, *J. Math. Phys.* **31** (1990) 1955.

[8] W. Gautschi, *SIAM Rev.* **9** (1967) 24; M. Znojil, K. Sandler and M. Tater, *J. Phys.* **A 18** (1985) 2541.

[9] S. N. Biswas, K. Datta, R. P. Saxena, P. K. Srivastava and V. S. Varma, *Phys. Rev.* **D 4** (1971) and *J. Math. Phys.* **14** (1973) 1190.

[10] V. Singh, S. N. Biswas and K. Datta, *Phys. Rev.* **D 18** (1978) 1901 and J. Math. Phys. **23** (1982) 1323; A. V. Turbiner, *Comm. Math. Phys.* **118** (1988) 467.

[11] E. Magyari, *Phys. Lett.* **A 81** (1981) 279.

[12] G. P. Flessas, *J. Phys.* **A 15** (1982) L1; G. P. Flessas and G. S. Anagnostatos, *J. Phys.* **A 15** (1982) L537; R. N. Chaudhuri, *J. Phys.* **A 16** (1983) 209.

[13] A. Hautot and A. Magnus, *J. Comput. Appl. Math.* **5** (1979) 3; M. Znojil, *Phys. Rev.* **D 26** (1982) 3750; D. Masson, *J. Math. Phys.* **24** (1983) 2074; M. Tater, *J. Phys.* **A 20** (1987) 2483.

[14] R. N. Chaudhuri and B. Mukherjee, *J. Phys.* **A 17** (1984) 3327; J. Killingbeck, *J. Phys.* **A 18** (1985) L1025; R. N. Chaudhuri, *Phys. Rev.* **D 31** (1985) 2687.

[15] J. Killingbeck, *Phys. Lett.* **A 115** (1986) 301; M. Znojil, *Phys. Lett.* **A 116** (1986) 207.

[16] A. Hautot, *Phys. Rev.* **D 33** (1986) 437.

[17] M. Znojil, *Phys. Rev.* **D 34** (1986) 1224.

[18] M. Znojil, *J. Math. Phys.* **29** (1988) 1433.

[19] C. A. Ginsburg, *Phys. Rev. Lett.* **48** (1982) 839.

[20] J. Oliver, *Numer. Math.* **11** (1968) 349 and **12** (1968) 459.

[21] D. A. Estrin, F. M. Fernandez and E. A. Castro, *Phys. Lett.* **A 130** (1988) 330.

[22] M. Znojil, *Comm. JINR Dubna* **E4-89-19** (1989), unpublished.

Waveguides and Contact Interactions

Quantum waveguides : energy bounds and critical thickness

Pavel Exner

Nuclear Physics Institute, CSAS
25068 Řež near Prague, Czechoslovakia

Several new results on discrete spectrum in bent quantum waveguides are presented. Nearly universal lower bounds to bound state energies are derived. The critical thickness is found for small bending angles. In the general case, its asymptotic behaviour with respect to the bending angle is shown to be governed by local smoothness properties of the boundary.

The recently discovered fact that a quantum particle living in a bent tube with Dirichlet boundary conditions can have bound states provided the boundary is smooth and the cross section is small enough [1,2] represents not only a mathematically interesting result but also a source of various physical applications - see, e.g. [3]. It opens a new field of research freight with unsolved problems. In this talk, we are going to discuss two of them. Before doing that, however, we review briefly the results obtained earlier.

1. PRELIMINARIES

Consider a *non-straight tube* Ω in $\mathbb{R}^n$, $n = 2,3$, of a fixed cross section generated by a smooth curve Γ. It is assumed to be *non-selfintersecting*, however, we bear in mind also the possibility that a nearly flat screw-type tube in $\mathbb{R}^3$ may be modeled by a winding strip on a multiple-sheeted Riemannian surface replacing the plane. In the case $n = 3$ we concentrate mostly on the circular tubes, though a wider class specified below can be treated without substantial modifications. The curve Γ will be then axis of the tube; in the two-dimensional case one can choose it alternately as one of the boundaries.

Each tube Ω is then characterized by its thickness $d \equiv 2a$ and the curve Γ specified by the curvature $\gamma(s)$ for $n=2$, and by curvature $\gamma(s)$ and torsion $\tau(s)$ for $n=3$; here s means the arc length of Γ . We adopt the following assumptions :

(i) *smoothness* : $\gamma \in C^2$ for $n=2$, and $\tau \in C^1$ while γ is modulus of a C^2 function for $n=3$,

(ii) *smoothness at infinity* : $\gamma, \gamma', \gamma''$, and also τ, τ' for $n=3$ are bounded,

(iii) *regularity of the boundary* : $a\|\gamma\|_\infty < 1$, or $\inf \gamma(s) < d^{-1}$ if $n=2$ and one of the boundaries is chosen as Γ ; it is fulfilled automatically for a simply bent strip, $\gamma(s) \ge 0$,

(iv) *injectivity* : Ω is non-selfintersecting (with the license made above for $n=2$).

The object of our interest is the operator describing free quantum motion in the tube Ω with infinitely hard walls, i.e., Dirichlet Laplacian [4],

$$H := -\Delta_D^\Omega \quad \text{on} \quad L^2(\Omega) \ ,$$

in particular, its discrete spectrum. The essence of the result mentioned in the introduction is that $\sigma_{disc}(H)$ is non-empty provided d is small enough and Ω is (more or less) asymptotically straight ; recall that we have excluded the trivial case of a straight strip. To be more specific, assume

(v) $\gamma(s) = O(|s|^{-1-\varepsilon})$ as $|s| \to \infty$ for n=2, and $\gamma(s) = O(|s|^{-2-\varepsilon})$, or alternately, $\gamma(s), \tau(s) = O(|s|^{-1-\varepsilon})$ as $|s| \to \infty$ for n=3.

Then we have

1.1 Theorem : Assume (i)-(v). Then $\sigma_{ess}(H) = [E_\infty, \infty)$, where $E_\infty = (\pi/2a)^2$ for n=2 and $E_\infty = j_{0,1}^2 a^{-2}$ with $j_{0,1}$ = 2.405... for n=3. Furthermore, there is $a_0 > 0$ such that for all $a \in (0, a_0)$, H has an eigenvalue in $[0, E_\infty)$.

Proof can be found in [1] for n=2 and in [2,5] for n=3 ; it is based on straightening the tube with the use of the natural curvilinear coordinates (for n=3 one has to employ the frame rotating with angular velocity τ' with respect to the Frenet triad) combined with minimax estimates. ∎

1.2 Remark : The argument modifies easily for the non-circular tubes obtained by moving a fixed compact connected set along the curve rotating it in the transversal plane in the above mentioned way.

Next we adopt stronger decay assumptions on the curvature and torsion, namely

(vi) $\gamma(s) = O(|s|^{-3/2-\varepsilon})$ as $|s| \to \infty$ for n=2, and $\gamma(s) = O(|s|^{-3-\varepsilon})$, or alternately, $\gamma(s), \tau(s) = O(|s|^{-3/2-\varepsilon})$ as $|s| \to \infty$ for n=3.

It allows us to prove a necessary and sufficient condition for existence of the bound states :

1.3 Theorem : Assume (i)-(vi), then $\sigma_{disc}(H) \neq \emptyset$ *iff*

$$\int_{\mathbb{R}\times[-a,a]} \left\{ -\frac{\gamma^2}{(1+u\gamma)^2} + \frac{u^2\gamma'^2}{(1+u\gamma)^4} \right\} \cos^2(\pi u/d)\, ds\, du < 0$$

for n=2, while for n=3 the condition reads

$$\int_{\mathbb{R}\times[0,a]} \left\{ -\frac{\gamma^2}{(1-\varrho^2\gamma^2)^{3/2}} + \frac{\varrho^2}{(1-\varrho^2\gamma^2)^{5/2}} \left[\gamma^2\tau^2 + \gamma'^2\,\frac{3+2\varrho^2\gamma^2}{1-\varrho^2\gamma^2} - \frac{32\varrho\gamma^2\gamma'\tau}{3\pi(1-\varrho^2\gamma^2)^{3/2}} \right]\right\} J_0(j_{0,1}a^{-1})^2\, \varrho\, d\varrho\, ds < 0$$

Proof is based on a lemma [1,5] which generalizes the well-known criterion [6,7] to the case of Schrödinger operators on a cylinder; one proves it combining the standard Birman-Schwinger argument with the decomposition according to transversal modes.The lemma together with minimax estimates (finer than in the previous theorem) and an integration by parts (for n=3) give the result. ∎

Another minimax estimate combined with transversal-mode decomposition and the known trick for one-dimensional Schrödinger operators [8,9] yield an upper bound on the number of bound states including multiplicity.

1.4 Proposition : Assume (i)-(v) and

$$a < \frac{c_n}{1+c_n\|\gamma\|_\infty},$$

where

$$c_2 := \min\left\{ \frac{\pi}{\|\gamma\|_\infty}, \left[\frac{\pi^2}{2\|\gamma''\|_\infty}\right]^{1/3}, \left[\frac{\pi}{\sqrt{5}\|\gamma\|_\infty}\right]^{1/2} \right\},$$

$$c_3 := \min\left\{ \frac{\sqrt{10}}{\|\gamma\|_\infty}, \left[\frac{5}{\|\gamma\tau'-\gamma''\|_\infty+\|2\gamma'\tau+\gamma\tau'\|_\infty}\right]^{1/3}, \left[\frac{\sqrt{2}}{\|\gamma\tau\|_\infty+\|\gamma'\|_\infty}\right]^{1/2} \right\}.$$

Then

$$N(H) \le 1 + \frac{1}{8}\left(\frac{1+a\|\gamma\|_\infty}{1-a\|\gamma\|_\infty}\right)^2 \frac{\int_{\mathbb{R}^2} \gamma(s)^2 |s-t|\, \gamma(t)^2\, ds\, dt}{\int_{\mathbb{R}} \gamma(s)^2\, ds}$$

Notice that in the three-dimensional case, the torsion enters this formula only through the range of its validity. Furthermore, if one chooses for n=2 one of the boundaries as Γ, then the preintegral factor in the estimate can be replaced by $(1/8)(1+d\|\gamma_+\|_\infty)^2(1-d\|\gamma_-\|_\infty)^{-2}$; for a simply bent (outward) strip it is slightly better than the original one.

2. LOWER BOUNDS TO BOUND STATE ENERGIES

Our present knowledge of where the bound state eigenvalues are situated is rather scarce. For thin tubes perturbation theory may be used [2,5] while in some simple cases with a non-smooth boundary the answer can be found numerically in a way adapted from the classical waveguide theory [10], however, any further result is welcome. In this section, we are going to derive a family of lower bounds based on a recent result of Ashbaugh and Benguria [11] who proved a conjecture on the ratio of eigenvalues for a Dirichlet Laplacian formulated by Payne, Polya and Weinberger in the fifties [12,13]; their paper finished a long line of gradually strengthening estimates [14-22].

Consider again a tube Ω fulfilling (i)-(v). We add two Dirichlet boundaries perpendicular to Γ at a distance l to both sides of a fixed point of Γ ; in this way Ω decomposes into the central region Ω_l and the tails $\Omega_l^{(\pm)}$, and the corresponding Hamiltonian is the orthogonal sum of three Dirichlet Laplacians

$$H_l := H_l^{(-)} \oplus H_{\Omega_l} \oplus H_l^{(+)} .$$

Since the central region is compact the respective part of the spectrum is discrete, $\sigma(H_{\Omega_l}) = \{\lambda_k(l)\}_{k=1}^{\infty}$. Due to the minimax principle the functions $\lambda_k(.)$ are non-increasing so the limits $\lambda_k := \lim_{l\to\infty} \lambda_k(l)$ exist ; we claim that those of them lying below E_∞ form the appropriate part of $\sigma(H)$. A simple limiting argument [23] shows that $H_l \longrightarrow H$ as $l \longrightarrow \infty$ in the strong resolvent sense so one has to check only that the "tail" operators do not contribute to $\sigma(H)$ below E_∞. This can be done using the procedure from proof of Theorem 1.1 ; see [23] for details.

We know from the previous section that for small enough d , $\sigma_{disc}(H) = \{ E_1<E_2<\ldots<E_N \} \neq \emptyset$. Suppose first we have just one bound state with energy E_1 , and consider the case n=2. In view of the described limiting argument $\lambda_1(l) \longrightarrow E_1$ while $\lambda_2(l) \longrightarrow E_\infty$ as $l \longrightarrow \infty$. However, the

PPW-inequality implies $\lambda_1(l)/\lambda_2(l) \le 2.539$ for any l, and therefore

$$E_1 \ge \left[\frac{j_{0,1}}{j_{1,1}}\right]^2 (\pi/2a)^2 \approx 0.3939\ E_\infty .$$

In the three-dimensional case, we get similarly

$$E_1 \ge \left[\frac{\pi}{j_{3/2,1}}\right]^2 j_{0,1}^2\ a^{-2} \approx 0.4888\ E_\infty .$$

For $N > 1$ we get weaker estimates, $E_1 \ge 0.3939 \times 3^{N-1}\ E_\infty$ and $E_k/E_1 \le 2.539 \times 3^{k-2}$ for $k \ge 3$ in the two-dimensional case, and analogous relations for $n=3$. If $k \ge 3$, the last bound can be improved by a factor of nearly 2 using the results of [17] ; see [23] for details.

The obtained bounds are certainly rough but on the other hand, they are nearly universal depending on the shape of Ω through the number of bound states only (an inspection of Proposition 1.4 shows that a tube with a simple and not very extended bend has typically just one bound state - cf.[2]). What is even more important is that the above sketched argument can be used even if the considerations of the previous section does not apply, for instance, for tubes with a non-smooth boundary. One has only to check that the tail operators do not contribute; in some cases it may be trivial, e.g., if the tails are straight half-cylinders for l large enough.

3. FINITENESS OF THE CRITICAL THICKNESS

For simplicity, we shall consider the case of a planar strip, n=2, only. First we introduce a parameterization in the set of all admissible tubes. We choose one of the boundaries as the reference curve Γ ; without loss of generality we may suppose that $\sup \gamma(s) \ge -\inf \gamma(s)$. Then one can choose $\alpha, \varrho > 0$ such that

$$\gamma(s) = \frac{\alpha}{\varrho}\, \phi(s/\varrho)$$

where the function ϕ satisfies $-1 < \phi(x) \le \phi(0) = 1,\ x \in \mathbb{R}$, and the normalization condition

$$\int_{\mathbb{R}} \phi(x)\, dx = \pi\ .$$

In other words, Γ has minimum curvature radius ϱ/α at $s=0$ and the total bending $\beta := \int_{\mathbb{R}} \gamma(s)\, ds = \pi\alpha$ (its existence is ensured by the assumptions (i) and (v)). Substituting to the criterion given by Theorem 1.3 we find that a bound state exists *iff* the integral

$$I(\alpha,\varkappa) := \int_{\mathbb{R}} \int_0^1 \left[-\alpha^2 \frac{\phi(x)^2}{(1+\varkappa y\phi(x))^2} + \frac{\varkappa^2 y^2 \phi'(x)^2}{(1+\varkappa y\phi(x))^4} \right] \sin^2 \pi y\, dxdy$$

is *negative* ; here we have introduced $\varkappa := \alpha d/\varrho$ as the thickness normalized to the minimum curvature radius.

A simple consequence of the above stated condition is that *there is no bound state for a fixed* $\varkappa$ *and* α *small enough,* i.e., for a tube which is only slightly bent.

3.1 Example : Let $\phi(x) = (1+x^2)^{-1}$. The integration over x is then easily performed and the equation $I(\alpha,\varkappa) = 0$ can be solved numerically. It is more convenient, however, to solve it as an equation for the critical angle which gives

$$\alpha(\varkappa)^2 = \left[\int_0^1 \frac{1}{2}\, \varkappa^2 y^2 \,\frac{\sin^2 \pi y}{(1+\varkappa y)^{5/2}}\, dy \right] \left[\int_0^1 \frac{\sin^2 \pi y}{(1+\varkappa y)^{3/2}}\, dy \right]^{-1} ;$$

it is obvious that the solution is unique. One can check numerically that the function $\alpha(\varkappa)$ is strictly monotonous so the inverse $\varkappa(\alpha)$ exists and it is growing with respect to α. Moreover, it is not difficult to find the asymptotic behaviour,

$$\varkappa(\alpha) = \alpha \left[\frac{12\pi^2}{3\pi^2 - 2} \right]^{1/2} + O(\alpha^2)$$

for small α and

$$\varkappa(\alpha) \approx 5.71\, \alpha^2$$

as $\alpha \to \infty$; it shows, in particular, that the lower bounds to critical thickness given in [1] are far from being precise.

Now we want to do the same for a general ϕ . In this case, of course, one cannot perform generally the integration "along the tube" but the equation $I(\alpha,\varkappa) = 0$ solves easily giving

$$\alpha(\varkappa) = \left\{ \frac{\int_{\mathbb{R}} \int_0^1 \frac{\varkappa^2 y^2 \phi'(x)^2}{(1+\varkappa y\phi(x))^4} \sin^2 \pi y \, dx \, dy}{\int_{\mathbb{R}} \int_0^1 \frac{\phi(x)^2}{(1+\varkappa y\phi(x))^2} \sin^2 \pi y \, dx \, dy} \right\}^{1/2} .$$

In particular, for small $\varkappa$ we get from here

$$\alpha(\varkappa) = \varkappa \left(\frac{1}{3} - \frac{1}{2\pi^2} \right)^{1/2} \frac{\|\phi'\|}{\|\phi\|} + O(\varkappa^2)$$

so for small α the critical thickness depends also linearly on the bending angle. In the general case, one wants to solve the following problems :

(a) finiteness of $\varkappa(\alpha)$ for any bending angle, which means to check $\lim_{\varkappa\to\infty} \alpha(\varkappa) = \infty$,

(b) uniqueness of $\varkappa(\alpha)$; it requires that the critical angle is strictly monotonous, $\alpha'(\varkappa) > 0$.

Let us remark first that $I(\alpha,\varkappa)$ is decreasing with respect to α so uniqueness is guaranteed as long as the curve $\alpha(\varkappa)$ does not leave the region of the $(\varkappa,\alpha)$-plane where $I(\alpha,\varkappa)$ is growing with respect to $\varkappa$. Calculating the derivative we get after a simple estimate

$$\frac{d}{d\varkappa} I(\alpha,\varkappa) > 0 \qquad \text{if} \quad \varkappa \le 1 ,$$

i.e., the critical thickness is defined as a (continuously growing) function of α provided its value does not exceed the minimum curvature radius. Later we get a better result.

3.2 Remark : Of course, one can always define $\varkappa(\alpha)$, e.g., as the infimum (supremum) of those $\varkappa$ for which $I(\alpha,\varkappa)$ is positive (negative) at a given α. The problem (b) requires to exclude the situations when we make a strip over a given Γ broader, with a bound state disappearing in the continuum, then appearing and disappearing again.

In order to proceed further, we perform first the "transversal" integration in the formula for $I(\alpha,\varkappa)$. Using the standard combination of integral sines and cosines,

$$f(z) = \sin z \,\mathrm{ci}(z) - \cos z \,\mathrm{si}(z) \ ,$$

$$g(z) = -\cos z \,\mathrm{ci}(z) - \sin z \,\mathrm{si}(z) \ ,$$

we introduce

$$G(z) := f\left(2\,\frac{\pi}{z}\right) - f\left(2\,\frac{\pi}{z} + 2\pi\right) \ ,$$

$$F(z) := \frac{\pi}{3(1+z)} + \left(1 - \frac{2\pi^2}{3z^2}\right) G(z)$$

$$- 2\,\frac{\pi}{z}\left[g\left(2\,\frac{\pi}{z}\right) - g\left(2\,\frac{\pi}{z} + 2\pi\right)\right] \ ;$$

then we have

$$\alpha(\varkappa) = \left\{ \frac{\int_{\mathrm{supp}\,\phi} \frac{\phi'(x)^2}{\phi(x)^4}\, F(\varkappa\phi(x))\, dx}{\int_{\mathrm{supp}\,\phi} G(\varkappa\phi(x))\, dx} \right\}^{1/2} .$$

It seems that the numerator contains a strong singularity

but in fact it cancels because the functions F,G behave around the point $z=0$ as

$$F(z) = \frac{z^4}{6\pi}\left[1 - \frac{3}{2\pi^2}\right] + \frac{z^5}{2\pi}\left[1 - \frac{3}{\pi^2}\right] + O(z^6) ,$$

$$G(z) = \frac{z^2}{2\pi} - \frac{z^3}{2\pi} + \frac{z^4}{2\pi}\left[1 - \frac{3}{2\pi^2}\right] + O(z^5) ,$$

in particular, we recover the above derived asymptotic behaviour of $\alpha(\varkappa)$ for small $\varkappa$.

Now we need to know more about the functions F,G. They are both positive and growing in $(0,\infty)$: for $G(z)$ it follows from the integral representation

$$G(z) = \int_0^\infty \frac{1-e^{-2\pi t}}{t^2+1}\, e^{-2\pi t/z}\, dt$$

while for $F(z)$ one can check it numerically. Furthermore,

$$\lim_{z\to\infty} F(z) = \lim_{z\to\infty} G(z) = \mathrm{Si}(2\pi) \approx 1.42 .$$

Using these properties we are going to show that *the asymptotic behaviour of* $\alpha(\varkappa)$ *for large* $\varkappa$ *is determined by the behaviour of* ϕ *around its zero points, both at infinity and at finite zero points if any exist.* We shall not attempt a general formulation, instead we illustrate the assertion with two types of power-like asymptotics. Suppose that $\mathrm{supp}\,\phi = \cup_{j=1}^N (a_j,b_j)$ with $b_j \le a_{j+1}$, $N < \infty$, and assume

(A_β) if $a_1 = -\infty$ or $b_N = \infty$, let the function ϕ fulfill $|\phi(x)| = C|x|^{-\beta}(1+O(|x|^{-1})$ as $x \longrightarrow \mp\infty$ (due to (vi), $\beta > 3/2$),

(B_δ) in any finite zero point x_0 (= a_j, b_j) assume that $|\phi(x)| = C|x-x_0|^{\delta}(1+O(|x-x_0|)$ as $x \longrightarrow x_0\pm$ (for a simple bend, $\phi(x) \geq 0$, one has $\delta > 2$ from (i) while in the general case one needs only $\delta > 0$ if, of course, $b_j = a_{j+1}$).

For a given ϕ, we denote $M(a) := \{ x \in \text{supp}\,\phi : \phi(x) \geq a \}$ and $N(a) := \text{supp}\,\phi \setminus M(a)$. Then it is easy to see that for each $\varkappa, K, j$ there is just one right (left) maximal neighbourhood $(a_j, a_j(K,\varkappa))$ or $(b_j(K,\varkappa), b_j)$ belonging to $N(K/\varkappa)$. The idea is to estimate the integrals appearing in the formula for $\alpha(\varkappa)$ separately on these neighbourhoods where one can use the asymptotics and on the rest of the support.

To get a lower bound to $\alpha(\varkappa)$, e.g., one has to estimate the denominator and numerator from above and below, respectively. We use the following rough estimates

$$G(z) \leq \begin{cases} z^2/2\pi & \ldots \quad z \leq 2.99 \\ \mathrm{Si}(2\pi) & \ldots \quad z \geq 2.99 \end{cases}$$

and

$$F(z) \geq \begin{cases} 0 & \ldots \quad z < 27.06 \\ \mathrm{Si}(2\pi)/2 & \ldots \quad z \geq 27.06 \end{cases}$$

(similar "reversed" estimates can be used to get an upper bound to $\alpha(\varkappa)$). Applying it to the mentioned integrals, we arrive at the following

3.3 Conclusion : The lower asymptotic bound of $\alpha(\varkappa)$ for $\varkappa \to \infty$ is governed, roughly speaking, by the "fastest" zero of the numerator and the "slowest" zero of the denominator; the upper bound has the same behaviour differing in numerical constants only.

Instead of making the formulation more precise, we present several

3.4 Examples : (a) infinite support with no internal zero points, asymptotics at the two ends with the *same* β :

$$\alpha(\varkappa) \geq \frac{\sqrt{\beta}\ (80.8)^{1/2\beta}}{45.6} \left[\frac{C_a^{-1/\beta}+C_b^{-1/\beta}}{C_a^{1/\beta}+C_b^{1/\beta}} \right]^{1/2} \varkappa^{1-1/\beta}\ [1+O(\varkappa^{-1/\beta})]$$

where the constants C_a, C_b refer to the assumed asymptotics at $-\infty$ and ∞, respectively,

(b) semiinfinite support with the asymptotics (B_δ) at $a_1 > -\infty$ and (A_β) at ∞, and no internal zeros ; then one has

$$\alpha(\varkappa) \geq \frac{\delta}{45.6} \left[\frac{2\beta-1}{\beta(2\delta+1)} \right]^{1/2} \left[\frac{C_a}{27.1} \right]^{1/2\delta} \left[\frac{2.99}{C_b} \right]^{1/2\beta} \times$$

$$\varkappa^{1+1/2\delta-1/2\beta}\ [1+O(\varkappa^{-1/\beta})+O(\varkappa^{-1/\delta})]$$

(c) the same as (a) but now with internal zeros among which just one (at some $b_j = a_{j+1}$) has a minimum asymptotic power δ ; then

$$\alpha(\varkappa) \geq \frac{\delta}{45.6}\left[\frac{2\beta-1}{(2\delta+1)}\left[\left(\frac{C_{b_j}}{27.1}\right)^{1/\delta}+\left(\frac{C_{a_{j+1}}}{27.1}\right)^{1/\delta}\right]\times\right.$$

$$\left.\left[\left(\frac{C_{a_1}}{2.99}\right)^{1/\beta}+\left(\frac{C_{b_N}}{2.99}\right)^{1/\beta}\right]^{-1}\right]^{1/2} \varkappa^{1+1/2\delta-1/2\beta} \times$$

$$[1+O(\varkappa^{-1/\beta})+O(\varkappa^{\lambda})] \quad ,$$

where $\lambda := -1/2\delta+1/2\delta'$ with δ' being the next-to-minimum asymptotic power, and $\lambda := -1/\delta$ when there is no other internal zero.

Hence the finiteness of the critical thickness is ensured in the considered cases since $\alpha(\varkappa) \geq \text{const}\ \varkappa^{1/3}$, the worst case being that with infinite support of ϕ with no internal zero points and the decay at infinity as slow as possible. Then the critical thickness (defined as in Remark 3.2) fulfills $\varkappa(\alpha) \leq C\,\alpha^3$ for some C what we wanted to show.

3.5 Remarks : (a) If the criterion expressed by Theorem 1.3 could be proven without the assumption (vi) (in the spirit of [7]), then the "coulombic" decay of the curvature, $\beta=1$, will be the borderline for a finite critical thickness.
(b) Other types of zero points can be treated in a similar way, e.g., if ϕ behaves as $\exp(-x^{\gamma})$ around a zero point, then the corresponding contribution to the asymptotics of $\alpha(\varkappa)$ is $\sim \varkappa\,(\ln \varkappa)^{1/\gamma}$.

The main lesson drawn from the above considerations is that a *local* change of the boundary can affect drastically the asymptotic behaviour : making the boundary locally flat or even changing the sign of the curvature (in the limits allowed by the assumption (iii)) one gets the function $\varkappa(\alpha)$ with a much slower growth at infinity. However, the importance of this conclusion must not be exaggerated since it is the *size* of the deformed region determines *when* the modified asymptotics will start to dominate ; the smaller it is the larger bending angles are needed, so tiny deformations of the boundary can be felt only for the coiled stripes mentioned in the introduction.

In conclusion, let us return to the monotonicity problem. One calculates easily

$$\frac{d}{d\varkappa} I(\alpha,\varkappa) = -\frac{2\pi}{\varkappa^2}\int_{\mathrm{supp}\,\phi}\left[-\alpha^2 B(\varkappa\phi(x)) + \frac{\phi'(x)^2}{\phi(x)^4} A(\varkappa\phi(x))\right] dx$$

with

$$A(z) := F(z) - \frac{1}{2} zF'(z) ,$$

$$B(z) := G(z) - \frac{1}{2} zG'(z) ;$$

we have mentioned in the opening of this section that the uniqueness of $\varkappa(\alpha)$ is ensured as long as the derivative is positive. The functions A, B are both smooth and tend to $\mathrm{Si}(2\pi)$ as $z \to \infty$. However, B is positive and growing for $z > 0$ while A assumes negative values for small z ; it has just one zero point in $(0,\infty)$ which can be determined numerically as $z = 2.086\ldots$

Consequently, the critical thickness is (under the assumptions (i)-(vi)) a uniquely determined monotonously growing function of the bending angle provided it does not exceed the value 2.086 times the minimum curvature radius. It is clear that this result as well as the asymptotic analysis presented above are far from being optimal, and a general answer to the problems (a),(b) has yet to be found.

ACKNOWLEDGMENT

The results presented in this lecture were obtained in collaboration with Mark Ashbaugh and Pierre Duclos ; the author is sincerely grateful to both of them.

REFERENCES

1. P.Exner, P.Šeba, J.Math.Phys.30, 2574 (1989).
2. P.Exner, in *Order, Disorder and Chaos in Quantum Systems*, Operator Theory : Advances and Applications, vol.46, Birkhäuser, Basel 1990 ; p.65.
3. P.Exner, P.Šeba, P.Šťovíček, Phys.Lett.A150,179 (1990)
4. M.Reed, B.Simon : *Methods of Modern Mathematical Physics IV* ; Academic Press, New York 1978 ; Sec.XIII.15
5. P.Duclos, P.Exner : *Curvature-induced bound states in quantum waveguides in two and three dimensions*, to appear
6. B.Simon, Ann.Phys.97, 279 (1976).
7. R.Blanckenbecler, M.L.Goldberger, B.Simon, Ann.Phys. 108, 69 (1977).
8. M.Klaus, Ann.Phys.108, 288 (1977).
9. R.G.Newton, J.Operator Theory 10, 119 (1983).
10. P.Exner, P.Šeba, P.Šťovíček, Czech.J.Phys.B39, 1181 (1989).
11. M.S.Ashbaugh, R.Benguria : *Proof of the Payne-Polya-Weinberger conjecture*, to appear
12. L.E.Payne, G.Polya, H.F.Weinberger, C.R.Acad.Sci. 241, 917 (1955).
13. L.E.Payne, G.Polya, H.F.Weinberger, J.Math.and Phys. 35, 289 (1956).

14. J.J.A.M.Brands, Arch.Rat.Mech.Anal.16, 265 (1964).
15. H.L.de Vries, Z.Naturforsch.22a, 152 (1967).
16. C.J.Thompson, Stud.Appl.Math.48, 281 (1969).
17. P.Marcellini, J.Diff.Eqs.37,438 (1980).
18. G.N.Hille, M.H.Protter, Indiana Univ.Math.J.29, 523 (1980).
19. G.Chiti, Bul.Un.Mat.Ital.(5) 18-A, 144 (1981).
20. G.Chiti, SIAM J.Appl.Math.Anal.14, 1163 (1983).
21. J.D.Kuttler, V.G.Sigillito, SIAM Rev.29, 163 (1987).
22. M.H.Protter, SIAM Rev.29, 185 (1987).
23. M.S.Ashbaugh, P.Exner, Phys.Lett A150, 183 (1990).

Spectral analysis for optical fibres and stratified fluids

S. DeBièvre†

Université Paris 7
LPTM, Tour Centrale 3ème étage
2 Place Jussieu, 75251 Paris, Cedex 05

ABSTRACT

The operator $\hat{H} = -c^2(x,y)\rho(x,y)\nabla_z \cdot \rho(x,y)^{-1}\nabla_z$, $z=(x,y)$, $x \in \mathbf{R}^k$, $y \in \mathbf{R}^m$, appears in the study of the propagation of acoustical waves in stratified fluids $(m=1,\ k=2)$, as well as of electromagnetic waves in layered dielectric media $(m=1,\ k=2)$, and in optical fibres $(m=2,\ k=1$, and $\rho(x,y)=1)$. The density $\rho(x,y)$ and signal speed $c(x,y)$ are (possibly long range) perturbations of a density $\rho_0(y)$ and speed $c_0(y)$, i.e. $(\rho(x,y)-\rho_0(y))$, $(c(x,y)-c_0(y)) \to 0$ as $|z| \to \infty$. Positive commutator methods along with Virial techniques are used to show that in many cases of physical interest the spectrum of $\hat{H}$ is purely absolutely continuous. This means in particular that the point spectrum of $\hat{H}$ is empty. In addition, a limiting absorption principle for $\hat{H}$ is established: it is proven that the resolvent $(\hat{H}-\kappa)^{-1}$ of $\hat{H}$ has a norm limit as an operator between $\mathcal{L}^2(\mathbf{R}^{k+m},(1+z^2)^{\frac{\alpha}{2}}(c^2\rho)^{-1}d^nz)$ and $\mathcal{L}^2(\mathbf{R}^{k+m},(1+z^2)^{-\frac{\alpha}{2}}(c^2\rho)^{-1}d^nz)$ for $\alpha > \frac{1}{2}$ as $\kappa \in \mathbf{C} \setminus \overline{\mathbf{R}^+}$ approaches the real axis from above or from below.

§1. Introduction

We study the spectral properties of the operators

$$\hat{H} \;=\; -\,c^2(z)\rho(z)\nabla_z\rho^{-1}(z)\nabla_z\,, \tag{1.1}$$

arising in the study of wave propagation in perturbed stratified acoustic and dielectric media ($z \in \mathbf{R}^n$, $\nabla_z \equiv (\partial_{z_1}, \cdots \partial_{z_n})$). Our main results, Theorems 1.1 and 1.2, are the establishment of the limiting absorption principle, and the absence of point spectrum for $\hat{H}$, under suitable conditions on ρ and c, to be spelled out shortly. For more details and for proofs of the theorems, we refer to [DBP1] and [DBP2].

The operator $\hat{H}$ in (1.1) can be realized as a self-adjoint operator on the Hilbert space $\mathcal{L}^2(\mathbf{R}^n,(c^2\rho)^{-1}d^nz)$, but for our purpose it will be more convenient to work with the unitarily equivalent operator on $\mathcal{L}^2(\mathbf{R}^n, d^nz)$, given by

$$H \;=\; -\,f(z)\nabla_z g(z)^2\nabla_z f(z)\,, \tag{1.2}$$

where

$$f(z) \;=\; (c(z)^2\rho(z))^{\frac{1}{2}}\,, \quad g(z) \;=\; \rho(z)^{-\frac{1}{2}}\,. \tag{1.3}$$

We first list our conditions on ρ and c in terms of conditions on f and g.

---

† Aangesteld Navorser, NFWO 1989-1990

To start, we remark that we think of H as being a perturbation of an operator H_0 (see (H4) and (H5) below), given by

$$H_0 \;=\; -\; f_0\, \nabla_z {g_0}^2 \nabla_z f_0 \;, \tag{1.4}$$

where f_0 and g_0 are functions of $y \in \mathbf{R}^m$ alone, and where we wrote

$$z = (x, y) \;\in\; \mathbf{R}^k \times \mathbf{R}^m \;=\; \mathbf{R}^n \;. \tag{1.5}$$

For $\theta \subset \mathbf{R}^p$ open, $p \in \mathbf{N}$, we use the notation

$$W^{\alpha,\infty}(\theta) = \{\phi \in \mathcal{L}^\infty(\theta)|\ \partial_z^{\overline{\alpha}}\phi \in \mathcal{L}^\infty(\theta),\ |\overline{\alpha}| \;\le\; \alpha\}, \tag{1.6}$$

where $\partial_z^{\overline{\alpha}} \equiv \partial_{z_1}^{\alpha_1}\partial_{z_2}^{\alpha_2}\ldots\partial_{z_p}^{\alpha_p}$, $|\overline{\alpha}| = \alpha_1 + \alpha_2 + \cdots + \alpha_p$ for any multi-index $\overline{\alpha}$. The conditions we need on f_0, g_0, f, and g are

(H1)
(a) $f_0, g_0 \in \mathcal{L}^\infty(\mathbf{R}^m)$ are $\mathbf{R}^+$-valued and $inf\ f_0 > 0$, $inf\ g_0 > 0$;
(b) $f, g \in \mathcal{L}^\infty(\mathbf{R}^n)$ are $\mathbf{R}^+$-valued and $inf\ f > 0$, $inf\ g > 0$.

(H2) $\exists \Omega \subset \mathbf{R}^m$, compact, such that, with $\Omega^c = \mathbf{R}^m \setminus \Omega$
(a) $g_0\big|_{\Omega^c}$, $f_0\big|_{\Omega^c} \in W^{3,\infty}(\Omega^c)$;
(b) $y \cdot \nabla_y g_0\big|_{\Omega^c}$, $y \cdot \nabla_y f_0\big|_{\Omega^c} \in W^{2,\infty}(\Omega^c)$;
(c) $(y \cdot \nabla)^2 c_0\big|_{\Omega^c}, (y \cdot \nabla)^2 \rho_0\big|_{\Omega^c} \in W^{1,\infty}(\Omega^c)$;
(d) $y \cdot \nabla_y g_0\big|_{\Omega^c} \to 0$, $\nabla_y(y \cdot \nabla_y g_0)\big|_{\Omega^c} \to 0$ as $|y| \to \infty$;
$y \cdot \nabla_y f_0\big|_{\Omega^c} \to 0$ as $|y| \to \infty$.

(H3) $g_0 \in W^{1,\infty}(\mathbf{R}^m)$.

For the perturbed media, we need, in analogy with (H2):

(H4) $\exists \Theta \subset R^n$, compact, $\Theta \subset R^k \times \Omega$, so that
(a) $cc_0^{-1}\big|_{\Theta^c}, \rho\rho_0^{-1}\big|_{\Theta^c} \in W^{3,\infty}(\Theta^c)$;
(b) $(x \cdot \nabla)cc_0^{-1}\big|_{\Theta^c}, (x \cdot \nabla)\rho\rho_0^{-1}\big|_{\Theta^c} \in W^{2,\infty}(\Theta^c)$;
$(y \cdot \nabla)cc_0^{-1}\big|_{\mathbf{R}^k \times \Omega^c}, (y \cdot \nabla)\rho\rho_0^{-1}\big|_{\mathbf{R}^k \times \Omega^c} \in W^{2,\infty}(\mathbf{R}^k \times \Omega^c)$;
(c) $(x \cdot \nabla)^2 cc_0^{-1}\big|_{\Theta^c}, (x \cdot \nabla)^2 \rho\rho_0^{-1}\big|_{\Theta^c} \in W^{1,\infty}(\Theta^c)$;
$(y \cdot \nabla)^2 cc_0^{-1}\big|_{\mathbf{R}^k \times \Omega^c}, (y \cdot \nabla)^2 \rho\rho_0^{-1}\big|_{\mathbf{R}^k \times \Omega^c} \in W^{1,\infty}(\mathbf{R}^k \times \Omega^c)$,
$(x \cdot \nabla)(y \cdot \nabla)cc_0^{-1}\big|_{\mathbf{R}^k \times \Omega^c}, (x \cdot \nabla)(y \cdot \nabla)\rho\rho_0^{-1}\big|_{\mathbf{R}^k \times \Omega^c} \in W^{1,\infty}(\mathbf{R}^k \times \Omega^c)$.

(H5)
(a) $(f - f_0)$ and $(g - g_0) \to 0$, as $|z| \to \infty$;
(b) $\nabla(gg_0^{-1})\big|_{\Theta^c} \;\to\; 0$ as $|z| \to \infty$;
(c) $(x \cdot \nabla)(gg_0^{-1})\big|_{\Theta^c} \;\to\; 0$, $(x \cdot \nabla)(ff_0^{-1})\big|_{\Theta^c} \;\to\; 0$ as $|z| \to \infty$;
$(y \cdot \nabla)(gg_0^{-1})\big|_{\mathbf{R}^k \times \Omega^c} \;\to\; 0$, $(y \cdot \nabla)(ff_0^{-1})\big|_{\mathbf{R}^k \times \Omega^c} \;\to\; 0$ as $|z| \to \infty$;
(d) $\nabla(x \cdot \nabla)(gg_0^{-1})\big|_{\Theta^c} \;\to\; 0$ as $|z| \to \infty$;
$\nabla(y \cdot \nabla)(gg_0^{-1})\big|_{\mathbf{R}^k \times \Omega^c} \;\to\; 0$ as $|z| \to \infty$.

(H6) $g \in W^{3,\infty}(\mathbf{R}^n)$, $(z \cdot \nabla g) \in W^{2,\infty}(\mathbf{R}^n)$.

The complicated appearance of the conditions is due to the fact that we want to permit discontinuities in the functions f and g. We then establish the following results.

Theorem 1.1: Suppose (H1a) and (H2) hold. Then
(a) $\sigma_{pp}(\hat{H}_0^0)$ is a discrete set without accumulation points, except possibly zero and ∞;
(b) $\forall\lambda \notin \sigma_{pp}(\hat{H}_0^0) \cup \{0\}$, $\forall\alpha > \frac{1}{2}$, the norm limits

$$\lim_{\mu\downarrow 0} < z >^{-\alpha} (\hat{H}_0 - \lambda \pm i\mu)^{-1} < z >^{-\alpha} \equiv < z >^{-\alpha} (\hat{H}_0 - \lambda \pm i0)^{-1} < z >^{-\alpha} \quad (1.7)$$

where $< z > = (1 + z^2)^{\frac{1}{2}}$, exist; the limits are uniform for $\lambda \in \Lambda$ compact, $\Lambda \cap (\sigma_{pp}(\hat{H}_0^0) \cup \{0\}) = \emptyset$. With this definition, $< z >^{-\alpha} (\hat{H}_0 - \kappa)^{-1} < z >^{-\alpha}$ is Hölder continuous of order θ as a map from $\hat{\mathcal{H}}_0 = \mathcal{L}(R^n, (c_0^2\rho_0)^{-1}d^n z)$ to $\hat{\mathcal{H}}_0$ on

$$\{\kappa \mid \mathrm{Im}\,\kappa \lesseqgtr 0 \text{ and } \mathrm{Re}\,\kappa = \lambda \notin \sigma_{pp}(\hat{H}_0^0) \cup \{0\}\}, \quad \text{for all} \quad 0 < \theta < \frac{2}{3}(\alpha - \frac{1}{2}).$$

If moreover (H1b), (H3), (H4) and (H5) hold, then
(c) $\sigma_{pp}(\hat{H}_0^0)$ is empty;
(d) $\sigma_{pp}(\hat{H})$ is a discrete set without accumulation points, except possibly zero and ∞;
(e) $\forall\lambda \notin \sigma_{pp}(\hat{H}) \cup \{0\}$ the conclusions of (b) above hold with $\hat{H}_0$ replaced by $\hat{H}$ and $\hat{\mathcal{H}}_0$ by $\hat{\mathcal{H}} = \mathcal{L}^2(R^n, (c^2\rho)^{-1}d^n z)$.

Theorem 1.2: Let (H1) - (H6) hold. Then the pure point spectrum of $\hat{H}$ is empty.

We prove Theorem 1.1 using positive commutator techniques that have proven very effective in the study of N-body Schrödinger operators ([CFKS], [M], [PSS]). The central part of the proof consists of showing a Mourre estimate for $\hat{H}_0^0, \hat{H}_0$ and $\hat{H}$.

Theorem 1.1 improves previously known results in several directions. In the case $m = 1$ a complete proof of the limiting absorption principle for $\hat{H}_0$ was given only in the special case when $\rho_0 \equiv 1$ and c_0 is piecewise constant ([We1],[We2],[We3]). In [BDG], more general ρ_0 and c_0 are considered, i.e. ρ_0, c_0 satisfy (H1a), are constant outside a compact interval, $\rho_0 \in C^1(R)$, $\nabla\rho_0$ is bounded, and c_0 is piecewise continuous with finitely many jump discontinuities. (Note that those conditions imply, but are not implied by (H2)). However, the proof of the limiting absorption principle in [BDG] requires some implicit information on the ξ-dependence of the spectrum of the operators

$$\hat{H}_0^\xi = c_0^2|\xi|^2 - c_0^2\rho_0\nabla_y \cdot \rho_0^{-1}\nabla_y \,. \quad (1.8)$$

The positive commutator method used in this paper does not require any such information and, as a result, allows for a complete proof of the limiting absorption principle for $\hat{H}_0$, using only (H1) and (H2). Note also that the conditions above do not restrict the rate at which ρ_0 and c_0 approach their limiting values at infinity. This means $\hat{H}_0^\xi$ is allowed to have point spectrum accumulating at the bottom of its essential spectrum, and that thresholds can accumulate. Moreover $\hat{H}_0^\xi$ could have imbedded eigenvalues. Remark that the weight $(1 + |z|^2)^{-\frac{\alpha}{2}}$, $\alpha > \frac{1}{2}$, used in Theorem 1.1 is an improvement over the weight

$(1+|x|^2)^{-\frac{a}{2}}\,(1+|y|^2)^{-\frac{a}{2}}$ used in [BDG]. For the case $m > 1$, we are not aware of previous rigorous results in the literature.

To prove Theorem 1.2, we adapt the commutator methods of [FH] (see [CFKS] for a textbook treatment) to the present case. As a first and crucial step in the proof of Theorem 1.2 we establish

Theorem 1.3: Suppose (H1) - (H5) hold. Let ψ be an $\mathcal{L}^2$-eigenfunction of H. Then

$$e^{\alpha|z|}\psi \in \mathcal{L}^2(\mathbf{R}^n, d^n z)\ , \quad \forall\alpha > 0\ . \tag{1.7}$$

Based on one's experience with the Schrödinger equation, one might think this result should immediately imply Theorem 1.2. This is no longer true for the operators appearing in (1.1) and we are forced to add the rather strong condition (H6). Previous results on the absence of point spectrum, for $y \in \mathbf{R}$ and $g \equiv 1$, are found in [We3] and [We4].

We note that for H_0 a much stronger result than Theorem 1.1 can be obtained very simply:

Theorem 1.4: Let (H1a) hold. Then the pure point spectrum of H_0 is empty.

Previous results on the absence of pure point spectrum for H_0 were established for the case $g_0 \equiv 1$ and with $y \in \mathbf{R}$.

References

[CFKS] H.L. Cycon, R.G. Froese, W. Kirsch, B. Simon, *Schrödinger Operators*, Springer-Verlag, Berlin, 1975.

[BDG] M. Ben-Artzi, Y. Dermenjian, J-C. Guillot, Acoustic Waves in Perturbed Stratified media, *Comm. P.D.E.*, Vol. 14(4), 1989, pp. 479-517.

[DBP1] S. DeBièvre, D.W. Pravica, Spectral analysis for optical fibres and stratified fluids I: The limiting absorption principle, *Journal of Functional Analysis*, to appear.

[DBP2] S. DeBièvre, D.W. Pravica, Spectral analysis for optical fibres and stratified fluids II: Absence of eigenvalues, preprint 1990.

[FH] R. Froese, I. Herbst, Exponential bounds and absence of positive eigenvalues for N-body Schrödinger operators, *Commun. Math. Phys.*, 87, 1982, pp.429-447.

[M] E. Mourre, Absence of singular continuous spectrum for certain self-adjoint operators, *Commun. Math. Phys.*, Vol. 78, 1981, pp. 391-408.

[PSS] P. Perry, I.M. Sigal, B. Simon, Spectral Analysis of N-body Schrödinger Operators, *Ann. Math.*, Vol. 114, 1981, pp. 519-567.

[We1] R. Weder, Spectral and Scattering Theory in perturbed stratified fluids, *J. Math. Pures et Appl.*, Vol. 64, 1985, pp. 149-173.

[We2] R. Weder, Spectral and Scattering Theory in perturbed stratified fluids II, *J. Diff. Eqns.*, Vol. 65, 1986, pp. 109-131.

[We3] R. Weder, The limiting absorption principle at thresholds, *J. Math. Pures et Appl.*, Vol. 67, 1988, pp. 313-338.

[We4] R. Weder, Absence of eigenvalues of the acoustic propagator in deformed wave guides, *Rocky Mountain Journal of Mathematics*, Vol. 18, No. 2, 1988, pp.495-503.

Derivation of the Optical Potential in a Model for Low-Energy Neutron Scattering

R. Figari
Dipartimento di Scienze Fisiche
Mostra d'Oltremare Pad. 19
80125 Napoli (Italy)

A. Teta
Istitut für Mathematik
Ruhr-Universität Bochum
D-4630 Bochum 1 (FRG)

Abstract

We propose a simple model for low-energy neutron scattering from condensed matter and compute the corresponding optical potential. The result is based on the analysis of a Schrödinger hamiltonian with N δ-sphere interactions of radius N^{-1} and suitable renormalized strength in the limit $N \to \infty$.

In this note we discuss an application to neutron optics of a result on the asymptotic behaviour of the Schrödinger operator with many (and suitable renormalized) δ-sphere interactions.

Neutron optics is the branch of nuclear physics devoted to the study of the scattering of slow neutrons from condensed matter, producing typical optical phenomena like reflection, diffraction, adsorbtion and so on (we refer to [1],[2] for a comprehensive treatment of the whole subject).

We shall limit our attention to the case in which the target contains a large number of nuclei and

$$\lambda_{neu.} \gg d \gg r_0 \tag{1}$$

where we have denoted by $\lambda_{neu.}$ the wavelength of the incident neutron, by d the mean internuclear distance and by r_0 the range of the interaction between the neutron and the nuclei of the target. Moreover the nuclei are sufficiently heavy to be considered fixed in their equilibrium positions.

The above conditions are satisfied, with a good approximation, in a variety of experimental situations (e.g. in the scattering of cold neutrons from liquids or amorphous solids typically one has $\lambda_{neu.} \gg 1$ Å, d is few Å and $r_0 \simeq 10^{-4}$ Å).

It is reasonable that, if (1) holds, the wavelength of the incoming particle "sees" only an effective potential, i.e. the motion of the neutron is well described by a one-body Schrödinger hamiltonian

$$H_{eff} = -\Delta + U \tag{2}$$

where U, appearing in the literature as the "optical potential", depends on the density of the distribution of the nuclei and some low-energy property of the neutron-nuclei interaction (to simplify the notation we put $\hbar = 1$ and $m_{neu.} = 1/2$).

In general U can be complex, with an imaginary part describing absorption of the neutrons by the target.

From a theoretical point of view a central task is to derive an explicit expression for U starting from a given distribution of the nuclei and a model of neutron-nuclei interaction.
Such a goal, which is the object of the theory of dispersion, has been studied in the physical literature since long time (see references in [2]). Different derivations of the optical potential have been given, using models of increasing complexity in order to take into account various effects like multiple scattering or correlations in the positions of the nuclei in the target. Nevertheless, as far as we know, only zero-range potentials were used to build up models for neutron-nuclei interaction.

Here we present a rigorous derivation of the optical potential for a model with a short-range interaction between neutron and nuclei. More precisely let $y_1, \ldots, y_N \in R^3$ be the positions of the nuclei of the target, distributed according to some density ρ, with

$$\rho > 0\,, \quad \rho \in L^1(R^3)\,, \quad \|\rho\|_{L^1(R^3)} = 1$$

and such that the following assumptions are satisfied

A_1

$$\lim_{N\to\infty} \frac{1}{N}\sum_{i=1}^{N} \phi(y_i) = \int_{R^3} \phi(x)\rho(x)\, dx$$

for any continuous function $\phi : R^3 \to R$.

A_2 There exist $C_1 > 0$ and $\nu > 0$ such that

$$min_{i\neq j} |y_i - y_j| > C_1 N^{-1+\nu} .$$

A_3 There exists C_2, $0 < C_2 < \infty$, such that

$$\frac{1}{N^2} \sum_{i,j=1,i\neq j}^{N} \frac{1}{|y_i - y_j|^2} < C_2 .$$

A_4

$$\lim_{N\to\infty} \frac{1}{N^3} \sum_{i,j=1,i\neq j}^{N} \frac{1}{|y_i - y_j|^4} = 0 .$$

Notice that $A_1, \ldots, A_4$ are satisfied with probability increasing to 1 if the points y_i, $i = 1, \ldots, N$, are independently randomly chosen according to the common density ρ. Obviously, they are also satisfied by a periodic array of decreasing mesh inside a finite region Ω with $\rho(x) = |\Omega|^{-1}$ if $x \in \Omega$ and $\rho(x) = 0$ otherwise.

Moreover we assume that the potential between the neutron and the nucleus in y_i, $i = 1, \ldots, N$, is given by a δ-sphere interaction with center in y_i, radius N^{-1} and strength $N\gamma_i$, where $\gamma_i \in R$. This means that the model hamiltonian for the neutron is

$$H_N = -\Delta + \sum_{i=1}^{N} N\gamma_i\, \delta(|\cdot - y_i| - N^{-1}) . \tag{3}$$

It is easily seen that the hamiltonian (3) is a well defined selfadjoint operator in $L^2(R^3)$ which is bounded below (in fact, using standard trace embeddings, the δ-sphere potential is shown to be a small form perturbation with respect to $-\Delta$).

Notice that the range of the neutron-nuclei interaction is chosen to be N^{-1} and the average internuclear distance in the target, by assumptions $A_1, \ldots, A_4$, is of order $N^{-1/3}$. If the wavelength of the incident neutron is taken of order 1 then, for large N, our model hamiltonian describes a situation in which condition (1) is satisfied.

Our result, summarized in the theorem below, shows that in the limit $N \to \infty$ H_N reduces to an effective hamiltonian of the type (2).

Theorem 1. Let the assumptions $A_1, \ldots, A_4$ be satisfied. Let $\gamma_i = \gamma(y_i)$ for some function γ such that, outside a set of $\rho(x)dx$-measure zero, γ^{-1} is continuous and $dist(R(\gamma), O^-) > C > 0$ where $R(\gamma)$ is the range of the function γ and O^- is the set of the negative odd integers.
There exists a $\lambda_1 > 0$ such that for any $\lambda > \lambda_1$ and for any $f \in L^2(R^3)$

$$\lim_{N\to\infty} \| (H_N + \lambda)^{-1} f - (-\Delta + \frac{4\pi\rho}{1+\gamma^{-1}} + \lambda)^{-1} f \|_{L^2(R^3)} = 0 . \qquad (4)$$

The proof can be found in [3], [4]. Here we shall limit ourselves to some comments on the result.

An explicit computation (see e.g. [5]) gives that quantity $(1+\gamma^{-1})^{-1}$ is the scattering length associated to a δ-sphere potential of radius 1 and strength γ. Theorem 1 then says that the optical potential appearing in the limit operator in (4) is nothing but 4π times the density of scattering length of the neutron-target interaction.

Introducing a complex strength for the δ-sphere interaction in (3), it is possible to generalize the optical potential to complex values without any essential change in the proof.

Conceivably the result expressed in theorem 1 is true for a very general class of neutron-nuclei interactions given by suitable rescaled potentials of arbitrary shape. Work in such more general context is in progress.

A result similar to theorem 1, in the context of the homogeneization theory, was obtained recently by Brillard ([6]) using epi-convergence techniques. He considered interactions supported by surfaces of arbitrary shape distributed in a periodic array with positive definite values of the strength of the interaction. Theorem 1 generalizes Brillard's result to attractive interactions, a typically important case for applications in non relativistic quantum mechanics.

References

1. S.W. Lovesey, *Theory of Neutron Scattering from Condensed Matter*, (Claredon Press, Oxford, 1984).
2. V.F. Sears, *Neutron Optics*, (Oxford University Press, Oxford, 1989).
3. R. Figari, H. Holden, A. Teta, *J. Stat. Phys.* **51** (1988) 205-214.
4. R. Figari, A. Teta, SBF 237 - Preprint Nr. 95, Institut für Mathematik, Ruhr-Universität Bochum, October 1990.
5. J.P. Antoine, F. Gesztesy, J. Shabani, *J. Phys.* **A20** (1987) 3687-3712.
6. A. Brillard, *Math. Mod. and Num. Anal.* **22** (1988) 187-216.

Three particles in a lattice : a model of interaction and dynamical equations

Yu.A.Kuperin, B.S.Pavlov

Leningrad University, Soviet Union

In this paper the three–body scattering problem on the lattice is considered, assuming that interaction between particles is of a resonance nature; that is, each two particles at small distances form a compound, which has some inner degrees of freedom. This sort of interaction is typical for the theory of solids. Thus an electron and a hole form a compound state-exciton, which has a hydrogen–like spectrum of excitations.

1. Basic construction and notations

Let A_α , $\alpha = 1,2,3$ be selfadjoint (s.a.) operators, which act in the Hilbert spaces $\mathcal{H}_\alpha^{ex}$. We use the realization of A_α in form of constant Jacobian matrices, $\mathcal{H}_\alpha^{ex} \simeq l_2(\mathbb{Z}_\alpha^3)$.

$$(A_\alpha u^\alpha)_s = \sum_{|s-t|\leq 1} a_{s-t}^\alpha u_t^\alpha \, , \, a_{-s}^\alpha = \bar{a}_s^\alpha \, , s,t \in \mathbb{Z}_\alpha^3$$

In the Hilbert space $\mathcal{G}^{ex} = \bigotimes_{\alpha=1}^{3} \mathcal{H}_\alpha^{ex}$ we consider the s.a. operator H^{ex} – the "tensor sum" of A_α – the free three–body Hamiltonian:

$$H^{ex} := \sum_\alpha A_\alpha \times I_{\hat{\alpha}}$$

Here we denote $\hat{\alpha} = \beta\gamma$, $\beta,\gamma = 1,2,3$, $\beta \neq \gamma$, $\alpha \neq \beta,\gamma$, and $I_{\hat{\alpha}} = I_\beta \times I_\gamma$ is the tensor product of units in $\mathcal{H}_\beta^{ex}$, $\mathcal{H}_\gamma^{ex}$. We denote T_α^l the translation on the lattice $\mathbb{Z}_\alpha^3$, $(T_\alpha^l u)_s = u_{s_{l_\alpha}}$ and by $L_{\hat{\alpha}}, \hat{\alpha} = 12, 23, 31$ - lattice planes

$$L_{\hat{\alpha}} = \{l_\alpha, l_\beta, l_\gamma : l_\beta = l_\gamma\} \, .$$

The "equidomoid" $\{l_\alpha, l_\beta, l_\gamma : l_\alpha = l_\beta = l_\gamma\}$ will be denoted by L_0.
Together with H^{ex} we consider the free pair Hamiltonians

$$H_{\hat{\alpha}}^{ex} = I_\beta \times A_\gamma + A_\beta \times I_\gamma \, , \, \hat{\alpha} = \beta\gamma \, , \tag{1}$$

which act in $\mathcal{H}_{\hat{\alpha}}^{ex} = \mathcal{H}_\beta^{ex} \otimes \mathcal{H}_\gamma^{ex}$. The Hamiltonians H^{ex} , $H_{\hat{\alpha}}^{ex}$ we call exterior ones.

We introduce the pair compound Hilbert space $\mathcal{B}_{\hat{\alpha}} = l_2(\mathbb{Z}^3_{\hat{\alpha}}) \otimes \mathcal{L}_{\hat{\alpha}}$ and the inner compound Hamiltonian $B_{\hat{\alpha}} = \text{diag}\ \{b_{\hat{\alpha}}\}$, $b_{\hat{\alpha}}$ is s.a. in $\mathcal{L}_{\hat{\alpha}}$ where $\mathcal{L}_{\hat{\alpha}}$ is an arbitrary Hilbert space. At the moment we do not demand any special properties from them, but in the further development we assume them to be finite dimensional. Actually we use in our constructions the aggregates

$$H^{in}_{\alpha} = A_{\alpha} \times I_{\hat{\alpha}} + I_{\alpha} \times B_{\hat{\alpha}} \ , \tag{2}$$

which are s.a. operators in $\mathcal{G}^{in}_{\hat{\alpha}} = l^2(\mathbb{Z}^3_{\alpha}) \times \mathcal{B}_{\hat{\alpha}}$. We call them the inner Hamiltonians. Actually they are the compound Hamiltonians in the laboratory coordinate system.

At last we introduce also the three–body compound Hamiltonian H^{in}_0 which is supposed to be s.a. and diagonal $H^{in}_0 = \text{diag}\ \{H_0\}$ in the Hilbert space $l^2(\mathbb{Z}_3) \otimes \mathcal{G}^{in}_0$, where H_0 is an arbitrary s.a. operator acting in $\mathcal{G}^{in}_0$. To save space we use the joint multiindex $a = \{\alpha, 0\}$ and $\hat{a} = \{\hat{\alpha}, 0\}$. Actually the role of the unperturbed ("background") Hamiltonian (with no interaction between channels) is played by the orthogonal sum

$$H = H^{ex} \oplus \sum_{a=0}^{3} H^{in}_a$$

which is a s.a. operator in $\mathcal{G}^{ex} \oplus \sum_{a=0}^{3} \mathcal{G}^{in}_a$. The interaction will be switched on by means of the extension theory.

2. Restriction of the background Hamiltonian and its s.a. extension

Usually when constructing solvable models in continuous case (see [1], [2]) one deals with densely defined symmetric operators. But still the problem of attaching a matrix to the Laplacian demands accurate treating of the non–densely defined case.

As for lattice models, the non–densely defined operators appear in most situations. In these situations the domain of the restricted operator, $A \to A_0$, is described usually (see [3]) by previous fixing of deficiency subspaces $\mathcal{M}_i$, $\mathcal{M}_i \in \mathcal{D}_A$:

$$\mathcal{D}_0 = \left\{ \frac{1}{A - iI} v^{\perp} \, , \, v^{\perp} \perp \mathcal{M}_i \right\} .$$

If $\mathcal{M}_i$ forms a non–zero angle with $\mathcal{M}_{-i} = \dfrac{A + iI}{A - iI} \mathcal{M}_i = U^+ \mathcal{M}_i$, then the adjoint operator can be correctly defined on the sum $\mathcal{M}_i + \mathcal{M}_{-i}$ by the usual rule [3]:

$$(A_0^* + iI)\theta = 0 \, , \ \theta \in \mathcal{M}_i \ ,$$
$$(A_0^* - iI)\theta = 0 \, , \ \theta \in \mathcal{M}_{-i} \ .$$

Starting from an orthonormal basis $\{\theta_s\}$, $\theta_s \in \mathcal{M}_i$, we can construct the basis $U^+\theta_s$ in $\mathcal{M}_{-i}$ and then the following basis in $\mathcal{M}_i + \mathcal{M}_{-i}$:

$$W_s^+ = \frac{U^* + I}{2}\theta_s = \frac{A}{A-i}\theta_s \,,$$
$$W_s^- = \frac{U^* - I}{2i}\theta_s = \frac{1}{A-i}\theta_s \,.$$

It is easy to see that the adjoint operator combines the elements $W_s^\pm$ in the following way:

$$A_0^* W_s^+ = -W_s^- \,,$$
$$A_0^* W_s^- = W_s^+ \,.$$

For elements $u_{1,2} = \sum_s \xi_{1,2}^{s+} W_s^+ + \sum_s \xi_{1,2}^{s-} W_s^- = \frac{A}{A-iI}\xi_{1,2}^+ + \frac{1}{A-iI}\xi_{1,2}^-, \xi^\pm = \sum_s \xi_{1,2}^\pm \theta_s \in \mathcal{M}_i$ the boundary form is easily calculated:

$$J(u-1, u_2) = \langle A_0^* u_1, u_2 \rangle - \langle u_1, A_0^* u_2 \rangle = \langle \xi_1^-, \xi_2^+ \rangle - \langle \xi_1^+, \xi_2^- \rangle \,. \tag{3}$$

The construction of a selfadjoint extension of A_0 is equivalent to the choosing of a neutral (Lagrangian) plane of the boundary form (3) in $\mathcal{M}_i + \mathcal{M}_{-i}$. Some of those planes correspond to the boundary conditions with Hermitian operators $\Gamma : \mathcal{M}_i \longrightarrow \mathcal{M}_i$:

$$\xi^- = \Gamma \xi^+;$$

the others can be produced from them by J-unitary transformation in $\mathcal{M}_i + \mathcal{M}_{-i}$, conserving the boundary form. So the domain of the selfadjoint extension A_Γ is

$$\mathcal{D}_\Gamma = \left\{ U : U = \frac{1}{A-iI} v^\perp + \frac{A}{A-iI}\xi^+ + \frac{1}{A-iI}\Gamma\xi^+ \right\} \,, \tag{4}$$

and acting of A_Γ on U is given by the formula

$$A_\Gamma U = \frac{A}{A-iI} v^\perp - \frac{1}{A-iI}\xi^+ + \frac{A}{A-i}\Gamma\xi^+ \,.$$

Thus the resolvent of A_Γ is easily calculated as a solution of the equation

$$A_\Gamma U - zU = f \,. \tag{5}$$

This equation is equivalent to the next equation in the deficiency space $\mathcal{M}_i$

$$\Gamma\xi = P_{\mathcal{M}_i} \frac{I + zA}{A - zI}\xi + P_{\mathcal{M}_i} \frac{A - iI}{A - zI} f \,.$$
$$\xi = \left[\Gamma - P_{\mathcal{M}_i} \frac{I + zA}{A - zI} |_{\mathcal{M}_i} \right]^{-1} P_{\mathcal{M}_i} \frac{A - iI}{A - zI} f \,. \tag{6}$$

where $P_{\mathcal{M}_i}$ is the orthoprojector onto deficiency subspace $\mathcal{M}_i$. From (6) follows the fundamental Krein's formula for resolvents of s.a. extensions:

$$(A_\Gamma - zI)^{-1} = (A - zI)^{-1} + \frac{A + iI}{A - zI}\left[\Gamma - P_{\mathcal{M}_i}\frac{I + zA}{A - Z}|_{\mathcal{M}_i}\right]^{-1} P_{\mathcal{M}_i}\frac{A - iI}{A - zI} \, .$$

In our case we start with fixing of deficiency subspaces in the exterior channel $\mathcal{G}^{ex}$. Planning to switch on the resonant interaction between pairs of particles, we restrict first the free pair Hamiltonian $H^{ex}_{\hat{\alpha}}$, specifying its deficiency subspace $\mathcal{M}^{\hat{\alpha}}_i$. Namely, we construct the basis in $\mathcal{M}^{\hat{\alpha}}_i$ by translating a fixed element by T^l_α ,

$$\theta_{\hat{\alpha}} = \frac{1}{H^{in}_{\hat{\alpha}} + iI}\rho_0 \quad , \quad \rho_0 = \delta_{l_\beta 0} \cdot \delta_{l_\gamma 0} \tag{7}$$

along the plane. The domain $\mathcal{D}^{\hat{\alpha}}_0$ of the restricted Hamiltonian $H^{ex}_{\hat{\alpha}0}$ consists of all vectors

$$\tilde{u} = \frac{1}{H^{ex}_\alpha - iI}v^\perp \quad \subset \quad l_2(\mathbb{Z}^3_\beta) \times l_2(\mathbb{Z}^3_\gamma) \, ,$$

satisfying the orthogonality condition $(H^{ex}_\alpha - iI)\tilde{u} = v^\perp \perp \mathcal{M}^{\hat{\alpha}}_i$. This condition is equivalent to the following,

$$\langle \tilde{u}, T^l_\alpha \rho \rangle = \tilde{u}_{ll} = 0 \quad , \ \forall l \in \mathbb{Z}^3_\alpha \, .$$

Thus the domain $\mathcal{D}^{\hat{\alpha}}_0$ consists of all vectors from $\mathcal{D}(H^{ex}_{\hat{\alpha}})$, which vanish on the plane $L_{\hat{\alpha}}$. The described restriction of $H^{ex}_{\hat{\alpha}}$ generates the corresponding restriction of the free Hamiltonian $(H^{ex})^{\hat{\alpha}}_0 = A_\alpha \times I_{\hat{\alpha}} + I_\alpha \times H^{ex}_{\hat{\alpha}0}$. We calculate the corresponding deficiency subspace $\mathcal{M}_{\hat{\alpha}}$.

Lemma 1.

$$\mathcal{M}_{\hat{\alpha}} = \bigvee_{l_\alpha, l_\beta = l_\gamma} \left\{ \frac{1}{H^{ex} - iI}\delta_{l_\alpha 0}\delta_{l_\beta 0}\delta_{l_\gamma 0} \right\} \, .$$

Proof. The domain $\mathcal{D}_{0\hat{\alpha}}$ of $(H^{ex})^{\hat{\alpha}}_0$ consists of all vectors $\tilde{u}_{\hat{\alpha}}$ from $\mathcal{D}(A_\alpha) \otimes \mathcal{D}(H^{ex}_{\hat{\alpha}0})$, which vanish at the plane $L_{\hat{\alpha}} = (l_\alpha, l_\beta, l_\gamma; l_\beta = l_\gamma)$:

$$0 = \tilde{u}_{\hat{\alpha}}(l_\alpha, l_\beta, l_\gamma)L_{\hat{\alpha}} =< \tilde{u}_{\hat{\alpha}}, \delta_{l_\beta l_\gamma}\delta_{l_\beta}\delta_{l_\beta}$$

Writing $\tilde{u}_{\hat{\alpha}}$ in usual form

$$\tilde{u}_{\hat{\alpha}} = \frac{1}{H^{ex} - iI}v^\perp \quad , \ v^\perp \in \mathcal{M}_{\hat{\alpha}}$$

we get for each l_α and $l_\beta = l_\gamma$:

$$0 = \langle v^\perp, \frac{1}{H^{ex} - iI}\delta_{l_\beta l_\gamma}$$

Note, that the described restriction is not densely defined, which causes generally some troubles with the construction of s.a. extensions (see [3]). But fortunately in our case the deficiency subspaces $\mathcal{M}_{\hat{\alpha}}$ and $\bar{\mathcal{M}}_{\hat{\alpha}}$, corresponding to the spectral points $i, -i$ form a positive angle. This can be deduced from the fact, that the Cayley transform of the free three-body Hamiltonian in Fourier representation $\frac{\sum_\alpha a_\alpha(P_\alpha)+i}{\sum_\alpha a_\alpha(p_\alpha)-i}$ can't be written as a function of quasimomentum $p_\beta + p_\gamma$ of the pair $\hat{\alpha}$. Using the positivity of this angle we could proceed in construction of the s.a. extension just formally (see [4]).
Analogously the other deficiency subspaces $\mathcal{M}_{\hat{\beta}}, \mathcal{M}_{\hat{\gamma}}$, associated with the planes L_β, L_γ can be constructed. The final restriction of H^{ex} we get by summation of deficiency subspaces

$$H^{ex} \longrightarrow H_0^{ex} \ ; \ \mathcal{D}_0^{ex} = \{\tilde{u} : u|_{L_{\hat{\alpha}}} = 0, \ \alpha = 1,2,3\} \ .$$

Note, that the elements of this domain vanish on the equidomoid $L_0 = L_{\hat{1}} \cap L_{\hat{2}} \cap L_{\hat{3}}$ automatically. The joint deficiency subspace $\mathcal{M}^{ex} = \mathcal{M}_i^{ex}$ associated with $L = \cup_\alpha L_{\hat{\alpha}}$ has an empty intersection with the deficiency subspace $\bar{\mathcal{M}}^{ex} = \mathcal{M}_{-i}^{ex}$. It simplifies the construction of s.a. extension.
Actually one can start with a description of a domain of formally adjoint operator $(H_0^{ex})^*$

$$\left\{ u := \tilde{u} + \frac{H^{ex}}{(H^{ex})^2+1}\rho^+ + \frac{1}{(H^{ex})^2+1}\rho^- \right\}, \sup \rho^\pm \subset L.$$

The corresponding boundary form can be calculated in terms of boundary values $\rho^\pm$ only (see [5])

$$\langle (H_0^{ex})^* u_1, u_2\rangle - \langle u_1, (H_0^{ex})^* u_2\rangle = \left\langle \frac{1}{(H^{ex})^2+1}\rho_1^-, \rho_2^+ \right\rangle_L - \left\langle \rho_1^+, \frac{1}{(H^{ex})^2+1}\rho_2^- \right\rangle_L . \tag{7}$$

It is useful to notice that the vector $\frac{1}{(H^{ex})^2+1}\rho^-|_L$ is just a renormalized value of the element $u, u \in \mathcal{D}((H_0^{ex})^*)$ on L:

$$u_L^{ren} = u - \frac{H^{ex}}{(H^{ex})^2+1}\rho^+|_L = \frac{1}{(H^{ex})^2+1}\rho^-|_L \ .$$

Thus the boundary form (4) is

$$\langle u_{1L}^{ren}, \rho_2^+ \rangle_L - \langle \rho_1^+, u_{2L}^{ren} \rangle \tag{8}$$

Notice, that ρ^+ and u_L^{ren} are defined correctly due to the fact that the angle between $\mathcal{M}_i^{ex}, \mathcal{M}_{-i}^{ex}$ is positive.
The restriction of the inner Hamiltonian $H_{\hat{\alpha}}^{in}$ will be also performed in terms of the two-body compound $b_{\hat{\alpha}}$. We restrict first $b_{\hat{\alpha}} \longrightarrow b_{\hat{\alpha}0}$ so that $\Theta_{\hat{\alpha}}, \Theta_{\hat{\alpha}} \in \mathcal{L}_{\hat{\alpha}} |\Theta_{\hat{\alpha}}| = 1,$

becomes a deficiency element, supposing Θ_α and $\frac{b_{\hat\alpha}-i}{b_{\hat\alpha}+1}\Theta_{\hat\alpha}$ are linearly independent. Then despite the fact of nondense domain of $(b_{\hat\alpha})_0$:

$$\mathcal{D}^{\hat\alpha}_{0_b} = \frac{1}{b_{\hat\alpha} - iI} v^\perp , v^\perp \perp \Theta_{\hat\alpha}$$

the formal adjoint $(H^{in}_{\hat\alpha 0})^*$ can be defined on

$$(\mathcal{D}^{\hat\alpha}_{0b})^* = \left\{ u : u = \frac{1}{b_{\hat\alpha} - iI} v^\perp_{\hat\alpha} + \frac{b_{\hat\alpha}}{b_{\hat\alpha} - iI} \xi^+_{\hat\alpha} + \frac{1}{b_{\hat\alpha} - iI} \xi^-_{\hat\alpha} \right\}$$

in the standard manner, where $\xi^\pm_{\hat\alpha}$ are the linear combination of deficiency elements $\Theta_{\hat\alpha}$, translated along the lattice $L_{\hat\alpha}$:

$$\xi^\pm_{\hat\alpha} = \sum (\xi^\pm_{\hat\alpha})^l \cdot T_l \Theta_{\hat\alpha} ,$$

and $v^\perp$ fulfills the orthogonality condition

$$\langle (v^\perp)_{l_\alpha, l_{\hat\alpha}}, \Theta_{\hat\alpha} \rangle_{\hat\alpha} = 0 .$$

The boundary form of the formal adjoint operator $(H^{in}_{\hat\alpha})^*_0$

$$(H^{in}_{\hat\alpha})^*_0 u = H^{in}_\alpha \frac{1}{\mathcal{B}_{\hat\alpha} - iI} (v^\perp_{\hat\alpha} + \xi^-_{\hat\alpha}) + \frac{A_\alpha \mathcal{B}_{\hat\alpha}}{\mathcal{B}_{\hat\alpha} - iI} \xi^+_{\hat\alpha} - \frac{1}{\mathcal{B}_{\hat\alpha} - iI} \xi^+_{\hat\alpha}$$

can be calculated similarly to [5]:

$$\langle (H^{in}_{\hat\alpha})^* u_1, u_2 \rangle - \langle u_1, (H^{in}_{\hat\alpha})^* u_2 \rangle = \langle \xi^-_{1_{\hat\alpha}}, \xi^+_{2_{\hat\alpha}} \rangle - \langle \xi^+_{1_{\hat\alpha}}, \xi^-_{2_{\hat\alpha}} \rangle . \tag{9}$$

Finally, the three-body compound Hamiltonian $H^{in}_0 = \text{diag}\ \{b_0\}$ will be restricted in a similar way:

$$b_0 \longrightarrow (b_0)_0 \quad : \quad \mathcal{D}_0 = \left\{ u : u = \frac{1}{b_0 - iI} \Theta_0 \right\} ,$$

$$\mathcal{D}^*_0 = \frac{1}{b_0 - iI} v^\perp + \frac{b_0}{b_0 - iI} \xi^+_0 + \frac{1}{b_0 - iI} \xi^-_0 ,$$

where $\langle (v^\perp)_l, \Theta_0 \rangle = 0$, $\xi^\pm_0 = \sum_{l \in Z_3} (\xi^\pm_0)^l T_l \Theta_0$. The corresponding boundary form is

$$\langle \xi^-_{10}, \xi^+_{20} \rangle - \langle \xi^+_{10}, \xi^-_{20} \rangle . \tag{10}$$

In the boundary forms (8), (9) we can write the scalar products in $l_2(Z^3_\alpha) \times l_2(Z^3_{\hat\alpha})$ and in $l_2(Z^3_0)$, correspondingly, taking the Fourier coefficients $(\xi^\pm_a)^l$ instead of $\xi^\pm_a$, $a = 0, 1, 2, 3$.

The boundary form of the restricted background Hamiltonian is just a sum of (7) (8) (9). The selfadjoint extension of it is defined by Lagrangian plane in the joint deficiency space $(\mathcal{H}+\bar{\mathcal{H}}) \oplus \sum_{\hat{\alpha}=1} (\mathcal{H}_{\hat{\alpha}} + \bar{\mathcal{H}}_{\hat{\alpha}}) \oplus (\mathcal{H}_0 + \bar{\mathcal{H}}_0)$, where this sum vanishes. From many possibilities of choosing the plane we take the quasilocal boundary condition, connecting locally the boundary values separately on the planes $L_{\hat{\alpha}}$

$$\begin{pmatrix} (u^{ren})_{l_\alpha, l_{\hat{\alpha}}} \\ \xi^-_{\hat{\alpha}_{l_\alpha, l_{\hat{\alpha}}}} \end{pmatrix} = \Gamma_{\hat{\alpha}} \begin{pmatrix} (\rho)_{l_\alpha, l_{\hat{\alpha}}} \\ \xi^+_{\hat{\alpha}_{l_\alpha, l_{\hat{\alpha}}}} \end{pmatrix}_{l_\alpha \neq l_{\hat{\alpha}}} \tag{11}$$

and on the equidomoid

$$\begin{pmatrix} (u^{ren})_{l,l,l} \\ (\xi_0^-)_{l_0} \end{pmatrix} = \Gamma_0 \begin{pmatrix} \rho_{l,l,l} \\ (\xi_0^+)_l \end{pmatrix} \tag{12}$$

Here $\Gamma_{\hat{\alpha}}, \Gamma_0$ are Hermitian matrices with elements γ^a

$$\Gamma_a = \begin{pmatrix} \gamma^a_{ll} & \gamma^a_{li} \\ \gamma^a_{il} & \gamma^a_{ii} \end{pmatrix} \quad , \quad a = \alpha, 0 \; .$$

Another sort of lagrangian planes, which also corresponds to s.a. extensions can be obtained by I–unitary transformations, preserving boundary forms, say

$$(\xi^-, \xi^+) \longrightarrow (-\xi^+, \xi^-) \; .$$

We shall not discuss this possibility.

3. Resolvent and dynamical equations

The general form of resolvent for extensions is given by Krein–formula, see [6], which replaces inverting of the operator by inverting of some reduced operators, acting in the boundary spaces. This procedure can be peformed explicitely for two–body problems, see [7], but in the three–body case it gives some sort of integral equations, which are equivalent to Faddeev equations for corresponding densities, which play the role of Faddeev components (see [8]).
To construct the resolvent, we must find the solution of adjoint equations

$$((H^{ex})_0^* - zI)\, u^{ex} = f^{ex} \tag{13}$$

$$((H^{in}_{\hat{\alpha}})_0^* - zI)\, u^{in}_{\hat{\alpha}} = f^{in}_{\hat{\alpha}} \tag{14}$$

$$((H^{in}_0)_0^* - zI)\, u^{in}_0 = f^{in}_0 \tag{15}$$

submitted to the boundary conditions (11), (12). It is easy to suggest the Ansätze for u^{ex}

$$u^{ex} = \frac{1}{H^{ex} - zI} f^{ex} + \sum_{a=0}^{3} \frac{1}{H^{ex} - zI} \xi^{ex}_\alpha \; . \tag{16}$$

Here ξ_α^{ex} is localized on $L_{\hat\alpha}\backslash L_0$ and ξ_0^{ex} on L_0. Similarly we can write Ansatz for u_0^{in}

$$u_0^{in} = \frac{1}{H_0^{in} - zI} f_0^{in} + \frac{1}{H_0^{in} - zI}\xi_0^{in} , \tag{17}$$

where $\xi_0^{in} = \sum_{l\in Z^3} (\xi_0^{in})^l T^l (b_0 + iI)\Theta_0^{in}$.

In both cases the boundary values can be easily calculated.

Lemma 2. The equations (13), (15) have the solutions of the following form

$$u^{ex} = \frac{1}{H^{ex} - \lambda I} f^{ex} + \frac{H_0^{ex} + iI}{H^{ex} - zI}\cdot \sum_a \xi_a^{ex} ,$$

$$u_0^{in} = \frac{1}{H_0^{in} - \lambda I} f^{in} + \frac{H_0^{in} + iI}{H_0^{in} - zI}\cdot \xi_0^{in} .$$

Here $\xi_a^{ex} = \frac{1}{H^{ex}+iI}\rho_a$; ρ_a is localized on $L_\alpha\backslash L_0$, ρ_0 localized on ρ_0; $\xi_0^{in} = \sum (\xi_0^{in})^l T^l \Theta_0^{in}$, $\rho = \sum_{a=0}^{3} \rho_a$.

The boundary values of u^{ex}, u_0^{in} are

$$u_a^{ex} \sim \begin{cases} \rho_a \\ u_{l_a}^{ren} = \frac{1}{H^{ex}-iI} P_a \left[\frac{HzH^{ex}}{H^{ex}-z}\frac{\rho}{H^{ex}+iI} + \frac{H^{ex}-iI}{H^{ex}-zI} f^{ex}\right] \end{cases} , \quad a = 0,1,2,3 \tag{18}$$

$$u_0^{in} \sim \begin{cases} (\xi_0^{in})^l \\ \langle \frac{1+b_0 z}{b_0 - z}\Theta_0, \Theta_0\rangle (\xi_0^{in})^l + \langle \frac{b_0^{in}-iI}{H_0^{in}-zI}(f_0^{in})^l, \Theta_0\rangle \end{cases} \tag{19}$$

Notice, that the renormalized value of the solution u^{ex} on the plane $L_\alpha\backslash L_0$ or L_0 can be calculated just by restriction onto them the agregate

$$\left\{ u^{ex} - \frac{H^{ex}}{(H^{ex})^2 - 1}\rho \right\}$$

The calculation concerning the two–body part of the inner hamiltonian are a bit more delicate. Mimicking the argument of [5], we compute the formal solution of (14), starting with the Ansatz

$$u_{\hat\alpha}^{in} = \frac{1}{b_{\hat\alpha} - iI} v_{\hat\alpha}^{in} + \sum_{\substack{l = l_\alpha, l_{\hat\alpha} \\ l_\alpha \neq l_{\hat\alpha}}} \frac{b_{\hat\alpha}}{b_{\hat\alpha} - iI} T^l \Theta_{\hat\alpha} (\xi_{\hat\alpha}^{+})^l + \sum_{\substack{l = l_\alpha, l_{\hat\alpha} \\ l_\alpha \neq l_{\hat\alpha}}} \frac{1}{b_{\hat\alpha} - iI} T^l \Theta_{\hat\alpha} (\xi_{\hat\alpha}^{-})^l . \tag{20}$$

According to general scheme [5] applying of $(H_{\hat\alpha}^{ex})_0^* = A_\alpha \times i_{\hat\alpha} + I_\alpha \times$ diag $\{b_{\hat\alpha}\}_0^*$

to (20) gives the following terms

$$\{(H^{ex}_{\hat\alpha})^*_0 - zI\}\,\frac{1}{b_{\hat\alpha} - iI}v^{in}_{\hat\alpha} = \frac{H^{ex} - zI}{b_{\hat\alpha} - iI}v^{in}_{\hat\alpha}$$

$$\{(H^{ex}_{\hat\alpha})^*_0 - zI\}\,\frac{1}{b_{\hat\alpha} - iI}T^l\Theta_h a = \frac{H_{\hat\alpha} - zI}{b_{\hat\alpha} - iI}T^l\Theta_{\hat\alpha}$$

$$\{(H^{ex}_{\hat\alpha})^*_0 - zI\}\,\frac{b_{\hat\alpha}}{b_{\hat\alpha} - iI}T^l\Theta_h a\cdot\xi = A_\alpha\cdot\xi\cdot\frac{b_{\hat\alpha}}{b_{\hat\alpha} - iI}T^l\Theta_{\hat\alpha} - \frac{1}{b_{\hat\alpha} - iI}T^l\Theta_{\hat\alpha}\cdot\xi$$
$$- z\frac{b_{\hat\alpha}}{b_{\hat\alpha} - iI}T^l\Theta_{\hat\alpha}\cdot\xi$$

since $(b_{\hat\alpha})^*_0\frac{b_{\hat\alpha}}{b_{\hat\alpha}-iI}\Theta_{\hat\alpha} = -\frac{1}{b_{\hat\alpha}-iI}\Theta_{\hat\alpha}$.

Lemma 3. Let us denote

$$\sum_{\substack{l=l_\alpha,l_{\hat\alpha}\\ l_\alpha\neq l_{\hat\alpha}}} T^l\Theta_0(\xi_{\hat\alpha})^l = \xi_{\hat\alpha}$$

Then the Ansatz for the solution of the equation (14) can be taken in form

$$u^{in}_{\hat\alpha} = \frac{1}{H^{in}_{\hat\alpha} - zI}f_{\hat\alpha} + \frac{\mathcal{B}_{\hat\alpha} + iI}{H^{in}_{\hat\alpha} - zI}\xi^{in}_{\hat\alpha}\,. \tag{21}$$

The boundary values of this Ansatz are

$$\begin{aligned}\xi^+_{\hat\alpha} &= \xi^{in}_{\hat\alpha} + \rho^f_{\hat\alpha}\\ \xi^-_{\hat\alpha} &= P^{in}_{\hat\alpha}\frac{1 + (z - A_\alpha\times i_{\hat\alpha})\mathcal{B}_{\hat\alpha}}{H^{in}_{\hat\alpha} - zI}\xi^{in}_{\hat\alpha} + P^{in}_{\hat\alpha}\frac{\mathcal{B}_\alpha - iI}{H^{in}_\alpha - zI}\end{aligned} \tag{22}$$

Proof. Starting from ansatz (20), we apply the operator $(H^{in}_{\hat\alpha})^*_0 - zI$ to it, then $(H^{in}_{\hat\alpha} - zI)^{-1}(\mathcal{B}_{\hat\alpha} - iI)$ to the result. It gives

$$v + \xi^-_\alpha = \frac{1 + (z - A_\alpha\times i_{\hat\alpha})\mathcal{B}_{\hat\alpha}}{H^{in}_{\hat\alpha} - zI}\xi^+_\alpha + \frac{\mathcal{B}_{\hat\alpha} - iI}{H^{in}_{\hat\alpha} - zI}f^{in}_{\hat\alpha}$$

Projecting $P^{in}_{\hat\alpha}$ onto $\mathcal{H}^{in}_{\hat\alpha}$, $\mathcal{H}^{in}_{\hat\alpha} \sim L_{\hat\alpha}\backslash L_0$, gives

$$\xi^-_{\hat\alpha} = P^{in}_{\hat\alpha}\frac{1 + (z - A_\alpha\times i_{\hat\alpha})\mathcal{B}_{\hat\alpha}}{H^{in}_{\hat\alpha} - zI}\xi^+_{\hat\alpha} + P^{in}_{\hat\alpha}\frac{\mathcal{B}_\alpha - iI}{H^{in}_{\hat\alpha} - zI}f_{\hat\alpha}\,,$$

and inserting into (20) results in (21) with $\xi^{in}_{\hat\alpha} = \xi^+_{\hat\alpha}$. Combining the calculated boundary values (15), (18), (19), (22) with the boundary conditions (11), (12) we get the system of equations for $\rho^{ex}_a, \xi^{in}_0, \xi^{in}_{\hat\alpha}$

$$\begin{pmatrix} \frac{1}{H^{ex}-iI}P_{\hat\alpha}\frac{H^{ex}-iI}{H^{ex}-zI}f + \frac{1}{H^{ex}-iI}P_{\hat\alpha}\frac{1+zH^{ex}}{H^{ex}-zI}\cdot\frac{1}{H^{ex}-iI}\rho^{ex} \\ P^{in}_{\hat\alpha}\frac{1+(z-A_\alpha\times i_{\hat\alpha})\mathcal{B}_{\hat\alpha}}{H^{in}_{\hat\alpha}-zI}\xi^{in}_{\hat\alpha} + P^{in}_{\hat\alpha}\frac{\mathcal{B}_\alpha-iI}{H^{in}_\alpha-zI}f_{\hat\alpha}\end{pmatrix}^{l,l_\alpha} = \Gamma_{\hat\alpha}\begin{pmatrix}\rho^{ex}_\alpha\\ \xi^{in}_{\hat\alpha}\end{pmatrix}^{l,l_\alpha}_{\substack{\alpha=1,2,3\\ l_\alpha\neq l_{\hat\alpha}}} \tag{23}$$

$$\begin{pmatrix} \frac{1}{H^{ex}-iI}P_0\frac{H^{ex}-iI}{H^{ex}-zI}f + \frac{1}{H^{ex}-iI}P_0\frac{1+zH^{ex}}{H^{ex}-zI}\cdot\frac{1}{H^{ex}-iI}\rho^{ex} \\ \langle\frac{\mathcal{B}_0-iI}{b_0-zI}f^{in}_0,\Theta_0\rangle + \langle\frac{1+b_0z}{b_0-z}\Theta^{in}_0,\Theta^{in}_0\rangle\xi^{in}_0\end{pmatrix}^{l} = \Gamma_0\begin{pmatrix}\rho^{ex}_0\\ \xi^{in}_0\end{pmatrix}^{l} \tag{24}$$

Note, that renormalized values of u_a^{ex} can be received by restriction of the expression

$$\left\{\frac{1}{H^{ex}-zI}f+\frac{1}{H^{ex}-zI}\rho-\frac{H^{ex}_{ex}}{(H^{ex})^2+1}\rho\right\}$$

onto planes $L_\alpha\backslash L_0$ or L_0 correspondingly in (23) and (24). Using this fact we get the system of dynamical equations which are equivalent to the Faddeev equations. Using the fact, that the second equation in every system (23), (24) can be resolved in two–body terms or even in one–body terms,

$$\xi^{in}_{\hat\alpha}=\left\{\gamma^{\hat\alpha}_{ii}I-P^{in}_{\hat\alpha}\frac{1+(z-A_\alpha\times i_{\hat\alpha})\mathcal{B}_{\hat\alpha}}{H^{in}_{\hat\alpha}-zI}\right\}^{-1}\cdot P^{in}_{\alpha}\frac{\mathcal{B}_\alpha-iI}{H^{in}_{\hat\alpha}-zI}f_{\hat\alpha}$$
$$-\left\{\gamma^{\hat\alpha}_{ii}I-P^{in}_{\hat\alpha}\frac{1+(z-A_\alpha\times i_{\hat\alpha})\mathcal{B}_{\hat\alpha}}{H^{in}_{\hat\alpha}-zI}\right\}^{-1}\gamma^{\hat\alpha}_{ie}\rho^{ex}_{\hat\alpha}\ ;$$
$$\xi^{in}_{0}=\left\{\gamma^{0}_{ii}I-\left(\frac{1+b_0z}{b_0-zI}\Theta^{in}_0\Theta^{in}_0\right)\right\}^{-1}\langle\frac{b_0-iI}{b_0-zI}f^{in}_0,\Theta_0\rangle$$
$$-\left\{\gamma^{0}_{ii}I-\langle\frac{1+b_oz}{b_0-z}\Theta^{in}_0,\Theta^{in}_0\rangle\right\}^{-1}\gamma^{0}_{ie}\rho^{ex}_0$$

We can exclude the inner variables from the first equations, transforming them into equations for the densities $\rho^{ex}_\alpha,\rho^{ex}_0$:

$$\left\{\frac{1}{H^{ex}-zI}-\frac{H^{ex}}{(H^{ex})^2+1}\right\}\rho^{ex}+\frac{1}{H^{ex}-zI}f^{ex}|_{\Gamma_\alpha\backslash\Gamma_0}$$
$$=\left[\gamma^{\hat\alpha}_{ee}-\gamma^{\hat\alpha}_{ei}\left\{\gamma^{\hat\alpha}_{ii}I-P^{in}_{\hat\alpha}\frac{1+(z-A_\alpha\times i_\alpha)\mathcal{B}_{\hat\alpha}}{H^{in}_{\hat\alpha}-zI}\right\}^{-1}\gamma^{\hat\alpha}_{ie}\right]_\alpha\rho^{ex}_{\hat\alpha}\qquad(25)$$
$$+\left\{\gamma^{\hat\alpha}_{ii}I-P^{in}_{\hat\alpha}\frac{1+(z-A_\alpha\times i_\alpha)\mathcal{B}_{\hat\alpha}}{H^{in}_{\hat\alpha}-zI}\right\}^{-1}P^{in}_{\alpha}\frac{\mathcal{B}_\alpha-iI}{H^{in}_{\hat\alpha}-zI}f_{\hat\alpha}$$

$$\left\{\frac{1}{H^{ex}-zI}-\frac{H^{ex}}{(H^{ex})^2+1}\right\}\rho^{ex}+\frac{1}{H^{ex}-zI}f^{ex}|_{\Gamma_0(l_1=l_2=l_3=l)}$$
$$=\left[\gamma^{0}_{ee}-\gamma^{0}_{ei}\left\{\gamma^{0}_{ii}-\langle\frac{1+b_0z}{b_0-z}\Theta_0,\Theta_0\rangle\right\}^{-1}\gamma^{0}_{ie}\right]_0\rho^{l,l,l}_0\qquad(26)$$
$$+\left[\gamma^{0}_{ee}-\gamma_{ei}\left\{\gamma^{0}_{ii}I-\langle\frac{1+b_0z}{b_0-z}\Theta_0,\Theta_0\rangle\right\}^{-1}\gamma^{0}_{ie}\right]\langle\frac{b_0-iI}{b_0-zI}(f^{in}_0)^l,\Theta_0\rangle\ .$$

The operators in the square brackets on the left sides of the last two equations play crucial role in future development. Note, that in special case, when $\gamma^0_{ii}+\langle b_0\Theta_0,\Theta_0\rangle=0$ the bracket $[\quad]_0$ in last equation (26) has linear asymptotics at infinity in z–plane.

$$\gamma_{ee}-\gamma^{0}_{ei}\left\{\gamma^{0}_{ii}-\langle\frac{1+b_0z}{b_0-z}\Theta_0,\Theta_0\rangle\right\}^{-1}\gamma^{0}_{ie}\cong\gamma_{ee}-\frac{|\gamma_{ei}|^2z+0(1)}{\langle(1+b_0^2)\Theta_0,\Theta_0\rangle}\ .$$

Similary, if $\gamma_{ii}^{\hat{\alpha}} + \langle b_{\hat{\alpha}}\Theta_{\hat{\alpha}}, \Theta_{\hat{\alpha}}\rangle = 0$, the square bracket in the first equation can be reduced to the following form

$$\gamma_{ee}^{\hat{\alpha}} + \frac{\gamma_{ei}^{\hat{\alpha}}|^2 \cdot (A_\alpha - z)}{\langle(1 + b_{\hat{\alpha}}^2)\Theta_0, \Theta_0\rangle} + 0(1)$$

If we move the terms with selfinteraction from left side to the right one, then the operator factor in front of $\rho_{\hat{\alpha}}$ in the first equation becomes

$$\gamma_{ee}^{\hat{\alpha}} + \frac{|\gamma_{ei}^{\hat{\alpha}}|^2(A_\alpha - z)}{\langle(1 + b_{\hat{\alpha}}^2)\Theta_{\hat{\alpha}}, \Theta_{\hat{\alpha}}\rangle} - \left\{\frac{1}{H^{ex} - zI} - \frac{H^{ex}}{(H^{ex})^2 + 1}\right\}|_{L_\alpha} + \ldots$$

and in the second

$$\gamma_{ee}^{0} + \frac{|\gamma_{ei}^{0}|^2 z}{\langle(1 + b_{0}^2)\Theta_0, \Theta_0\rangle} - \left\{\frac{1}{H^{ex} - zI} - \frac{H^{ex}}{(H^{ex})^2 + 1}\right\}|_{L_0} + \ldots$$

Both of them become big for large z. At the same time on the left-hand-side side the discrete integral operators

$$\mathcal{U}_{ab} \sim \left\{\frac{1}{H^{ex} - zI} - \frac{H^{ex}}{(H^{ex})^2 + 1}\right\} \qquad (b \longrightarrow a)$$

remain which act from the space of densities, defined on L_b into the class of densities, defined onto L_a as bounded operators, $a, b = 0, 1, 2, 3$. These operators have the same structure as the corresponding terms in Faddeev's equations. The operator $\phi \times \phi$ matrix

$$\mathcal{U} = \{\mathcal{U}_{ab}\} \quad , \quad \mathcal{U}_{aa} = 0$$

can be proved to be compact in the space of densities. It will be analyzed in a separate publication. Here we finish with the formulation of the main result.

Theorem. Let two–body compound Hamiltonians $b_{\hat{\alpha}}$ and three–body compound Hamiltonian fullfill the conditions

$$\gamma_{ii}^{0} + \langle b_0\Theta_0, \Theta_0\rangle = 0; \gamma_{ii}^{\hat{\alpha}} + \langle b_{\hat{\alpha}}\Theta_0, \Theta_0\rangle = 0 \ , \tag{27}$$

which guarantee the appearence of the term linear with respect to the energy in equations (25), (26).

Then the spectrum of corresponding three–body Hamiltonian is bounded.

Proof. Actually, to perform the proof, it is sufficient to mention, that the operator–matrix M defines uniformely bounded operator for large z. Thus the equation for densities can be solved for those z just by perturbation methods, using the fact that they contain constant terms, linearly depending on z.

It is interesting, that the conditions (27) are of importance also in continuous case (see [5]), where they guarantee the triviality of the high–energy asymptotics of two–body scattering matrices. One can suppose, that the same connection exists also in the lattice case.

4. Conlusion

1. Notice, that the Fredholm character of the system (25,26) can be proved on the basis of the condition $\mathcal{U}_{aa} = 0$.
2. The structure of the equations and the character of boundary conditions (11), (12) opens possibility to generalize the described approach for many-particle systems. The corresponding system of equations would be analogous to the Faddeev–Jakubobsky–system.
3. Despite the fact that our exposition is concerned with the particles in the lattice, it is clear, that the described approach is valid for any abstract quantum–mechanical situation. Thus, it is valid for analysis of many–body problems in different laboratory coordinate systems.

Acknowledgment

One of the authors (B.P.) is grateful to Professor S. Albeverio, and The University of Ruhr, Bochum, for hospitality.

References

[1] *F. Berezin, L. Faddeev:* DAN UdSSR 137, N 5, (1961)
[2] *S. Albeverio, F. Gestezy, R. Høegh–Krohn, H. Holden:* Solvable models in quantum mechanics. Springer, (1988)
[3] *B. Pavlov:* Teor. Mat. Fiz. 59, 3 (1984)
[4] *P. Exner, P. Šeba (ed.):* Applications of selfadjoint extensions in quantum physics, Springer (1989), p. 3–11
[5] *B. Pavlov:* Matem. Sbornik 136, N 2, (1988)
[6] *M. Krein:* DAN UdSSR 52, N 8, (1946)
[7] *B. Pavlov:* Vestnik Leningr. Univ. 15 (1987)
[8] *S. Merkuriev, L. Faddeev:* Quantum scattering theory for many–body systems. Nauka (1985)

Time-dependent Systems, Berry Phase, Quantum Chaos

Schrödinger evolution equations and associated smoothing effect

KENJI YAJIMA

Department of Pure and Applied Sciences
University of Tokyo
3-8-1 Komaba, Meguroku, Tokyo 153, Japan

Abstract. We shall review author's recent results on the existence, uniqueness, regularity and the smoothing property of a unitary propagator $U(t,s): u(s) \to u(t)$ in $\mathsf{H} = L^2(\mathbf{R}^n)$ for time dependent Schrödinger equations in an external electromagnetic field. Potentials are assumed to be real and are possibly unbounded at infinity $|x| \to \infty$ and scalar potential can have local singularities.

1. Introduction, Assumptions and Theorems.

In this lecture we report author's recent works [**29**],[**30**] and [**31**], improving some results in [**31**], on the Cauchy problem for time dependent Schrödinger equations

$$i\frac{\partial u}{\partial t} = H(t)u = (\sum_{j=1}^{n} \frac{1}{2}(\frac{1}{i}\frac{\partial}{\partial x_j} - A_j(t,x))^2 + V(t,x))u, \tag{1.1}$$

where the real potentials $V(t,x)$ and $A(t,x) = (A_1(t,x), \cdots, A_n(t,x))$ are possibly unbounded at infinity $|x| \to \infty$ and $V(t,x)$ can have local singularities. We shall be concerned with the existence, uniqueness, regularity and the smoothing property of a unitary propagator $U(t,s): u(s) \to u(t)$ in $\mathsf{H} = L^2(\mathbf{R}^n)$ solving (1.1).

The existence of a unique dynamics associated with (1.1) is of fundamental importance in quantum mechanics. When $H(t) = H$ is t-independent, this is equivalent to the selfadjointness of H, via celebrated Stone's theorem, and various satisfactory criteria have been known since Kato's fundamental paper [**14**] (see e.g. [**5**], [**23**]). When $H(t)$ is genuinely time dependent, on the other hand, the situation has been by far less satisfactory. We recall three common methods, among others, for solving (1.1). One of them is to apply to (1.1) the abstract theory of evolution equations (cf. e.g. [**15**]). The method has been widely and successfully employed by many authors. The theory, however, has been originally designed for solving parabolic or hyperbolic equations and, when applied to Schrödinger equations, it often imposes rather severe restrictions on the potentials, which exclude many concrete and important physical examples out of its scope (cf. [**30**]). Another one, which we call the method of semi-classical approximation, is to construct the propagator $U(t,s)$ directly in the form of oscillatory integral operator using, so called, semi-classical approximations. This method was introduced by Fujiwara([**9**] and [**10**]) for Schrödinger equations and extended by Kitada-Kumanogo[**22**], which, however, requires the potentials be C^∞ in the spatial variables. The third one is to study (1.1) through its equivalent integral equation using the perturbation technique.

Synthesizing the last two methods mentioned above, we first establish the following existence and uniqueness theorem. For an interval I, $L^{\ell,\theta}(I) = L^\theta(I, L^\ell(\mathbf{R}^n))$ and $L^{\ell,\theta}_{loc}(I) = L^\theta_{loc}(I, L^\ell(\mathbf{R}^n))$. $|\cdot|$ denotes the norm of matrices as operators on $\mathbf{C}^n$ as well as the Euclidean norm of vectors. $\partial_x = (\partial/\partial x_1, \cdots, \partial/\partial x_n)$ and for a multi-index $\alpha = (\alpha_1, \cdots, \alpha_n)$, $\partial_x^\alpha = (\partial/\partial x_1)^{\alpha_1} \cdots (\partial/\partial x_n)^{\alpha_n}$. We sometimes write $\partial/\partial x_j = \partial_{x_j}$.

ASSUMPTION (A): (1) For $j = 1, \cdots, n$, $A_j(t,x)$ is a real function of $(t,x) \in \mathbf{R}^{n+1}$ such that $\partial_x^\alpha A_j(t,x) \in C^1(\mathbf{R}^{n+1})$ for any α. There exists $\epsilon > 0$ such that

$$(1.2) \qquad |\partial_x^\alpha B(t,x)| \le C_\alpha (1+|x|)^{-1-\epsilon}, \qquad |\alpha| \ge 1,$$

$$(1.3) \qquad |\partial_x^\alpha A(t,x)| + |\partial_x^\alpha \partial_t A(t,x)| \le C_\alpha, \quad |\alpha| \ge 1, \quad (t,x) \in \mathbf{R}^{n+1},$$

where $B(t,x)$ is the strength tensor of the magnetic field, i.e. the skew symmetric matrix with (j,k)-component $B_{jk}(t,x) = (\partial A_k/\partial x_j - \partial A_j/\partial x_k)(t,x)$.
(2) $V(t,x)$ is a real function of $(t,x) \in \mathbf{R}^{n+1}$ such that $V \in L^{p,\alpha}(\mathbf{R}^1) + L^{\infty,1}(\mathbf{R}^1)$ for some $p > n/2$ with $p \ge 1$ and $\alpha = 2p/(2p-n)$.

For $k = 0, 1, \cdots$, $\Sigma(k) = \{u \in L^2(\mathbf{R}^n) : \sum_{|\alpha|+|\beta| \le k} \|x^\beta \partial_x^\alpha u\|^2 = \|u\|^2_{\Sigma(k)} < \infty\}$ and $\Sigma(-k)$ is the dual space of $\Sigma(k)$. For Banach spaces X and Y, $B(X,Y)$ stands for the set of bounded operators from X to Y, $B(X) = B(X,X)$. We use C_* to indicate the strong continuity of operator valued functions, e.g. $C_*(I^2, B(X,Y))$ is the set of strongly continuous $B(X,Y)$-valued functions on I^2.

THEOREM 1. *Let Assumption (A) be satisfied,* $q = 2p/(p-1)$ *and* $\theta = 4p/n$. *Then,* $H(t) \in B(\mathbf{H} \cap L^q(\mathbf{R}^n), \Sigma(-2))$ *and its part in* $\mathbf{H}$ *is selfadjoint for a. e.* $t \in \mathbf{R}^1$. *There exists a unique propagator* $\{U(t,s) : t,s \in \mathbf{R}^2\}$ *with the following properties.*
(1) $U(t,s)$ *is unitary in* $\mathbf{H}$ *with* $U(t,s)U(s,r) = U(t,r)$.
(2) $U(\cdot,\cdot,h) \in C_*(\mathbf{R}^2, B(\mathbf{H}))$.
(3) For any $s \in \mathbf{R}^1$ *and* $f \in \mathbf{H}$, $U(\cdot,s)f \in L^{q,\theta}_{loc}(\mathbf{R}^1)$ *and with constant* $C > 0$ *independent of* s, f *and compact intervals* I,

$$(1.4) \qquad \|U(\cdot,s)f\|_{L^{q,\theta}(I)} \le C(1+|I|)^{1/\theta} \|f\|.$$

(4) If $f \in \mathbf{H}$, $u(t) = U(t,s)f$ is $\Sigma(-2)$-valued absolutely continuous in t and satisfies the equation (1.1) in $\Sigma(-2)$ at almost every $t \in \mathbf{R}^1$.

In what follows, the exponents $p > n/2$, $q = 2p/(p-1)$ and $\alpha = 2p/(2p-n)$ are reserved to denote those appeared in Assumption (A) and Theorem 1.

Several remarks are in order.

REMARK(A): If $G(t,x)$ is real smooth and T is the gauge transformation:

$$Tu(t,x) = \exp(-iG(t,x))u(t,x), \tag{1.5}$$

then T transforms (1.1) into the one with $A + \partial_x G$ and $V - \partial_t G$, in place of A and V respectively. In particular, by taking $G(t,x) = \int_0^t V(s,x)ds$, smooth scalar potentials may be eliminated from (1.1) by a gauge transformation by changing A to $A + \int_0^t \partial_x V(s,x)ds$. This is the reason why we assumed in Assumption (A) that $V(t,x)$ is bounded at infinity in a certain L^p norm, which allows us to regard (1.1) as a perturbed equation of

$$i\frac{\partial u}{\partial t} = H_0(t)u = \sum_{j=1}^{n} \frac{1}{2}(\frac{1}{i}\frac{\partial}{\partial x_j} - A_j(t,x))^2 u. \tag{1.6}$$

It should be also clear that, if $V_0(t,x)$ is a real function such that $\partial_x^\alpha V_0(t,x)$ is continuous in (t,x) for all $|\alpha| \geq 0$ and $|\partial_x^\alpha V_0(t,x)| \leq C_\alpha$, $|\alpha| \geq 2$, then, Theorem 1 remains to hold for (1.1) with $V(t,x) + V_0(t,x)$ in place of $V(t,x)$, the latter being assumed to satisfy Assumption (A). Similar remark will apply to theorems to follow, though we shall not mention it explicitly any more.

REMARK(B): (1.4) implies that for every $f \in \mathbf{H}$ and $s \in \mathbf{R}$, $U(t,s)f \in \mathbf{H} \cap L^q(\mathbf{R}^n)$ for *a.e.* $t \in \mathbf{R}^1$ and manifests the smoothing effect of the propagator $U(t,s)$ in the sense that it improves L^p-smoothness. This is an improvement of Strichartz' inequality

$$(\int\{\int |e^{it\Delta}f(x)|^p dx\}^{\theta/p}dt)^{1/\theta} \leq C\|f\|, \quad f \in L^2(\mathbf{R}^n) \tag{1.7}$$

for $0 \leq 2/\theta = n(1/2 - 1/p) < 1$ (cf. [**11**]). Note that (1.7) is an estimate global in time and manifests simultaneously the local decay property (as $|t| \to \infty$) of $\exp(it\Delta)f$ (which, incidentally, implies the existence of the restriction of Fourier transforms of $L^{p',\theta'}$-functions on the quadratic surface $\eta = \xi_1^2 + \cdots + \xi_n^2$, cf. [**26**]). This is in contrast to (1.4) where the right hand side increases with the length of the time interval. This is, of course, natural because bound states exist in general which never decay as $t \to \pm\infty$.

REMARK(C): If $V(t,x) = V(x)$ satisfies (2) of Assumption (A) with some $p_0 > n/2$, it does so with any $n/2 < p \leq p_0$. Hence $U(\cdot,s)f \in L^{\ell,\theta}_{loc}(\mathbf{R}^1)$ for any $2 \leq \ell < 2n/(n-2)$ and $1 \leq \theta \leq 4\ell/n(\ell-2)$.

REMARK(D): The L^p-smoothing property of the propagator is expected to hold, at least in a weaker form, for Schrödinger equations with faster increasing scalar potentials. In fact, if $H = -d^2/dx^2$ on $[0,1]$ with Dirichlet conditions at $x = 0, 1$, which is an extreme case that $V_0(x) = \infty$ for $x \in (-\infty, 0) \cup (1, \infty)$ in the terminology of Remark (a), it can be easily checked by an explicit computation using Fourier series expansions that $\exp(-itH)u \in L^4_{loc}(\mathbf{R}^1, L^4([0,1]))$, for $u \in L^2([0,1])$ and $\exp(-itH)$ *does* possess certain smoothing effect even in this extreme case. (This is a result of Zygmund [**33**].) In view of Zelditch's work ([**32**]), we note that faster increase of scalar potential implies increasing number of caustics for higher momentum which suggests stronger singularities in the integral kernel of the propagator $U(t,s)$. We suspect that the smoothing effect may be used for measuring the *strength of its singularity*.

REMARK(E): When $H(t) = H$, the L^p-smoothing property implies an interesting result on *change of phases* of Fourier coefficients in (generalized) eigenfunction expansions associated with H. The following is an obvious consequence of (1.4) and the fact that $\{u \in \mathsf{H} : (-i\partial_x - A(x))u \in \mathsf{H}^n\} \subset L^q(\mathbf{R}^n)$. We refer to Edward[**8**] for informations on related results for the ordinary Fourier series.

COROLLARY 2. *Let A and V be independent of t and satisfy Assumption (A) and let $H = H(t)$ be the selfadjoint operator in H defined by the Friedrichs extension:*

$$D(H) = \{u \in \mathsf{H} : (-i\partial/\partial x_j - A_j(x))u \in \mathsf{H},\ j = 1, \cdots, n,\ Hu \in \mathsf{H}\}.$$

Suppose that H has only pure point spectrum $\{\lambda_j\}_{j=1}^{\infty}$ and let $\{\phi_j\}_{j=1}^{\infty}$ be the associated complete set of orthonormal eigenfunctions of H. Then for any $\{c_j\} \in \ell^2$, we have $\sum_{j=1}^{\infty} \exp(-it\lambda_j)c_j\phi_j(x) \in L^\ell$, a.e. $t \in \mathbf{R}^1$ for $2 \le \ell < 2n/(n-2)$.

REMARK(F): The L^p-smoothing property of $e^{it\Delta/2}$ has wide applications in the study of *non-linear* Schrödinger equations $i\partial_t u = -(1/2)\Delta u + F(u)$ (cf. e.g. [**11**] and [**17**]). Theorem 1 (and Theorems 7 and 8 to follow) may be used to extend some of their results to more general non-linear Schrödinger equations $i\partial_t u = H(t)u + F(u)$, with $H(t)$ as in (1.1) (cf. [**7**]).

The propagator $U(t,s)$ has not only L^p-smoothing property but also the differentiability improving property, which is the content of the following two theorems. Because of a technical difficulty, we were forced to assume either $A(t,x)$ or $V(t,x)$ vanishes identically. If Theorems 3 and 4 are true under the conditions of Theorem 1 is still an open question.

THEOREM 3. *Let Assumption (A) be satisfied and $\mu > 1/2$. Suppose either $A(t,x) = 0$ or $V(t,x) = 0$. Then, there exists a constant $C_\mu > 0$ such that for $s \in \mathbf{R}^1$ and $T > 0$*

$$\int_{s-T}^{s+T} \|\langle x\rangle^{-\mu}\langle D\rangle^{1/2}U(t,s)f\|^2 dt \le C_\mu(1+T)\|f\|^2, \quad f \in \mathcal{S}(\mathbf{R}^n). \tag{1.7}$$

THEOREM 4. *Let Assumption (A) be satisfied,* $2 \le \ell \le q = 2p/(p-1)$, $0 \le 2\sigma + n(1/2 - 1/\ell) \le 2/\theta < 1$ *and* $\mu > 1/2$. *Suppose either* $A(t,x) = 0$ *or* $V(t,x) = 0$. *Then, there exists a constant* $C_{\ell\theta\sigma} > 0$ *such that for* $s \in \mathbf{R}^1$ *and* $T > 0$

$$(1.8) \quad \int_{s-T}^{s+T} \|\langle x\rangle^{-2\mu\sigma}\langle D\rangle^{\sigma} U(t,s)f(x)\|_{L^\ell}^{\theta} dt \le C_{\ell\theta\sigma}(1+T)\|f\|^{\theta}, \quad f \in \mathcal{S}(\mathbf{R}^n).$$

REMARK (G): As in Remark(c), if $V(t,x) = V(x)$, the condition $2 \le \ell \le q = 2p/(p-1)$ in Theorem 4 may be replaced by $2 \le \ell < 2n/(n-2)$.

It was Kato [**16**] who discovered that a certain unitary group in L^2 can improve the differentiability property: The solutions of the KdV equation

$$\partial_t u + \partial_x^3 u + a(u)\partial_x u = 0, \quad t > 0,\ x \in \mathbf{R}^1$$

with $u(0) = f \in L^2(\mathbf{R}^1)$ satisfy

$$\int_0^T \int_{-R}^{R} |\partial_x u(t,x)|^2 dx dt \le K(\|f\|, T, R)$$

if $\limsup_{|\lambda|\to\infty} |\lambda|^{-4} a(\lambda) \le 0$. Later it was found (cf. [**25**], [**4**] and [**19**]) that such property is common for unitary groups generated by dispersive differential equations with constant coefficients and, in particular, for the free Schrödinger propagator they proved

$$(1.9) \qquad \int_{\mathbf{R}^1}\int_{\mathbf{R}^n} |\phi(t,x)(1-\Delta)^{1/4} e^{it\Delta} u(x)|^2 dx dt \le C\|u\|^2, \quad u \in L^2(\mathbf{R}^n)$$

and for $1 \le q \le 2$, $r \ge 2$ and $\alpha < 1/r - n(1/q - 1/r)$,

$$(1.10) \quad [\int_{\mathbf{R}^1}\int_{\mathbf{R}^n} |\phi(t,x)(1-\Delta)^{\alpha/2} e^{it\Delta} u(x)|^r dx dt]^{1/r} \le C\|u\|_{L^q}, \quad u \in L^q(\mathbf{R}^n),$$

where $\phi \in C_0^\infty(\mathbf{R}^{n+1})$ (see also [**18**] and [**24**] where ϕ in (1.9) is replaced by $\langle x\rangle^{-1/2-\epsilon}$, $\varepsilon > 0$).

Theorem 3 and Theorem 4 are extensions of (1.9) and (1.10) (with $q = 2$) to differential equations with variable coefficients, however, by the same reason as in (1.4), the right hand sides of (1.7) and (1.8) increase linearly in time length T in contrast to (1.9) and (1.10) which imply also the local decay as $|t| \to \infty$ of $\exp(it\Delta)f$ in the local Sobolev norm. In this connection we should mention that Ben-Artz and Klainermann [**2**] have recently shown, in the case $H(t) = H$ and $A = 0$ and $|V(x)| \le C(1+|x|)^{-2-\epsilon}$, $\epsilon > 0$, that, if zero is not a resonance energy of H, the estimate global in time of type (1.9) is valid for u belonging to the continuous spectral subspace of H (cf. also [**4**]). We should emphasize that

Theorem 4 is an improvement of (1.10) (with $q = 2$) even when $A(t,x) \equiv 0$ and $V(t,x) \equiv 0$ as far as the smoothness is concerned. It uses different L^p-norms for space and time variables in the first place and when $\theta = p$, σ in (1.8) can be as big as $1/p - (n/2)(1/2 - 1/p)$ whereas in (1.10), $\sigma < 1/p - n(1/2 - 1/p)$.

We record two consequences of Theorem 3. One is the *maximal inequality* of Schrödinger type and the other is on the almost everywhere convergence of $U(t,s)f(x)$ as $t \to s$.

THEOREM 5. *Assume the condition of Theorem 3, $\gamma > 1/2$ and $\delta > 2\gamma + 1/2$. Then there exists a constant $C > 0$ such that*

$$\int \sup_{|t-s|<T} |\langle x\rangle^{-\delta} U(t,s)f(x)|^2 dx \le C\|f\|_{H^\gamma}^2, \quad f \in \mathcal{S}(\mathbf{R}^n). \tag{1.12}$$

THEOREM 6. *Let Assumption (A) be satisfied and $f \in H^\gamma(\mathbf{R}^n) \cap L^1(\mathbf{R}^n)$ with $\gamma > 1/2$. Then for any $s \in \mathbf{R}^1$*

$$\lim_{t\to s} U(t,s)f(x) = f(x), \quad a.e.\, x \in \mathbf{R}^n. \tag{1.13}$$

Note that, when $H(t) = H$, Theorem 6 gives a *summability theorem* for (generalized) eigenfunction expansions associated with H. For $\exp(it\Delta)$ Carleson [**3**] proved the summability theorem with $\gamma = 1/4$ when $n = 1$. This result is shown to be sharp by Dahlberg and Kenig [**6**]. Again for $n = 1$, Kenig and Ruiz[**20**] showed that the maximal inequality (1.9) is satisfied with $\gamma = 1/4$ and that the estimate is sharp. These one-dimensional results were subsequently extended by Sjölin [**25**] and Kenig-Ponce-Vega [**19**] to higher dimensions. It seems still open, however, if these estimates are sharp or not in general.

We close this section with three theorems on the regularity of the solutions of (1.1). The solution $u(t) = U(t,s)f$, $f \in \mathsf{H}$, obtained in Theorem 1 is a *weak solution* in the sense that it satisfies (1.1) only in $\Sigma(-2)$ for $a.e.t \in \mathbf{R}^1$, and it is natural to ask the following question: When data f is nice, say $f \in \Sigma(2)$, is $u(t) = U(t,s)f$ a *strong solution*, that is, does $u \in C^1(\mathbf{R}^1, \mathsf{H})$, $u(t) \in D(H(t))$ for every $t \in \mathbf{R}^1$ and satisfy (1.1) in H? This is equivalent to *finding an invariant subspace* $\mathcal{D} \subset D(H(t))$ for $U(t,s)$: $U(t,s)\mathcal{D} = \mathcal{D}$ such that $U(\cdot,\cdot)f \in C^1(\mathbf{R}^2, \mathsf{H})$ for $f \in \mathcal{D}$. This turns out to be a rather subtle and difficult question in general and we shall study this problem only under certain stronger conditions on $A(t,x)$ and $V(t,x)$.

When $V(t,x)$ is smooth in the spatial variables, the gauge transformation (1.5) eliminates it from (1.1) reducing it to the form (1.6). For (1.6) we can construct the propagator $U_0(t,s)$ in the form of oscillatory integral operator(OIO) by the method of semi-classical approximation and derive much information out of it. In particular, $\Sigma(2)$ is an invariant subspace for $U_0(t,s)$. We should recall (cf. [**13**]) that $H_0(t)$ is essentially selfadjoint on $C_0^\infty(\mathbf{R}^n)$ and its closure, which we denote by the same symbol, is identical with its maximal extension. Thus

$$\Sigma(2) \subset D(H_0(t)) = \{u \in \mathsf{H} : H_0(t)u \in \mathsf{H}\}, \qquad t \in \mathbf{R}^1.$$

In virtue of (1.5), the following theorem is an extension of Fujiwara [9],[10], Kitada [21] and Kitada-Kumanogo [22] to the case with general magnetic potentials.

THEOREM 7. *Let Assumption (A) be satisfied and $H(t) = H_0(t)$. Then the propagator $U(t,s) = U_0(t,s)$ of Theorem 1 satisfies the following additional properties.*
(1) $U_0(\cdot,\cdot) \in C_*(\mathbf{R}^2, B(\Sigma(k))) \cap C_*^1(\mathbf{R}^2, B(\Sigma(k), \Sigma(k-2)))$ for $k = 0, \pm 1, \cdots$.
(2) If $f \in \Sigma(2)$, then $U_0(t,s)f$ satisfies

$$i(\partial/\partial t)U_0(t,s)f = H_0(t)U_0(t,s)f \in C(\mathbf{R}^2, \mathsf{H}),$$

$$i(\partial/\partial s)U_0(t,s)f = -U_0(t,s)H_0(s)f \in C(\mathbf{R}^2, \mathsf{H}).$$

(3) $U_0(t,s)$ is an isomorphism of the Schwartz space $\mathcal{S}(\mathbf{R}^n)$.
(4) For a small $T > 0$, $U_0(t,s)$, $0 < |t-s| \le T$, can be written in the form of an OIO:

$$U_0(t,s)f(x) = (2\pi i(t-s))^{-n/2} \int e^{iS(t,s,x,y)} b(t,s,x,y) f(y) dy, \tag{1.14}$$

where $S(t,s,x,y)$ and $b(t,s,x,y)$ satisfy the following properties.
(4.a) $S(t,s,x,y)$ is a real solution of Hamilton-Jacobi equation

$$(\partial_t S)(t,s,x,y) + (1/2)((\partial_x S)(t,s,x,y) - A(t,x))^2 = 0. \tag{1.15}$$

It is C^1 in (t,s,x,y), C^∞ in (x,y) and satisfies the estimate

$$|\partial_x^\alpha \partial_y^\beta \{S(t,s,x,y) - \frac{(x-y)^2}{2(t-s)}\}| \le C_{\alpha\beta}, \qquad |\alpha + \beta| \ge 2. \tag{1.16}$$

(4.b) For any α and β, $\partial_x^\alpha \partial_y^\beta b(t,s,x,y)$ is C^1 in (t,s,x,y) and satisfies

$$|\partial_x^\alpha \partial_y^\beta b(t,s,x,y)| \le C_{\alpha\beta}, \qquad |\alpha + \beta| \ge 0. \tag{1.17}$$

When $V(t,x)$ is singular, we take a perturbation point of view and consider the case $H(t) = H_0(t) + V(t)$ with $H_0(t) = H_0$ first. We give a sufficient condition on $V(t,x)$ so that $D(H_0)$ becomes an invariant subspace. Recall that $p > n/2$ with $p \ge 1$ is fixed as in Assumption (A) and $q = 2p/(p-1)$.

ASSUMPTION (B): $V(t,x)$ satisfies $V \in C(\mathbf{R}^1, L^{\tilde{p}}(\mathbf{R}^n)) + C(\mathbf{R}^1, L^\infty(\mathbf{R}^n))$ and $\partial_t V \in L^{p_1,\alpha_1}(\mathbf{R}^1) + L^{\infty,1}(\mathbf{R}^1)$ for $\tilde{p} = max(p,2)$, $p_1 = 2np/(n+4p)$ $(n \ge 5)$, $p_1 > 2p/(p+1)$ $(n = 4)$, $p_1 = 2p/(p+1)$ $(n \le 3)$ and $\alpha_1 = 4p/(4p-n)$.

THEOREM 8. *Let Assumptions (A) and (B) be satisfied and $H_0(t) = H_0$ is independent of t. Then, $H(t)$ is essentially selfadjoint on $C_0^\infty(\mathbf{R}^n)$ and its closure, which will be denoted by the same symbol, has constant domain $D(H(t)) = D(H_0)$. The propagator $U(t,s)$ satisfies the following properties:*
(1) $U(\cdot,\cdot) \in C_*(\mathbf{R}^2, B(D(H_0))) \cap C_*^1(\mathbf{R}^2, B(D(H_0), \mathbf{H}))$.
(2) If $f \in D(H_0)$, then

$$i\partial_t U(\cdot\,,s)f = H(\cdot)U(\cdot\,,s)f \in C(\mathbf{R}^1, \mathbf{H}) \cap L_{loc}^{q,\theta}(\mathbf{R}^1), \tag{1.18}$$

$$i\partial_s U(\cdot\,,s)f = -U(\cdot\,,s)H(s)f \in C(\mathbf{R}^1, \mathbf{H})) \cap L_{loc}^{q,\theta}(\mathbf{R}^1), \tag{1.18}$$

where we have equipped $D(H_0)$ with graph norm and $\theta = \theta(q) = 4p/n$.

When $H_0(t)$ is genuinely t-dependent, $D(H_0(t))$ is in general wildly changing with t, and it is too ambitious to expect $U(t,s)D(H(s)) \subset D(H(t))$ even when $H(t) = H_0(t)$. However, we know by Theorem 7 that $\Sigma(2) \subset D(H_0(t))$ is an invariant subspace for $U_0(t,s)$, and this suggests we try to find conditions on $V(t,x)$ such that $\Sigma(2)$ be also an invariant subspace for $U(t,s)$. The next theorem shows that this is indeed the case, if $V(t,x)$ suitably decays at infinity. We write $\langle x\rangle = (1+|x|^2)^{1/2}$.

ASSUMPTION (C): $V(t,x)$ satisfies $V \in \langle x\rangle^{-3}C(\mathbf{R}^1, L^{\tilde{p}}(\mathbf{R}^n)) + \langle x\rangle^{-\beta}C(\mathbf{R}^1, L^\infty(\mathbf{R}^n))$ and $\partial_t V \in \langle x\rangle^{-2}L^{p_1,\alpha_1}(\mathbf{R}^1) + \langle x\rangle^{-\gamma}L^{\infty,1}(\mathbf{R}^1)$ for $\tilde{p} = max(p,2)$, $2 < \beta < 3$, $p_1 = 2np/(n+4p)$ $(n \geq 5)$, $p_1 > 2p/(p+1)$ $(n=4)$, $p_1 = 2p/(p+1)$ $(n \leq 3)$, $\gamma > 0$ and $\alpha_1 = 4p/(4p-n)$.

THEOREM 9. *Let Assumptions (A) and (C) be satisfied. Then,*
(1) $U(\cdot\,,\cdot\,) \in C_*(\mathbf{R}^2, B(\Sigma(2))) \cap C_*^1(\mathbf{R}^2, B(\Sigma(2), \mathbf{H}))$.
(2) If $f \in \Sigma(2)$, then $U(t,s)f$ satisfies (1.18) and (1.19) of Theorem 8.

The proofs of Theorems 1 and 7 to 9 can be found in [**30**] and those of Theorem 3 to 6 are given in [**31**] for the case $V(t,x) = 0$. Theorems 3 to 6 for the case $A(t,x) = 0$ with singular potentials are new materials here.

We shall review the outline of the proofs of Theorem 1 and 7 in section 2. In section 3 we shall prove Theorems 3 and 4 for the case $A(t,x) = 0$. Since the proofs of Theorems 5 and 6 for this case are identical with the corresponding ones in [**31**] for the case $V(t,x) = 0$, we omit it here.

§2. Strategy of the proof of Theorems 1 and 7.

Since the proof of Theorem 1 heavily relies on Theorem 7, we begin with a sketch of the proof of the latter theorem.

We denote the classical mechanical Hamiltonian associated with (1.12) by

$$H_0(t,x,\xi) = (1/2)\sum_{j=1}^n (\xi_j - A_j(t,x))^2$$

and by $(x(t,s,y,\eta),\xi(t,s,y,\eta))$ the solutions of Hamilton's equations

$$dx/dt = \partial H_0/\partial\xi, \quad d\xi/dt = -\partial H_0/\partial x \tag{2.1}$$

with the initial conditions $(x(s),\xi(s)) = (y,\eta)$. To make the dimension same, we introduce new variables and set

$$\tilde{x}(t,s,y,\eta) = x(t,s,y,\eta/(t-s)), \quad \tilde{\xi}(t,s,y,\eta) = (t-s)\xi(t,s,y,\eta/(t-s))$$

for $t \neq s$ and $\tilde{x}(t,t,y,\eta) = y+\eta, \quad \tilde{\xi}(t,t,y,\eta) = \eta$.

LEMMA 2.1. *$(\tilde{x}(t,s,y,\eta),\tilde{\xi}(t,s,y,\eta))$ exists globally for $t,s \in \mathbf{R}^1$ and, for every α and β, $\partial_y^\alpha\partial_\eta^\beta\tilde{x}$ and $\partial_y^\alpha\partial_\eta^\beta\tilde{\xi}$ are C^1 functions of (t,s,y,η). For $|t-s|\le 1$*

$$\begin{aligned} &|\partial_y^\alpha\partial_\eta^\beta(\partial\tilde{x}_j/\partial y_k - \delta_{jk})| + |\partial_y^\alpha\partial_\eta^\beta(\partial\tilde{x}_j/\partial\eta_k - \delta_{jk})| \\ &+ |\partial_y^\alpha\partial_\eta^\beta(\partial\tilde{\xi}_j/\partial y_k)| + |\partial_y^\alpha\partial_\eta^\beta(\partial\tilde{\xi}_j/\partial\eta_k - \delta_{jk})| \le C_{\alpha\beta}|t-s|, \end{aligned} \tag{2.2}$$

where δ_{jk} is Kronecker's delta;

$$|\tilde{\xi}(t,x,y,\eta)-\eta| + |\tilde{x}(t,s,y,\eta)-y-\eta| \le C|t-s|(1+|y|+|\eta|). \tag{2.3}$$

It can be deduced from (2.2) that there exists $T>0$ such that, for every fixed $|t-s|<T$ and $y\in\mathbf{R}^n$, $\mathbf{R}^n \ni \eta \to x = \tilde{x}(t,s,y,\eta)\in\mathbf{R}^n$ is a global diffeomorphism of $\mathbf{R}^n$. Write $\tilde{\eta}(t,s,y,x)$ for its inverse. Then

$$(x(\tau),\xi(\tau)) = (x(\tau,s,y,(t-s)^{-1}\tilde{\eta}(t,s,y,x)),\xi(\tau,s,y,(t-s)^{-1}\tilde{\eta}(t,s,y,x))) \tag{2.4}$$

is a unique solution of (2.1) such that $x(t)=x$ and $x(s)=y$. We denote by $S(t,s,x,y)$ the action integral along this path:

$$S(t,s,x,y) = \int_s^t \{(\partial_\xi H_0)(\tau,x(\tau),\xi(\tau))\cdot\xi(\tau) - H_0(\tau,x(\tau),\xi(\tau))\}d\tau.$$

LEMMA 2.2. *$S(t,s,x,y)$ satisfies the property (a) of Theorem 7, i.e. it satisfies Hamilton-Jacobi equation (1.15) and the estimate (1.16). Moreover,*

$$(t-s)(\partial_x S)(t,s,x,y) = \tilde{\xi}(t,s,y,\tilde{\eta}(t,s,y,x)), \tag{2.5}$$

$$(t-s)(\partial_y S)(t,s,x,y) = -\tilde{\eta}(t,s,y,x). \tag{2.6}$$

Note that (1.16) shows that $S(t,s,x,y)$ has the properties of phase functions of OIOs studied in [**1**]. We let $e(t,s,x,y)$ be the solution of the transport equation

$$\partial_t e + (\nabla_x S - A(t,x))\nabla_x e + \frac{1}{2}(\Delta_x S - \frac{n}{t-s} - div_x A(t,x))e = 0$$

associated with $S(t,s,x,y)$ with the initial condition $e(s,s,x,y)=1$. For $T>0$ small enough (to be specified below) we write $\Delta(T) = \{(t,s)\in\mathbf{R}^2 : |t-s|\le T\}$. $\mathcal{B}(\mathbf{R}^m) = \{f : \|f\|_{\mathcal{B}^k} = sup_{x\in\mathbf{R}^m}\sum_{|\alpha|\le k}|\partial_x^\alpha f(x)| < \infty,\ k=0,1,2,...\}$ is Fréchet space of bounded C^∞ functions with bounded derivatives.

LEMMA 2.3. *$e(t,s,\cdot,\cdot) \in C(\Delta(T), \mathcal{B}(\mathbf{R}^n \times \mathbf{R}^n)$ and satisfies*

$$|\partial_x^\alpha \partial_y^\beta (e(t,s,x,y) - 1)| \le C_{\alpha\beta}|t-s|. \tag{2.7}$$

Using S of Lemma 2.2 and e of Lemma 2.3, we define the (first order) semi-classical approximation $E(t,s)$ of the propagator by

$$E(t,s)f(x) = (2\pi i(t-s))^{-n/2} \int e^{iS(t,s,x,y)} e(t,s,x,y) f(y) dy.$$

When $a(t,s,x,y)$ (resp. $p(t,s,x,\xi,y)$) is continuous in $(t,s) \in \Delta(T)$ with values in $\mathcal{B}(\mathbf{R}^n \times \mathbf{R}^n)$ (resp. $\mathcal{B}(\mathbf{R}^n \times \mathbf{R}^n \times \mathbf{R}^n)$), we say $a \in Amp$ (resp. $p \in Sym_0$). $I(t,s,a)$ is the *OIO* with the phase $S(t,s,x,y)$ and amplitude $a(t,s,x,y)$:

$$I(t,s,a)f(x) = (2\pi i(t-s))^{-n/2} \int e^{iS(t,s,x,y)} a(t,s,x,y) f(y) dy \tag{2.8}$$

and $\Psi DO(t,s)$ is the set of all ν-pseudo-differential operators

$$p(t,s,x,\nu D,x)f(x) = (2\pi\nu)^{-n} \int e^{i(x-y)\xi/\nu} p(t,s,x,\xi,y) f(y) dy d\xi$$

with symbols $p \in Sym_0$, $\nu = t-s$.

Operators of the form (2.8) are of class Asada and Fujiwara[1] and the following lemma summarizes their properties to be used in what follows.

LEMMA 2.4. *There exists $T > 0$ such that for $(t,s) \in \Delta(T)$ the following statements are satisfied:*
(1) For $k = 0,1,\cdots$, there exists a constant C_k such that

$$\|I(t,s,a)f\|_{\Sigma(k)} \le C_k \|a(t,s,\cdot,\cdot)\|_{\mathcal{B}^{M(k)}(\mathbf{R}^{2n})} \|f\|_{\Sigma(k)}, \quad a \in Amp, \tag{2.9}$$

where $M(k) = 2n + k + 1$.
(2) $I(t,s,a)$ is a continuous operator on $\mathcal{S}(\mathbf{R}^n)$.
(3) For $a, b \in Amp$ and $p \in Sym_0$, there exist $c, d, f, g \in Amp$ and $q \in Sym_0$ such that

$$I(t,s,a)^* = I(s,t,c), \quad I(t,r,a)I(r,s,b) = I(t,s,d). \tag{2.10}$$

$$I(t,s,a)p(t,s,x,\nu D,x) = I(t,s,f), \quad p(t,s,x,\nu D,x)I(t,s,a) = I(t,s,g). \tag{2.11}$$

$$I(t,s,a)I(s,t,b) = q(t,s,x,\nu D,x). \tag{2.12}$$

Moreover c, d, f, g $\in$ Amp and q $\in$ Sym_0 in (2.10) to (2.12) depend continuously on a, b and p in the topology of $\mathcal{B}(\mathbf{R}^{2n})$ or $\mathcal{B}(\mathbf{R}^{3n})$.
(4) For subdivision $\Omega : s = s_0 < s_1 < \cdots < s_L = t$ and $a_1, a_2, \cdots, a_L \in Amp$ there exists $a \in Amp$ such that $I(t, s_{L-1}, a_L) \cdot I(s_{L-1}, s_{L-2}, a_{L-1}) \cdots I(s_1, s, a_1) = I(t, s, a)$. For $m = 0, 1, \cdots$,

$$\|a(t,s,\cdot,\cdot)\|_{\mathcal{B}^m(\mathbf{R}^{2n})} \le \kappa(m)^L \prod_{j=1}^{L} \|a_j(s_j, s_{j-1}, \cdot, \cdot)\|_{\mathcal{B}^{R(m)}(\mathbf{R}^{2n})}, \tag{2.13}$$

where constants $\kappa(m)$ and $R(m)$ are independent of L, subdivision Ω, functions $a_j \in Amp$, $1 \le j \le L$ and ν.

In what follows $T > 0$ will be always assumed to be sufficiently small so that Lemma 2.4 is satisfied. In virtue of standard semi-group theory, it suffices to prove Theorems 1 and 7 for $|t - s| < T$.

PROOF OF THEOREM 7: A simple computation shows that

$$\begin{aligned} iG(t,s)f(x) &\equiv (-i\partial_t + H_0(t))E(t,s)f(x) \\ &= -(2\pi i\nu)^{-n/2} \int e^{i\tilde{S}(t,s,x,y)/\nu}(1/2)\Delta_x a(t,s,x,y)f(y)dy, \end{aligned} \tag{2.14}$$

where $\tilde{S}(t,s,x,y) = (t-s)S(t,s,x,y)$ and $\nu = t - s$. In virtue of Lemma 2.4 and (2.7) it follows that $E(t,s)$ and $G(t,s)$ satisfy the following properties:
(1) For $t \ne s$, $E(t,s)$ and $G(t,s)$ are continuous operators both in $\Sigma(k)$ and $\mathcal{S}(\mathbf{R}^n)$, and

$$\|E(t,s)f\|_{\Sigma(k)} \le C\|f\|_{\Sigma(k)}, \quad \|G(t,s)f\|_{\Sigma(k)} \le C|t-s|\|f\|_{\Sigma(k)}. \tag{2.15}$$

(2) For $f \in \Sigma(k)$, $\|E(t,s)f - f\|_{\Sigma(k)} \to 0$ as $t \to s$ and if we set $E(t,t)f = f$ and $G(t,t)f = 0$, then $E(\cdot,\cdot)$ and $G(\cdot,\cdot) \in C_*(\Delta(T), B(\Sigma(k)))$.
(3) If $f \in \mathcal{S}(\mathbf{R}^n)$, then $E(t,s)f$ and $G(t,s)f$ are continuous in $\mathcal{S}(\mathbf{R}^n)$ for $(t,s) \in \Delta(T)$.

Using properties (1) to (3) above, we see that, if $U_0(t,s)$ is the propagator with the properties of Theorem 7, then (2.14) may be solved for $E(t,s)$ in the form:

$$E(t,s)f = U_0(t,s)f - \int_s^t U_0(t,r)G(r,s)f\,dr. \tag{2.16}$$

For constructing $U_0(t,s)$ we regard (2.16) as an operator equation for $U_0(t,s)$ and solve it by successive approximation:

$$U_0(t,s) = E(t,s) + (E * G)(t,s) + (E * G * G)(t,s) + \cdots, \tag{2.17}$$

where $(F * G)(t,s) = \int_s^t F(t,r)G(r,s)dr$. (2.15) implies

$$\|E * G * G * \cdots * G\|_{B(\Sigma(\ell))} \le C_\ell^k |t-s|^{2k} \frac{1}{\Gamma(2(k+1))}$$

and, for every $\ell = 0, 1, \cdots$, the series (2.17) converges in operator norm of $B(\Sigma(\ell))$ uniformly in $(t,s) \in \Delta(T)$ to define $U_0(\cdot,\cdot) \in C_*(\Delta(T), B(\Sigma(\ell)))$, which clearly satisfies (2.16), hence (1) to (3) of Theorem 7.

Moreover, by (2.13), we see that $(E * G * \cdots * G)(t,s) = I(t,s,c_{Nk})$ with $c_{Nk}(t,s,x,y)$ satisfying

$$\|c_k\|_{\mathcal{B}^m} \le C_m^{k+1} \frac{|t-s|^{2k}}{\Gamma(2k+1)}, \quad m = 0, 1, \cdots. \tag{2.18}$$

Thus, the sequence $b(t,s,x,y,h) = \sum_{k=0}^\infty c_{Nk}(t,s,x,y,h)$ converges in $\mathcal{B}(\mathbf{R}^{2n})$ uniformly in $\Delta(T)$ and $U_0(t,s) = I(t,s,b)$. This implies that $U_0(t,s)$ satisfies (4) of Theorem 7.

Proof of Theorem 1:

We first remark that the multiplication operator $V(t)$ with $V(t,x)$ is $H_0(t)$-form bounded with bound 0 and the Friedrich extension of $H(t) = H_0(t) + V(t)$ coincides with the part of $H(t) \in B(\mathbf{H} \cap L^q(\mathbf{R}^n), \Sigma(-2))$ in $\mathbf{H}$.

We solve (1.1) via the equivalent integral equation:

$$u(t) = U_0(t,s)f - i\int_s^t U_0(t,r)V(r)u(r)dr. \tag{2.19}$$

We set, for the parameter ℓ with $0 \le 2/\theta = n(1/2 - 1/\ell) < 1$, the function spaces

$$X(I,\ell) = C_b(I,\mathbf{H}) \cap L^{\ell,\theta}(I), \quad X^*(I,\ell) = L^{2,1}(I) + L^{\ell',\theta'}(I),$$

and solve (2.19) in $X(I,q)$, $q = 2p/(p-1)$ with a small interval I, where ℓ' and θ' are indices conjugate to ℓ and θ. We say $u \in X_{loc}(I,\ell)$ if $u \in X(I',\ell)$ for any compact subinterval I' of I. We need the following four lemmas. The first lemma is an immediate consequence of (2.9) and the complex interpolation theory.

Lemma 2.5. *Let $a \in Amp$. Then:*
(1) There exists a constant C depending only on n such that for any $2 \le \ell \le \infty$ and t, $s \in \Delta(T) \setminus \Delta$,

$$\|I(t,s,a)f\|_\ell \le C\|a(t,s,\cdot,\cdot)\|_{\mathcal{B}^{4n+1}} |t-s|^{-n(1/2-1/\ell)} \cdot \|f\|_{\ell'}. \tag{2.20}$$

(2) For $f \in L^{\ell'}(\mathbf{R}^n)$, $I(t,s,a)f \in C(\Delta(T) \setminus \Delta, L^\ell(\mathbf{R}^n))$.

The next lemma can be proved by using (2.20) and the Hardy-Littlewood-Sobolev inequality.

LEMMA 2.6. *Let $0 \leq 2/\theta(\ell) = n(1/2 - 1/\ell) < 1$ and let $Z(t,s)$ be $B(\mathbf{H})$-valued measurable function of $(t,s) \in \mathbf{R}^2$ which satisfies the following conditions:*
(a) For any $f \in \mathbf{H}$ and $t_0 \in \mathbf{R}^1$, $Z(t,s)f$ and $Z(s,t)^*f$ are continuous in $\mathbf{H}$ with respect to t at $t = t_0$ for almost all $s \in \mathbf{R}^1$.
(b) For $f \in L^{\ell'}(\mathbf{R}^n) \cap \mathbf{H}$ and $(t,s) \notin \Delta$, $Z(t,s)f \in L^{\ell}(\mathbf{R}^n)$ and

$$\|Z(t,s)f\|_{\ell} \leq C|t-s|^{-n(1/2-1/\ell)}\|f\|_{\ell'}$$

with constant $C > 0$ independent of t, s and f.
(c) For any $r \in \mathbf{R}^1$ and $(t,s) \notin \Delta$, $Z(r,t)^*Z(r,s)$ and $Z(t,r)Z(s,r)^*$ defined on $L^{\ell'}(\mathbf{R}^n) \cap \mathbf{H}$ can be extended to bounded operators from $L^{\ell'}(\mathbf{R}^n)$ to $L^{\ell}(\mathbf{R}^n)$ and

$$\|Z(r,t)^*Z(r,s)f\|_{\ell} \leq C|t-s|^{-n(1/2-1/\ell)}\|f\|_{\ell'},$$
$$\|Z(t,r)Z(s,r)^*f\|_{\ell} \leq C|t-s|^{-n(1/2-1/\ell)}\|f\|_{\ell'}.$$

Then:
(1) For any $f \in \mathbf{H}$ and s, $u(\cdot) = Z(\cdot,s)f \in L^{\ell,\theta}(\mathbf{R}^1)$ and with constant C independent of s and f,

$$\|Z(\cdot,s)f\|_{\ell,\theta} \leq C\|f\|.$$

(2) The integral operator $Ju(t) = \int_{-\infty}^{\infty} Z(t,s)u(s)ds$ is bounded from $X^*(\mathbf{R}^1,\ell)$ to $X(\mathbf{R}^1,\ell)$:

$$\|Ju\|_{X(\mathbf{R}^1,\ell)} \leq C\|u\|_{X^*(\mathbf{R}^1,\ell)}.$$

Since $U_0(t,s) = I(t,s,b)$ with $b \in Amp$, it obviously satisfies Lemma 2.5, hence the conditions (a) to (c) of Lemma 2.6. Accordingly, Lemma 2.6 implies the following

LEMMA 2.7. *Let $0 \leq 2/\theta = n(1/2 - 1/\ell) < 1$. Then there exists a constant $C > 0$ such that the following estimates are satisfied for any $R > 0$ and $a \in \mathbf{R}^1$.*
(1) For $f \in \mathbf{H}$, $U_0(\cdot,a)f \in L^{\ell,\theta}_{loc}(\mathbf{R}^1)$ and

$$\|U_0(\cdot,a)f\|_{L^{\ell,\theta}([a-R,a+R])} \leq C(1+R)^{1/\theta}\|f\|, \quad f \in \mathbf{H}.$$

(2) Let G_a be the integral operator defined by $G_a u(t) = \int_a^t U_0(t,s)u(s)ds$. Then G_a is a bounded operator from $X^*([a-R,a+R],\ell)$ to $X([a-R,a+R],\ell)$ and

$$\|G_a u\|_{X([a-R,a+R],\ell)} \leq C(1+R)^{2/\theta}\|u\|_{X^*([a-R,a+R],\ell)}.$$

If we define a Banach space for potentials by $M(I) = L^{p,\alpha}(I) + L^{\infty,1}(I)$ with the standard norm:

$$\|u\|_{M(I)} = inf\{\|V_1\|_{L^{p,\alpha}}(I) + \|V_2\|_{L^{\infty,1}}(I) : V = V_1 + V_2\},$$

where p and α are the exponents in Assumption (A), then Hölder's inequality implies that the multiplication operator V satisfies the following properties.

LEMMA 2.8. *Let $q = 2p/(p-1)$. Then:*

$$\|Vu\|_{X^*(I,q)} \le \|V\|_{M(I)} \cdot \|u\|_{X(I,q)}.$$

Setting $Q_s f(t) = U_0(t,s)f$, we write the integral equation (2.19) in the form

$$u(t) = Q_s f(t) + G_s V u(t). \tag{2.19'}$$

As in the proof of Theorem 3, we only have to consider in a small interval $I = [t_0 - T/2, t_0 + T/2]$. Lemma 2.7 and 2.8 show that $Q_s \in B(\mathsf{H}, X(I,q))$ and $G_s V \in B(X(I,q))$, both depending on s continuously in the strong topology of operator valued functions. Moreover, if we take $T > 0$ sufficiently small, $G_s V$ is a contraction on $X(I,q)$. It follows that (2.19') can be uniquely solved for $u \in X(I,q)$:

$$u = (1 - G_s V)^{-1} Q_s f = \sum_{k=0}^{\infty} (G_s V)^k Q_s f.$$

We define $U(t,s)$, for t, $s \in I$, by $U(t,s)f = u(t)$, that is, $U(t,s) = \Gamma_t(1 - G_s V)^{-1} Q_s$, where $\Gamma_t \in B(X(I,q), \mathsf{H})$ is the evaluation functional defined by $\Gamma_t u = u(t)$. Clearly $U(\cdot,\cdot) \in C_*(I^2, B(\mathsf{H}))$ and $u(\cdot) = U(\cdot,s)f \in L^{q,4p/n}(I)$ for any $f \in \mathsf{H}$, and, as can be easily shown from its construction, this is the desired propagator for (1.1) satisfying the properties on Theorem 1.

§3. Proof of Theorem 3 and 4 for the case $A(t,x) = 0$.

In virtue of the complex interpolation theory for weighted Sobolev spaces (cf. e.g. [**28**]), it is well known that

$$[W^{k_1,p_1}(\mathbf{R}^n, \langle x\rangle^{\mu_1}dx), W^{k_2,p_2}(\mathbf{R}^n, \langle x\rangle^{\mu_2}dx)]_\theta = W^{k_\theta,p_\theta}(\mathbf{R}^n, \langle x\rangle^{\mu_\theta}(x)dx), \tag{3.1}$$

where $k_\theta = (1-\theta)k_1 + \theta k_2$, $1/p_\theta = (1-\theta)/p_1 + \theta/p_2$ and $\mu_\theta = (1-\theta)\mu_1 + \theta\mu_2$, $0 \le \theta \le 1$. Hence Theorem 4 is a consequence of Theorem 3 and Theorem 1, (3).

We prove Theorem 3 for the case $A(t,x) = 0$. We assume $s = 0$ and suppress the index s. In this case $U_0(t,s) = \exp(i(t-s)\Delta/2)$ and (2.19) reads

$$u(t) = Qf(t) + GVu(t) = \exp(it\Delta/2)f - i\int_0^t \exp(i(t-s)\Delta/2)V(s)u(s)ds.$$

We let $F_\epsilon(\xi) = \langle\xi\rangle^{1/2}/(1+\epsilon\xi^2)$ and write, with a little abuse of notation, $F_\epsilon(D_j) = (1+D_j^2)^{1/4}/(1+\epsilon D_j^2)$, $\epsilon > 0$, and $\phi(t) = (1+t^2)^{-\mu/2}$, $t \in \mathbf{R}$. The commutators $[F_\epsilon(D), \langle x\rangle^{-\mu}]$ and $F_\epsilon(D_j)[\phi^2(x_j), F_\epsilon(D_j)]$ are uniformly bounded in H for $\epsilon > 0$, $j = 1, \cdots, n$. Hence we have

$$\|F_\epsilon(D)\langle x\rangle^{-\mu}v\|^2 \le \sum_{j=1}^n \|F_\epsilon(D_j)\langle x\rangle^{-\mu}v\|^2 \le \sum_{j=1}^n \|\phi(x_j)F_\epsilon(D_j)v\|^2 + C\|v\|^2;$$

$$\|\phi_j(x)F_\epsilon(D_j)v\|^2 = (F_\epsilon(D_j)\phi^2(x_j)F_\epsilon(D_j)v, v)$$
$$= (F_\epsilon(D_j)^2\phi^2(x_j)v, v) + O(\|v\|), \quad v \in \mathsf{H}.$$

It follows that with a constant $C > 0$ independent of $\epsilon > 0$

(3.2)
$$\int_0^T \|\langle x\rangle^{-\mu}F_\epsilon(D)u(t)\|^2 dt \le \sum_{j=1}^n \int_0^T \|\phi(x_j)F_\epsilon(D_j)u(t)\|^2 dt + C\|f\|^2$$
$$\sum_{j=1}^n \int_0^T (F_\epsilon(D_j)^2\phi^2(x_j)u(t), u(t))dt + O(\|f\|^2).$$

Here we used $\|u(t)\| = \|f\|$. Introducing the notation

$$g(t) = e^{-it\Delta/2}GVu(t) = \int_0^t e^{-is\Delta}V(s)u(s)ds$$

and

$$\Phi_j(T,x,D) = \int_0^T e^{-it\Delta/2}F_\epsilon(D_j)^2\phi^2(x_j)e^{it\Delta/2}dt,$$

we write the integral in the right of (3.2) in the form

$$(\Phi_j(T,x,D)f,f) + \int_0^T (\Phi_j'(t,x,D)f, g(t))dt \tag{3.3}$$
$$+ \int_0^T (\Phi_j'(t,x,D)g(t), f)dt + \int_0^T (\Phi_j'(t,x,D)g(t), g(t)),$$

where $\Phi_j'(t,x,D) = e^{-it\Delta/2}F_\epsilon(D_j)^2\phi^2(x_j)e^{it\Delta/2}$ is the t-derivative of $\Phi_j(t,x,D)$.
We use the following

LEMMA 3.1. *For every $j = 1,\cdots,n$ and $t \in \mathsf{R}$, the operator $\Phi_j(t,x,D)$, is a bounded operator of $L^\ell(\mathsf{R}^n)$ for every $1 < \ell < \infty$. Its operator norm is bounded uniformly in $|t| \le T$, $T > 0$, and $\epsilon > 0$.*

$$e^{it\Delta/2}\Phi_j(t,x,D)e^{-it\Delta/2} = -\Phi_j(-t,x,D). \tag{3.4}$$

PROOF: (3.4) is obvious and we prove the boundedness only. Since $F_\epsilon(D_j)^2$ and $e^{-it\Delta/2}$ commutes, we have

$$e^{-it\Delta/2}F_\epsilon(D_j)^2\phi^2(x_j)e^{it\Delta/2} = F_\epsilon(D_j)^2\phi^2(x_j + tD_j)$$

and $\phi^2(x_j + tD_j)$ is Weyl operator with the symbol $\phi^2(x_j + t\xi_j)$. It is esay to see that $(1 + \epsilon D_j^2)^{-1}$ has L^1 convolution integral kernel with uniform L^1-norm in

x_j-variable, hence, is a uniformly bounded operator in every $L^\ell(\mathbf{R}^n)$, $1 \le \ell \le \infty$. Take $\chi(\xi) \in C_0^\infty(\mathbf{R})$ such that $\chi(\xi) = 1$ near zero and split

$$(1+\xi_j^2)^{1/2} = (1+\xi_j^2)^{1/2}\chi(\xi_j) + (1+\xi_j^2)^{1/2}(1-\chi(\xi_j))\xi_j^{-1}\cdot\xi_j.$$

Here $(1+D_j^2)^{1/2}\chi(D_j)$ is obviously L^ℓ-bounded and the standard result of pseudo-differntial operators shows that $(1+D_j^2)^{1/2}(1-\chi(D_j))D_j^{-1}$ is also bounded in $L^\ell(\mathbf{R}^n)$, $1 < \ell < \infty$. (This is essentially the Hilbert-Transform.) Hence it suffices to prove the boundedness of $\int_0^T \phi^2(x_j+tD_j)dt$ and $\int_0^T D_j\phi^2(x_j+tD_j)dt$, the latter being the Weyl operator with the symbol

$$\int_0^T \phi^2(x_j+t\xi_j)\xi_j dt - (i/2)\int_0^T (\phi^2)'(x_j+t\xi_j)dt \tag{3.5}$$

Note that in general

$$\begin{aligned} g(x+tD)f(x) &= (2\pi)^{-n}\int e^{i(x-y)\cdot\xi}g((x+y)/2+t\xi)f(y)dyd\xi \\ &= e^{-ix^2/2t}(g(tD)e^{iy^2/2t}f)(x) \end{aligned} \tag{3.6}$$

It follows that $\phi^2(x_j+tD_j)$ and $(\phi^2)'(x_j+tD_j)$, hence the Weyl operator of which the symbol given by the second integral of (3.5) are uniformly bounded in L^ℓ for every $1 \le \ell \le \infty$. Writing the first integral of (3.5) as

$$\int_0^T \phi^2(x_j+t\xi_j)\xi_j dt = \int_{x_j}^{x_j+T\xi_j} \phi^2(t)dt = \Psi(x_j+T\xi_j) - \Psi(x_j)$$

with $\Psi(t) = \int_0^t \phi^2(s)ds$, we see by (3.6) that the Weyl operator with the symbol $\int_0^T \phi(x_j+t\xi_j)^2\xi_j dt$ is given by

$$\Psi(x_j+TD_j)f(x) - \Psi(x_j) = e^{-ix_j^2/2T}(\Psi(TD_j)e^{iy_j^2/2T}f)(x) - \Psi(x_j)f(x). \tag{3.7}$$

Since $\Psi(t)$ is bounded and its derivatives satisfies

$$|d^k\Psi/dt^k(t)| \le C_k\langle t\rangle^{-\mu+1-k}, \quad k = 1, 2, \cdots$$

the classical theorem for L^ℓ-multiplier theorem (cf. e.g. [**27**]) implies that $\Psi(TD_j)$ is L^ℓ-bounded for $1 < \ell < \infty$. This completes the proof of the lemma. ∎

By integration by parts we rewite (3.3) into the following form with $\Phi_j(t) = \Phi_j(t, x, D)$:

$$\begin{aligned} &(\Phi_j(T)f, f) + (\Phi_j(T)f, g(T)) + (\Phi_j(T)g(T), f) + (\Phi_j(T)g(T), g(T)) \\ &+ \int_0^T (\Phi_j(-t)e^{-it\Delta}f, (Vu)(t))dt + \int_0^T (\Phi_j(-t)(Vu)(t), e^{-it\Delta}f)dt \\ &+ \int_0^T (\Phi_j(-t)(Vu)(t), (GVu)(t))dt + \int_0^T (\Phi_j(-t)(GVu)(t), (Vu)(t))dt. \end{aligned}$$

Each term above can be estimated by $C\|f\|^2$ by applying Lemma 3.1 and the estimates

$$\|g(T)\| \leq C\|f\|; \tag{3.8}$$

$$\|e^{-it\Delta}f\|_{L^{4p/n}(I,L^q)} \leq C\|f\|, \quad \|Vu\|_{L^{4p/n}(I,L^q)^*} \leq C\|f\|; \tag{3.9}$$

and

$$\|GVu\|_{L^{4p/n}(I,L^q)} \leq C\|f\| \tag{3.10}$$

which are the results of Lemma 2.7 and 2.8. (3.8) and Lemma 3.1 show that the first four terms are bounded by $C\|f\|^2$; (3.9) and Lemma 3.1 show the following two terms are bounded by $C\|f\|^2$; and (3.10) implies that the last two terms are also bounded by $C\|f\|^2$. This completes the proof of Theorem 3.

REFERENCES

1. Asada, K. and D. Fujiwara, *On some oscillatory integral transformations in* $L^2(\mathbf{R}^n)$, Japan. J. Math. **4** (1978), 299-361.
2. Ben-Artzi, M. and S. Klainerman, *Decay and reguralіty for the Schrödinger equation*, preprint, Hebrew University of Jerusalem.
3. Carleson, L., *Some analytical problems related to statistical mechanics*, in "Euclidean Harmonic Analysis," Lecture Notes in Math. 799, Springer-Verlag, Berlin and New York, 1979, pp. 5-45.
4. Constantin, P. and J. C. Saut, *Local smoothing properties of dispersive equations*, J. of Amer. Math. Soc. **1** (1988), 413–439.
5. Cycon, H. L., R. G. Froese, W. Kirsch and B. Simon, "Schrödinger operators with application to quantum mechanics and global geometry," Springer, Berlin-Heidelberg, 1987.
6. Dahlberg, B. E. J. and C. E. Kenig, *A note on the almost everywhere behavior of solutions to the Schrödinger equation*, in "Lecture Notes in Math. 908," Springer-Verlag, Berlin and New York, 1982, pp. 205-208.
7. De Bouard, A., *Nonlinear Schrödinger equations with magnetic field*, preprint, Univ. Paris-Sud, 1990.
8. Edward, R. E., "Fourier series, a modern introduction. Vol. 2," Springer-Verlag, Berlin-Heidelberg-New York, 1982.
9. Fujiwara, D., *A construction of the fundamental solution for the Schrödinger equation*, J. d'Analyse Math. **35** (1979), 41–96.
10. Fujiwara, D., *Remarks on convergence of the Feynman path integrals*, Duke Math. J. **47** (1980), 559-600.

11. Ginibre, J. and G. Velo, *The global Cauchy problem for non-linear Schrödinger equation revisited*, Ann. Inst. H. Poincaré, Anal. Nonlinéare **2** (1985), 309–327.
12. Howland, J. S., *Stationary scattering theory for time dependent Hamiltonians*, Math. Ann. **207** (1974), 315–335.
13. Ikebe, T. and T. Kato, *Uniqueness of self-adjoint extension of singular elliptic differential operators*, Arch. Rat. Mech. Anal. **9** (1962), 77–92.
14. Kato, T., *Fundamental properties of Hamiltonian operators of Schrödinger type*, Trans. Amer. Math. Soc. **70** (1951), 195–211.
15. Kato, T., *Linear evolution equations of "hyperbolic" type*, J. Fac. Sci. Univ. Tokyo Sec. I **17** (1972), 241–258.
16. Kato, T., *On the Cauchy problem for the (generalized) Kortweg-de Vries equation*, Studies in Appl. Math. Adv. in Math. Supplementary Studies **18**, 93–128.
17. Kato, T., *On nonlinear Schrödinger equations*, Ann. Inst. H. Poincaré Phys. Theor. **46** (1987), 113–129.
18. Kato, T. and K. Yajima, *Some examples of smooth operators and the associated smoothing effect*, Reviews in Math. Physics, **1** (1990), 481–496.
19. Kenig, C. E. and G. Ponce and L. Vega, *Oscillatory integrals and regularity of dispersive equations*, preprint, Univ. Chicago, 1989.
20. Kenig, C. E. and A. Ruiz, *A strong type* (2, 2) *estimate for a maximal operator associated to the Schrödinger equation*, Trans. Amer. Math. Soc. **280** (1983), 239-246.
21. Kitada, H., *On a construction of the fundamental solution for Schrödinger equation*, J. Fac. Sci. Univ. Tokyo Sec. IA **27** (1980), 193–226.
22. Kitada, H. and H. Kumanogo, *A family of Fourier Integral operators and the fundamental solution for a Schrödinger equation*, Osaka J. Math. **18** (1981), 291–360.
23. Reed, M. and B. Simon, "Methods of modern mathematical physics, Vol.II. Fourier analysis and selfadjointness," Academic Press, New York, 1977.
24. Simon, B., *Best constants in some operator smoothness estimates*, preprint, Caltech, 1990.
25. Sjölin, P., *Regularity of solutions to the Schrödinger equation*, Duke Math. J. **55** (1987), 699-715.
26. Strichartz, R. S., *Restrictions of Fourier transforms to quadratic surfaces and decay of solutions of wave equations*, Duke Math. J. **44**, 705–714.
27. Taylor, M. E., "Pseudo-differential operators," Princeton Univ. Press, Princeton, New Jersey, 1981.
28. Triebel, H., "Interpolation theory, function spaces, differential operators," North Holland, Amsterdam, New York, Oxford, 1978.
29. Yajima, K., *Existence of solutions for Schrödinger evolution equations*, Commun. Math. Phys. **110** (1987), 415–426.
30. Yajima, K., *Schrödinger evolution equation with magnetic fields*, in press, J. d'Analyse Math..

31. Yajima, K., *On smoothing property of Schrödinger propagators*, in "Lecture Notes in Math. 1450," Springer-Verlag, Berlin and New York, 1990, pp. 20-35.
32. Zelditch, S., *Reconstruction of singularities for solutions of Schrödinger's equation*, Commun. Math. Phys. **90** (1983), 1–26.
33. Zygmund, A., *On Fourier coefficients and transforms of functions of two variables*, Studia Math. **50** (1974), 189–201.

MOVING POTENTIALS AND SCATTERING THEORY

Hagen Neidhardt
Karl-Weierstrass-Institut für Mathematik
Mohrenstr. 39, Berlin 1o86, Germany

ABSTRACT

For the explicitly time-dependent charge transfer model existence and completeness of the wave operators are shown.

1. Introduction

Time-dependent potentials may arise in quantum mechanics either under an influence from without or in the presence of moving classical particles in the system. The first case usually leads to explicitly time-dependent potentials [1-6], whereas the second case forces implicitly time-dependent potentials [7-9]. However, both cases are treated separately in the literature. In the following we try to combine both types of time-dependent potentials in one model and to establish the completeness of the wave operators generalizing the definition of the scattering subspace in a natural way. Since models of the second kind are usually called charge-transfer models we call our one the explicitly time-dependent charge transfer model.

In detail, in $\mathfrak{h} = L^2(\mathbb{R}^n)$, $n \geq 1$, we consider the Schrödinger equation

$$i\frac{\partial u}{\partial t} = H(t)u \equiv (H_0 + V(t))u, \qquad u|_{t=s} = u_0,$$

where $H_0 = -\frac{1}{2}\Delta$ is given as usual and $\{V(t)\}_{t\in\mathbb{R}^1}$ is a time-dependent perturbation of the form

$$V(t) = \sum_{j=1}^{N} V_j(t),$$

where the time-dependent perturbations $\{V_j(t)\}_{j=1}^{N}$, $t \in \mathbb{R}^1$, arise from time-dependent potentials q_j as follows:

$$(V_j(t)f)(x) = q_j(t, x - x_j(t)), \quad f \in \mathfrak{h}, \ t \in \mathbb{R}^1,$$

$x_j(.)$: $\mathbb{R}^1 \longrightarrow \mathbb{R}^n$, $j = 1,2,\ldots,N$.

In the following, by $C^1_{loc}(\mathbb{R}^m)$ and $C^1_{loc}(\mathbb{R}^m, \mathbb{R}^k)$, $m, k \geq 1$, we denote the sets of all functions defined on $\mathbb{R}^m$ with values in $\mathbb{R}^1$ and $\mathbb{R}^k$, respectively, whose first derivatives exist and are continuous.

Assumption P. The potentials q_j, $j = 1,2,\ldots,N$, belong to $C^1_{loc}(\mathbb{R}^{n+1})$ and satisfy the properties

$$|q_j(t,x)| \leq M_j(1+|x|)^{-1-\varepsilon}, \ (t,x) \in \mathbb{R}^{n+1}, \ \varepsilon > 0,$$

$$|x||\nabla q_j| \in L^\infty(\mathbb{R}^{n+1}), \ \lim_{|x|\to\infty} x\nabla q_j(t,x) = 0, \ t \in \mathbb{R}^1,$$

$$\dot{q}_j \in L^\infty(\mathbb{R}^{n+1}), \ \sup_{x\in\mathbb{R}^n} |q_j(t,x)| \in L^1(\mathbb{R}^1),$$

$j = 1,2,\ldots,N$, where we have used the notation $\dot{q}_j = \frac{\partial}{\partial t} q_j$.

The function $x_j(.)$: $\mathbb{R}^1 \longrightarrow \mathbb{R}^n$ can be regarded as a trajectory along which the potentials q_j move. Concerning the trajectories we assume the following.

Assumption T. The trajectories $x_j(.)$, $j = 1,2,\ldots,N$, belong to $C^1_{loc}(\mathbb{R}^1, \mathbb{R}^n)$ such that

$$\lim_{t\to\pm\infty} \frac{1}{t} x_j(t) = v_j^\pm, \quad j = 1,2,\ldots,N,$$

exist and, moreover,

$$\sup_{\pm t\geq 0} |x_j(t) - v_j^\pm t| < +\infty, \quad j = 1,2,\ldots,N,$$

$$\sup_{\pm t>0} |t\dot{x}_j(t) - v_j^{\pm}t| < +\infty, \quad j = 1,2,\ldots,N.$$

If $q_j \in C^1_{loc}(\mathbb{R}^{n+1})$, q_j, $|\nabla q_j|$ and $\dot{q}_j \in L^{\infty}(\mathbb{R}^{n+1})$ as well as $x_j(.) \in C^1_{loc}(\mathbb{R}^1,\mathbb{R}^n)$, $j = 1,2,\ldots,N$, by Proposition 2.2 and Remark 2.1 of [10] with the Schrödinger equation we can associate a unique propagator $\langle U(t,s)\rangle_{(t,s)\in\mathbb{R}^2}$ consisting of unitary operators and obeying the properties of Proposition 2.2 of [10]. Using this propagator the scattering states are defined as follows.

Definition. The state f belongs to the scattering subspace $\mathfrak{H}^{sc}_{\pm}(s)$, $s \in \mathbb{R}^1$, if for every $R > 0$ we have

$$\lim_{T\to\pm\infty} \frac{1}{T}\int_s^{s+T} dt \; \|F(|X| < R)e^{ix_j(t)P}\, U(t,s)f\|^2 = 0, \tag{1}$$

$j = 1,2,\ldots,N$, and if for every $\varepsilon_j^{\pm} > 0$ there exist $\eta_j^{\pm} > 0$ and $\tau_j^{\pm} > 0$ such that

$$\sup_{\pm t>\tau_j^{\pm}} \|F(|P - v_j^{\pm}| < \eta_j^{\pm})U(t,s)f\| < \varepsilon_j^{\pm}, \tag{2}$$

$j = 1,2,\ldots,N$.

By $F(.)$ we denote the spectral projection of the self-adjoint operator to the part of the spectrum as indicated in the parenthesis. By X and P we denote the commuting n-tuples $X = \langle X_1, X_2, \quad, X_n\rangle$ and $P = \langle -i\frac{\partial}{\partial x_1}, -i\frac{\partial}{\partial x_2}, \ldots, -i\frac{\partial}{\partial x_n}\rangle = -i\nabla$ of position and impulse operators, respectively.

Remark 1. (i) If the potentials q_j, $j = 1,2,\ldots, N$, are nonmoving, i.e. $x_j(t) \equiv 0$, $j = 1,2,\ldots,N$, and time-independent, i.e $q_j(t,x) = q_j(x)$, $j = 1,2,\ldots,N$, then condition (1) coincides with those of Ruelle [11] and Amrein-Georgescu [12]. Moreover, condition (2) is a consequence of condition (1) and Assumption P.

(ii) If the potentials q_j, $j = 1,2,\ldots,N$, are nonmoving but time-dependent, our definition of the scattering subspace coincides with Definition 5.1 of Kitada and Yajima [5]. See also [1,2]. As it has been pointed out by Kitada and Yajima the condition (2) is essential by a counter example given by Yafaev [3,4]. The same takes place in our case despite the fact that we have a slightly stronger condition (2) than Kitada and Yajima.

Therefore, it seems to us that the definition is a natural generalization of the definition of the scattering subspace to moving time-dependent potentials.

Theorem. *If the* Assumptions P *and* T *are satisfied, then the wave operators* $W_{\pm}(s) = \operatorname{s-lim}_{t\to\pm\infty} U(t,s)^{*} e^{-i(t-s)H_0}$ *exist and are complete, i.e* $\mathfrak{R}(W_{\pm}(s)) = \mathfrak{H}_{\pm}^{sc}(s)$, $s \in \mathbb{R}^1$.

Remark 2. (i) If the potentials q_j, $j = 1,2,\ldots,N$, are nonmoving and time-independent on account of Remark 1 (i), the problem coincides with the existence and completeness problem for short range potentials which is solved.

(ii) If the potentials q_j, $j = 1,2,\ldots,N$, are nonmoving but time-dependent, the problem was solved by Kitada and Yajima [5,6] even for long range potentials.

(iii) If the potentials are moving but time-independent a stronger asymptotic completeness result than in the theorem was proved by Yajima [7], Graf [8] and Wüller [9]. It can be shown that our completness relation follows for time-independent short range potentials from [7] or [8,9]. However, from [7-9] we obtain stronger assumptions concerning the trajectories $x_j(.)$, $j = 1,2,\ldots,N$, and the behavior of the potentials q_j, $j = 1,2,\ldots,N$, at infinity.

References

1. D.R.Yafaev, *Dokl. Akad. Nauk SSSR* **251** (1980) 812.
2. D.R.Yafaev, *Funkz. Analysis Pril.* **14** (1980) 91.
3. D.R.Yafaev, *Dokl. Akad. Nauk. SSSR* **243** (1978) 1150.

4. D.R.Yafaev, *Mat. Sbornik* **111** (1980) 187.
5. H.Kitada, K.Yajima, *Duke Math. J.* **49** (1982) 341.
6. Kitada, K.Yajima, *Duke Math. J.* **50** (1983) 1005.
7. K.Yajima, *Commun. Math. Phys.* **75** (1980) 153.
8. G.M.Graf, *Phase space analysis of the charge transfer model*, preprint, ETH-Zürich, Zürich 1988.
9. U.Wüller, *Geometric Methods in Scattering Theory of the Charge Transfer Model*, preprint, FU-Berlin, Berlin 1989.
10. H.Neidhardt, *Moving Potentials and the completeness of wave operators. Part I: The Propagator*, to appear in *Int. Eq. Oper. Theory*.
11. D.Ruelle, *Nuovo Cimento* **59** A (1969) 655.
12. W.O.Amrein, V.Georgescu, *Helv. Phys. Acta* **46** (1973) 653.

ON THE ADIABATIC THEOREM, LANDAU-ZENER FORMULA, AND THE GEOMETRY OF ISOSPECTRAL HAMILTONIANS

VOJKAN JAKŠIĆ
Department of Mathematics
California Institute of Technology
Pasadena, CA 91125

and

JAN SEGERT
Department of Mathematics
University of Missouri
Columbia, MO 65211

ABSTRACT

For a class of Hamiltonians with complex-analytic dependence on the time variable, we prove that the correction to the adiabatic approximation is exponentially small in the adiabatic parameter .

1. Introduction

The adiabatic theorem of quantum mechanics is concerned with the dynamics generated by time-dependent slowly changing Hamiltonians $H(s/\tau)$, where τ is the adiabatic parameter. Roughly, the theorem says that states do not evolve across the spectral gaps when $\tau \to \infty$. As long as τ is finite, there is a positive probability $p(s, s_0; \tau)$ that if the system was prepared in a state which lies in an isolated spectral band at time s_0, it has made a transition to a state outside the band at time s. Giving precise asymptotics of $p(s, s_0; \tau)$ has been a question of considerable mathematical and physical interest, with aplications to such diverse problems as the Zener diode, atomic scattering theory, the solar-neutrino problem, and the quantum Hall effect.

The first substantive result was obtained in 1928 by M. Born and V. Fock, who considered Hamiltonians that are nondegenerate matrices of finite dimension. They showed that $p(s, s_0; \tau) = O(1/\tau)$. Their results were extended by T. Kato in 1952, when he derived the same estimate for transition from a spectral band consisting of an isolated eigenvalue, without making any assumptions on the nature of the rest of the spectrum. In 1987, J. Avron, R. Seiler and L. Yaffe[1] showed that if $H(s)$ is k-times differentiable in a certain sense, then $p(s, s_0; \tau) = O(1/\tau^{k-1})$, assuming only that the spectral band stays separated from the rest of spectrum by a gap of finite size. However, physical heuristics due to L. Landau[4] and C. Zener[5] in 1932 indicate that $p(s, s_0; \tau)$ should be exponentially small in τ, providing that $s = \infty, s_0 = -\infty$, and that Hamiltonians are analytic in a certain sense.

We were able to prove the Landau-Zener conjecture in a number of special cases. A typical such result follows: Let $\mathrm{H}^1(a)$ be the Hardy class of operator-valued analytic functions $T(z)$ in the strip $\mathcal{S}(a) = \{z : -a < \mathrm{Im}(z) < a\}$, satisfying

$$\|T\|_{\mathrm{H}^1} = \sup_{-a<c<a} \int_{-\infty}^{+\infty} \|T(x + ic)\| dx < \infty.$$

Let $H(s) = W(s)HW^{-1}(s)$, where H is bounded and $W(s)$ is a unitary operator. Denote by g the size of the gap between the band and the rest of spectrum, and by $W'(s)$ the derivative $\partial W/\partial s$.

Theorem: Suppose that $W'(s) \in \mathrm{H}^1(a)$. Then there exists finite constant C_a such that

$$\lim_{s\to\infty} \lim_{s_0\to-\infty} p(s, s_0; \tau) \leq C_a \exp(-2\tau a g).$$

The exponential factor, $2\tau a g$, is exactly what one expects on the basis of the Landau-Zener argument. The above result is optimal in the sense in which the related Paley-Wiener theorem is optimal (see section 3).

We are also able to treat the case where H is unbounded, and various cases of time-dependent spectra, and in particular, the case when $H(s)$ has only discrete spectrum[3]. However, such results are more difficult to state, and we will omit that discussion here.

2. Sketch of the Proof

First, we will make some concepts from the previous section more rigorous. The Hamiltonians $H(s) = W(s)HW^{-1}(s)$ are isospectral. For an interval $[a, b] \subset \mathbf{R}$, let $\chi_{[a,b]}$ be the characteristic function. We denote the spectral projection by $P_{[a,b]}(s) = \chi_{[a,b]}(H(s))$. We call $P_{[a,b]}(s)$, or the corresponding interval $[a, b]$, a band in the spectrum. The band is separated from the rest of the spectrum by a gap of finite size g if $g = \mathrm{dist}(\sigma(H) \setminus [a, b], [a, b]) > 0$. After a (linear) reparametrization of the time variable, we are concerned with evolution $U_\tau(s, s_0)$ which solves (in scaled time) the equation

$$\frac{\partial}{\partial s} U_\tau(s, s_0) = \tau H(s) U_\tau(s, s_0), \qquad U_\tau(s_0, s_0) = 1.$$

Denote

$$p(s, s_0; \tau) = \|(1 - P_{[a,b]}(s)) U_\tau(s, s_0) P_{[a,b]}(s_0)\|^2. \tag{2.1}$$

$p(s, s_0; \tau)$ is the upper bound on the probability that the time evolution $U(s, s_0)\psi$ of a system prepared at time s_0 in a state ψ, which lies in the band $[a, b]$ (that is, $P_{[a,b]}(s_0)\psi = \psi$), lies outside the band. The strategy of the proof, which goes back to Kato, is to choose a unitary evolution that preserves the spectral bands and approximates the physical evolution as closely as possible. It is usually called the adiabatic evolution $W_A(s, s_0)$, and is the solution of

$$iW_A'(s,s_0) = (\tau H(s) + iW'(s)W^{-1}(s))W_A(s,s_0), \qquad W_A(s_0,s_0) = 1.$$

Such a choice is natural, but we will not enter into that issue here. W_A follows spectral bands exactly and if it is close to $U_\tau(s, s_0)$, we expect the wave operator $W_A^{-1}(s, s_0)U_\tau(s, s_0)$ to be close to 1. Denote

$$\Omega(s,s_0) = W_A(0,s_0)W_A^{-1}(s,s_0)U_\tau(s,s_0)W_A^{-1}(0,s_0),$$

and rewrite (2.1) as

$$p(s,s_0,\tau) = \|(1 - P_{[a,b]}(0))\,\Omega(s,s_0)\,P_{[a,b]}(0)\|^2.$$

If Ω is close to 1, p has to be close to 0. $\Omega(s, s_0)$ solves

$$\Omega'(s,s_0) = -\exp(i\tau sH)K(s)\exp(-i\tau sH)\Omega(s,s_0), \qquad \Omega(s_0,s_0) = 1, \qquad (2.2)$$

where $K(s) = W^{-1}(s)W'(s)$. The above equation determines the tunnelling probability. One rewrites (2.2) as a Volterra integral equation and writes the perturbative expansion of the solution. $W'(s) \in \mathrm{H}^1(a)$ iff $K(s) \in \mathrm{H}^1(a)$: Interchanging the line of integration in all terms of the perturbative expansion and inserting the projections, we get

$$\lim_{s\to\infty}\lim_{s_0\to-\infty} \|(1 - P_{[a,b]}(0))\Omega(s,s_0)P_{[a,b]}(0)\|^2 \le C_a \exp(-2\tau a g).$$

The Theorem follows.

3. Geometry

The geometric aspects are illustrated in a simple example. Choose a unitary representation of SU(2) on a Hilbert space for which the self-adjoint generators $J = \{J_1, J_2, J_3\}$ are *bounded*. The commutation relations of the generators are, for $a, b \in \mathbf{R}^3$

$$[a\cdot J, b\cdot J] = i(a\times b)\cdot J. \qquad (3.1)$$

In general, the triplet of self-adjoint operators $V = \{V_1, V_2, V_3\}$, which satisfies the commutation relation (3.1) we call a vector operator. We define Hamiltonians as

$$H(s) = H_I + \mathbf{k}(s)\cdot V \qquad (3.2)$$

where $\mathbf{k}(s) = (k_1(s), k_2(s), k_3(s))$ is a path on the unit sphere S^2, and $[a\cdot J, H_I] = 0$ for $a \in \mathbf{R}^3$. For example, any two-level system, after suitable reparametrization of the time variable, is of the above form. Hamiltonians (3.2) are isospectral: If $W(s)$ is the solution of

$$iW'(s) = (\mathbf{k}(s) \times \mathbf{k}'(s))W(s), \qquad W(0) = 1,$$

we have $H(s) = W(s)H(0)W^{-1}(s)$. $W'(s) \in \mathrm{H}^1(a)$ iff $k_i'(s) \in \mathrm{H}^1(a)$ for $1 \leq i \leq 3$: Under such easily verifiable conditions, the result of the Theorem is valid. Furthermore, the quantitity $K(s)$ in the equation (2.2) can be expressed (assuming boundary condition $\mathbf{k}(0) \cdot J = J_3$, $\mathbf{k}'(0) \cdot J = |\mathbf{k}'(0)| J_2$) as

$$K(s) = i|\mathbf{k}'(s)| \exp(-i\gamma(s)J_3)J_1 \exp(i\gamma(s)J_3),$$

where

$$\gamma(s) = \int_0^s \kappa_g(t)\, dl(t).$$

κ_g is the geodesic curvature of the path $\mathbf{k}$ and dl is a Riemannian length element. The result is particularly appealing for the two-level systems, where $H(s) = \mathbf{k}(s) \cdot \sigma$ and $\sigma = (\sigma_1, \sigma_2, \sigma_3)$ are Pauli matrices. In first-order perturbation theory, tunnelling probability is expressed as (taking limits $s \to \infty$, $s_0 \to -\infty$)

$$\hat{p}(\tau) = \left| \frac{1}{2} \int_{-\infty}^{+\infty} |\mathbf{k}'(s)| \exp(-i\gamma(s)) \exp(-2i\tau s)\, ds \right|^2. \tag{3.3}$$

The role of the Paley-Wiener theorem is now transparent. The integral (3.3) can be evaluated by residue calculus: we obtain that the tunnelling probability in the first-order approximation, in addition to the standard Landau-Zener expression, has another term (coming from $\gamma(s)$), independent of τ and of purely geometric origin. According to M. V. Berry[2], who obtained a similar result for the two-level system at about the same time, the geometric term may have experimental significance.

REFERENCES

1. J. Avron, R. Seiler, L. Yaffe: Adiabatic Theorems and Applications to the Quantum Hall Effect, *Commun. Math. Phys.* **110**, 110 (1987).
2. M. V. Berry: Geometric Amplitude Factors in Adiabatic Quantum Transitions, to appear in *Proc. Roy. Soc. Lond.*.
3. V. Jakšić, J. Segert: Exponential Approach to the Adiabatic Limit and the Landau-Zener Formula, submitted to *Commun. Math. Phys.*.
4. L. Landau: Zur Theorie der Energieübertragung bei Stössen, *Phys. Z. Sowjet.* **1**, 88 (1932).
5. C. Zener: Non-Adiabatic Crossing of Energy Levels, *Proc. Roy. Soc. London* A **137**, 696 (1932).

On the Relation between the Geometric Phase and the Schwinger Term

H. Grosse
Institut für Theoretische Physik
Universität Wien

Abstract

Transport of a quantum mechanical system along a closed loop of parameter space may yield a geometric phase. We discuss models for which nonintegrable phase factors are obtained from the adiabatic parallel transport.

After second quantizing such systems, we obtain a Schwinger term which indicates that representations of gauge groups will become, in general, only ray representations. Depending on the type of transformation a subtle relationship between these two obstructions are worked out.

We indicate finally how we may transport density matrices along loops in parameter space.

1 Introduction

Some time ago we were interested in supersymmetric quantum mechanics. During an early visit in Prague I met Jiri Blank and realized his broad interests. We had interesting discussions especially on representations of superalgebras, a subject to which he contributed together with his collaborators. When I met Jiri Blank last year I had no idea about his sickness. I was informed by friends shortly afterwards. I shall remember him very warmly and will dedicate my notes to his memory.

The main purpose of my contribution is to explain two somewhat unrelated notions and point out certain relationships: On the one hand, we shall discuss the occurrence of geometric phases in simple models and, on the other hand, some connections to the appearance of a Schwinger term will be shown.

Geometric phases were first obtained by Fock in the late twenties but run also under the name of Berry phases [1,2] who showed their occurrence for very simple and general settings. We consider a quantum mechanical problem with an Hamiltonian operator depending on various parameters $H(\lambda)$. Time evolution is then given by the time dependent Schrödinger equation $i\dot{\psi}_\lambda = H(\lambda)\psi_\lambda$, since λ may depend on time. We are used to deal usually with wave functions but know that expectation values depend only on the rays. To a vector ψ out of the Hilbert space $\mathcal{H}$ we may assign a ray $\{e^{i\alpha}\psi\}$ which is an element of the projective Hilbert space $\mathcal{P}(\mathcal{H})$ over $\mathcal{H}$. If we let ψ_λ depend on λ we may consider a line bundle with the parameter space as base space and fibres equal to the rays $\{e^{i\alpha}\psi_\lambda\}$. Next we imagine a motion along a closed loop in parameter space. The main question is how to lift the loop to the line bundle. We shall use the adiabatic approximation and obtain the Berry connection for the quantum mechanical problem. Afterwards we shall second quantize such problems.

In order to formulate the parallel transport for the above mentioned line bundle we start with the static problem for $H(\lambda)$. We assume a pure point spectrum and assume that no level crossings occur as we move along the curve C. Let $H(\lambda)|n(\lambda)\rangle = E_n(\lambda)|n(\lambda)\rangle$ be the eigenvalue problem for $H(\lambda)$. The adiabatic time evolution of an initial state $n(\lambda(0))$ is given by

$$|\psi(t)\rangle = \exp[-i\int_0^t d\tau E_n(\tau)]\exp[i\gamma_n(t)]|n(\lambda(t))\rangle. \tag{1.1}$$

We have splitted off a dynamical phase factor. The main question concerns the phase $\gamma_n(t)$. It is simple to derive the time evolution for $\gamma_n(t)$. We may expand the full solution $|\psi\rangle = \sum_n a_n|n\rangle$ and obtain

$$i\dot{a}_m + ia_m\langle m|\dot{m}\rangle + i\sum_{n\neq m} a_n \frac{\langle m|\dot{H}n\rangle}{E_n - E_m} = E_m a_m. \tag{1.2}$$

Within the adiabatic approximation we neglect the excitations to levels $m \neq n$ due to $\dot{H}$ and obtain

$$\dot{\gamma}_n = i\langle n|\frac{d}{dt}n\rangle. \tag{1.3}$$

We define a connection one-form $A_n = i\langle n|dn\rangle$ and may write the phase picked up along the curve C as

$$\gamma_n^C = \int_C A_n = \int_{S,\partial S=C} F_n \tag{1.4}$$

where the curvature two-form $F_n = dA_n$.

A purely geometric anholonomy measures the geometry of the parameter space and results from a nontrivial first Chern class.

A simple example is given by the interaction of a spin with a magnetic field $|\vec{B}| = 1$

$$H\psi = \vec{\sigma}\cdot\vec{B}\psi = \begin{pmatrix} \cos\vartheta & e^{-i\varphi}\sin\vartheta \\ e^{i\varphi}\sin\vartheta & -\cos\vartheta \end{pmatrix}\psi, \psi = \begin{pmatrix} e^{-i\varphi}\cos\vartheta/2 \\ \sin\vartheta/2 \end{pmatrix}. \tag{1.5}$$

ψ is eigenvector to eigenvalue 1 and is ill-defined at the north pole. The parameter space is S^2 and the singular point comes from the Dirac string. The connection form becomes $A = i\langle\psi|d\psi\rangle = \cos^2\frac{\vartheta}{2}d\varphi$ and the curvature $F = dA = -\frac{1}{2}\sin\vartheta d\vartheta\wedge d\varphi$. The phase obtained equals half the area surrounded by $\vec{B}$ while moving along C.

2 Models

Not many models with nontrivial geometric phases are known. We have looked to the "free" Dirac operator defined on the interval [3]. The Hilbert space is $\mathcal{H} = L^2([0,1],dx)\otimes \mathbf{C}^2$. As Hamiltonian we take $H = \alpha\frac{1}{i}\frac{d}{dx}$ where α denotes a σ-matrix. A unitary transformation allows to get rid of α and yields the Dirac equation $h_U\phi = \frac{1}{i}\frac{d}{dx}\phi$. We shall study the most general boundary conditions such that h_U becomes a self-adjoint operator: $\phi(1) = U\phi(0)$. U denotes a two times two unitary matrix mapping one deficiency space onto the other. From a plane wave ansatz for ϕ we obtain as solutions

$$\phi_n^j(x) = e^{i\lambda_n^j x}f_j, \qquad \lambda_n^j = 2\pi n + \mu_j, \qquad Uf_j = e^{i\mu_j}f_j, \qquad n\in\mathbf{N}, \qquad j = 1,2. \tag{2.1}$$

The spectrum of H_U consists of two points within the interval $[0, 2\pi)$ given by μ_j, where $\exp(i\mu_j)$ are the eigenvalues of U, and is periodic with period 2π. We distinguish three possibilities:

a) each spinor component is coupled to itself,

b) spinors are coupled at both ends to each other,

c) both cases mentioned before occur together and give mixed boundary conditions.

Our idea was to make the boundary conditions time-dependent. We will deal only with the adiabatic approximation although we note that the full problem is solvable as well. We write $U = T\,\mathrm{diag}\,(e^{i\mu_1}, e^{i\mu_2})T^\dagger$ where T denotes the $U(2)$ matrix diagonalizing U, and distinguish three different types of motions in our four-dimensional parameter space spanned by the set of $U(2)$ matrices.

1. We vary only $0 < \mu_1, \mu_2 < 2\pi$ on $U(1) \times U(1)$ but do not wind around the circles. It is easy to check that the connection form is independent of n and given by $A_j = -\frac{i}{2}d\mu_j$. A_j is therefore closed and no nontrivial phases occur.

2. We vary only μ_1, μ_2 but wind around one or both $U(1)$ manifolds. A somewhat surprising result is obtained. We indeed made a closed loop in parameter space but all levels moved upwards or downwards depending on the two winding numbers. It does not make sense therefore to compare two different eigenfunctions. We do not end up in the ray from which we started with.

3. We vary two parameters of T and fix μ_1 and μ_2. The manifold is again given by the two-sphere S^2, the line bundle is given by the Hopf fibration of the $SU(2)$ over S^2 with fibre S^1. That the resulting phase equals half the surrounded area comes as no surprise.

Apparent and avoided level crossings can be discussed in this model as well.

We have extended the above considerations to the Schrödinger interval on the line and treated the most general boundary conditions [4]. The deficiency indices turn out to be again (2.2) and all self-adjoint extensions for $H_0 = -d^2/dx^2$ on L^2-functions for which ψ'' is in $\mathcal{H}' = L^2([0,1], dx)$, ψ' is absolutely continuous and $\psi(0) = \psi(1) = 0$ and $\psi'(0) = \psi'(1) = 0$ are given by mixed boundary conditions of the form

$$\begin{pmatrix} \phi(1) \\ \phi'(1) \end{pmatrix} = \begin{pmatrix} a & z \\ \bar{z} & b \end{pmatrix} \begin{pmatrix} \phi(0) \\ \phi'(0) \end{pmatrix}, \qquad z = ce^{i\varphi}, \tag{2.2}$$

where a, b, c denote real parameters and $\varphi \in [0, 2\pi)$. The spectrum is this time determined by solutions of transcendental equations. For $a = b = 0$ a particular simple situation results. The curvature two-form for the parallel transport of the n-th level becomes highly nontrivial

$$d\langle n|dn\rangle = \frac{1 - k\cot k}{k^2}\frac{d\bar{z}\wedge dz}{(1 + z\bar{z})^2} \tag{2.3}$$

where k solves $(1 + z\bar{z})\cos k = z + \bar{z}$. Close to level crossings (2.3) yields the curvature form on S^2.

3 Second Quantization

We start from the one-particle Hilbert space $\mathcal{H}$ and go over for fermions to the Fock space $\mathcal{F} = \bigoplus_{n=0}^{\infty} \mathcal{H}^{(n)}$, where $\mathcal{H}^{(0)} = \mathbf{C}$, $\mathcal{H}^{(1)} = \mathcal{H}$ and $\mathcal{H}^{(n)}$ denotes the n-fold antisymmetric tensor product of $\mathcal{H}$. We intend to study representations of the C^*-algebra $\mathcal{A}$ of operators $a(f)$ and $a^\dagger(g)$ which fulfill the canonical anticommutation relations

$$\{a(f), a^\dagger(g)\} = \langle f|g\rangle, \qquad \{a(f), a(g)\} = 0, \tag{3.1}$$

where $\langle\cdot|\cdot\rangle$ denotes the scalar product in $\mathcal{H}$.

The Fock-Cook representation of $\mathcal{A}$ is obtained by demanding that all $a(f)$ annihilate the vacuum $\Omega = (1,0,0,\dots) \in \mathcal{F}$: $a(f)\Omega = 0$. Bounded operators $A \in \mathcal{B}(\mathcal{H})$ are transported to $\mathcal{F}$ by defining $\bar{\Gamma}(A) = A \otimes A \otimes \dots \otimes A$ on $\mathcal{H}^{(n)}$. For generators of strongly continuous groups $U_t = \exp(tB)$ we define $d\bar{\Gamma}(B) = B \otimes \mathbf{1} \otimes \dots \otimes \mathbf{1} + \dots + \mathbf{1} \otimes \dots \otimes B$ on $\mathcal{H}^{(n)}$ and obtain

$$\bar{\Gamma}(U_t) = e^{td\bar{\Gamma}(B)}, \qquad [d\bar{\Gamma}(A), d\bar{\Gamma}(B)] = d\bar{\Gamma}([A,B]), \qquad [d\bar{\Gamma}(A), a^\dagger(f)] = a^\dagger(Af). \tag{3.2}$$

(3.2) holds if the operators A and B are trace class.

The so obtained representation is of no big interest since the "Dirac sea" is not filled and the Hamiltonian would not be bounded from below.

We are therefore interested in quasifree states over the algebra $\mathcal{A}$, which are obtained from projection operators P_+ onto the positive spectral part of a Dirac operator H:

$$\omega_{P_+}(a(f_1)\dots a(f_N)a^\dagger(g_M)\dots a^\dagger(g_1)) = \delta_{NM} \det\langle f_i|P_+ g_j\rangle. \tag{3.3}$$

With the help of the GNS-construction we obtain a vacuum vector Ω, a Hilbert space $\mathcal{F}$ and a cyclic representation Π_{P_+} of $\mathcal{A}$ for which we define

$$\Pi_{P_+}(a(P_+f)) = B(P_+f), \qquad \Pi_{P_+}(a(CP_-f)) = D^\dagger(P_-f). \tag{3.4}$$

C denotes an antilinear involution like charge conjugation and makes $D^\dagger(P_-f)$ linearly depending on P_-f. According to the splitting of $\mathcal{H} = P_+\mathcal{H} \oplus P_-\mathcal{H}$ into positive and negative frequency modes we obtain a decomposition of the field operator $\Pi_{P_+}(a(f)) \equiv \psi(f)$ like

$$\psi(f) = B(P_+f) + D^\dagger(CP_-f). \tag{3.5}$$

We are interested in comparing two representations connected to two splittings $P_\pm$ and $\tilde{P}_\pm$. The transformation from B and $D^\dagger$ to operators $\tilde{B}$ and $\tilde{D}^\dagger$ is in general a Bogoliubov automorphism of the algebra $\mathcal{A}$. Whether this transformation is implementable in the representation with which we started with, is the main question. The Shale-Stinespring, Berezin, Friedrichs criterion asserts the implementability iff

$$\|P_+\tilde{P}_-\|_{HS} < \infty, \qquad \|P_-\tilde{P}_+\|_{HS} < \infty. \tag{3.6}$$

In order to transport operators from $\mathcal{H}$ to the physical Fock space representations we have to redefine the functors $d\bar{\Gamma}$ and $\bar{\Gamma}$. This is done through normal ordering with respect to the vacuum Ω. Let A be trace class; we define then

$$d\Gamma(A) =: d\tilde{\Gamma}(A) := d\tilde{\Gamma}(A) - \mathrm{Tr}\ P_-AP_-, \qquad \Gamma(e^A) = e^{d\Gamma(A)}. \tag{3.7}$$

We can answer now the question under which conditions a unitary transformation in $\mathcal{H}$ can be implemented in Π_{P_+}: The automorphism $\tau_\alpha a(f) = a(V_\alpha f)$, where V_α is a unitary representation of a group G and $\alpha \in G$, is implementable iff

$$\omega_{P_+} \cong \omega_{V_\alpha P_+ V_\alpha} \Leftrightarrow \exists \Gamma(V_\alpha). \tag{3.8}$$

The action of $\Gamma(V_\alpha)$ is given by

$$\Gamma(V_\alpha)\Pi_{P_+}(a(f))\Gamma^\dagger(V_\alpha) = \Pi_{P_+}(a(V_\alpha f)) \tag{3.9}$$

iff the two representations ω_{P_+} and $\omega_{V_\alpha P_+ V_\alpha}$ are equivalent. Equivalently

$$||P_\pm V_\alpha P_\mp||_{HS} < \infty \Leftrightarrow ||X_\alpha^\pm||_{HS} < \infty \tag{3.10}$$

with $X_\alpha^\pm = V_\alpha P_\pm V_\alpha^+ - P_\pm$. Both operators $X_\alpha^\pm$ actually fulfill a cocycle condition

$$V_\beta X_\alpha^\pm V_\beta^+ - X_{\alpha\beta}^\pm + X_\beta^\pm = 0. \tag{3.11}$$

If we require implementability of a one-parameter group of transformations $V_{t\Lambda} \cong 1 + tj_\Lambda$ with infinitesimal generators Λ we have to impose finiteness of the Hilbert-Schmidt norms

$$||P_+ j_\Lambda P_-||_{HS} < \infty, \qquad ||P_- j_\Lambda P_+||_{HS} < \infty. \tag{3.12}$$

$\Gamma(V_{t\Lambda})$ exists then as a strong one-parameter group of unitary operators and $d\Gamma(j_\Lambda)$ become smeared charges. The algebra of these operators is of particular interest. Let A, B first be trace class. Due to normal ordering the algebra (3.2) is changed to

$$[d\Gamma(A), d\Gamma(B)] = d\Gamma([A,B]) + S(A,B), \qquad S(A,B) = \mathrm{Tr}\,(P_- A P_+ B P_- - P_- B P_+ A P_-) \tag{3.13}$$

where $S(A,B)$ denotes the Schwinger term. We are allowed to extend (3.13) now to operators for which $\mathrm{Tr}\ P_- A P_+ A P_- < \infty$. Clearly, $S(A,B)$ cannot be absorbed anymore: This nontrivial cocycle implies that we are dealing with a ray representation of the group G. It fulfills the Hochschild condition

$$S(AB,C) + S(BC,A) + S(CA,B) = 0. \tag{3.14}$$

Introducing $F = P_+ - P_-$ with $F^2 = 1$ it can be rewritten in the form

$$4S(A,B) = \mathrm{tr}\ F[F,A][F,B] \tag{3.15}$$

which connects to Connes' noncommutative geometry [5]. We note that special cases of $S(A,B)$ give the central extension term of the Virasoro algebra and are the main reason for the boson-fermion correspondence.

4 Models

The above mentioned program can be completely worked out for the model of part two. We compare two representations of the CAR obtained from operators H_{U_1} and H_{U_2} and derive equivalence iff $U_1 = U_2$ [6]. Furthermore, chiral transformations

$$V_{\underline{\Lambda}} = \frac{1+\gamma_5}{2} e^{i\Lambda_+} + \frac{1-\gamma_5}{2} e^{i\Lambda_-} \tag{4.1}$$

turn out to be implementable iff $\Lambda_\pm(1) = \Lambda_\pm(0) + 2\pi n_\pm$ with $n_\pm \in \mathbf{N}$ for case a), $\Lambda_+(1) = \Lambda_-(1) + 2\pi n_+$, $\Lambda_+(0) = \Lambda_-(0) + 2\pi n_-$ for case b), and both conditions a) and b) have to hold for case c). Such models determine obviously certain quantum numbers "by themselves". We note that the difference of the integers $n_+ - n_-$ equals the index of the Bogoliubov transformation and is also equal to the winding number of $\Lambda_\pm$. In addition, chiral transformations may charge the vacuum. The occurrence of such sectors is due to the fact that $\Pi_1(S^1) = \mathbf{Z}$. Note that there is a relationship to the shifts of energy levels, if we wind around $U(1) \times U(1)$ in case 2). In order to have existence of generators of one-parameter groups connected to the unit element we have to have $n_\pm = 0$. The Schwinger term in this case has the well-known δ' singularity:

$$i[Q_{\underline{\Lambda}}, Q_{\underline{\mu}}] = \int_0^1 \frac{dx}{2\pi}(\Lambda'_-(x)\mu_-(x) - \Lambda'_+(x)\mu_+(x)). \tag{4.2}$$

Fourier coefficients of the currents fulfill a Kac-Moody algebra with central extension term.

Next we let everything depend on external parameters λ. If $\{\phi_n(\lambda)\}$ denote the set of eigenfunctions of the Dirac operator $h(\lambda)$ we get the projection operators

$$P_+(\lambda) = \sum_{n>0} |\phi_n(\lambda)\rangle\langle\phi_n(\lambda)|. \tag{4.3}$$

For explicit calculations we choose a section of the Fock bundle over parameter space $\lambda \to \phi_n(\lambda)$ with fixed n and introduce antihermitian operators

$$\Lambda_j(\lambda) = \sum_n |\partial_j\phi_n(\lambda)\rangle\langle\phi_n(\lambda)|, \qquad j = 1, 2, \ldots \tag{4.4}$$

which generate infinitesimal transformations from $\{\phi_n(\lambda)\}$ to $\{\phi_n(\lambda + \varepsilon e_j)\}$, where e_j denotes the j-th unit vector in parameter space. We note that the connection one-form A_n and the connection two-form F_n of equ. (1.4) can be rewritten in terms of $\Lambda_j(\lambda)$ as

$$A_n = i\langle\phi_n|\Lambda_j\phi_n\rangle d\lambda^j, \qquad F_n = -\frac{i}{2}\langle\phi_n|[\Lambda_j, \Lambda_k]\phi_n\rangle d\lambda^j \wedge d\lambda^k. \tag{4.5}$$

Second quantization yields, as described before, irreducible representations $\Pi_{P_+(\lambda)}$. We assume them all to be unitarily equivalent; therefore $\|P_+(\lambda)\Lambda_j(\lambda)P_-(\lambda)\|_{HS} < \infty$ has to be finite and all representations can be realized in one Fock space. We assume that $h(\lambda)$ has pure point spectrum, that no level crossings occur for our loop and no levels cross the energy zero. At level crossings the adiabatic approximation breaks down. If a level goes through zero we have to redefine the vacuum. Due to our assumptions, generators $Q_i(\lambda) = d\Gamma(\Lambda_i(\lambda))$ exist and are of the form

$$Q_j(\lambda) = \sum_{m,n} :\psi^\dagger(\phi_m(\lambda))\langle\phi_m|\Lambda_j\phi_n\rangle\psi(\phi_n(\lambda)):_{\Omega(\lambda)} \tag{4.6}$$

normal ordered with respect to $\Omega(\lambda)$ such that $\langle\Omega(\lambda)|Q_j(\lambda)\Omega(\lambda)\rangle_{\mathcal{F}} = 0$.

We are interested in the geometric phase picked up if $\Omega(\lambda)$ is transported along a closed loop. Consider the line bundle $\lambda \to \{e^{i\alpha}\Omega(\lambda)\}$ over parameter space and the connection form $A(\lambda) = i\langle\Omega|d\Omega\rangle_{\mathcal{F}} = i\langle\Omega|\partial_j\Omega\rangle_{\mathcal{F}} d\lambda^j$ with $\partial_j \equiv \partial/\partial\lambda_j$. Since $(1 + \varepsilon Q_j(\lambda))\Omega(\lambda)$ is an element out of the ray $\Omega(\lambda + \varepsilon e_j)$ we deduce for the parallel transport of the vacuum vector

$$\partial_j\Omega(\lambda) = (Q_j(\lambda) + \langle\Omega|\partial_j\Omega\rangle_{\mathcal{F}}\mathbf{1})\Omega(\lambda). \tag{4.7}$$

The c-number contribution in (4.7) can be fixed by taking the scalar product of both sides with $\Omega(\lambda)$. (4.7) allows to express the curvature form $F(\lambda) = dA(\lambda)$ as an expectation value of charge commutators

$$F(\lambda) = -\frac{i}{2}\langle\Omega|[Q_j, Q_k]\Omega\rangle_{\mathcal{F}} d\lambda^j \wedge d\lambda^k. \tag{4.8}$$

We use next the algebra (3.13) and obtain the expression of $F(\lambda)$ in terms of the Schwinger term [7]

$$F(\lambda) = -\frac{i}{2}S(\Lambda_j, \Lambda_k)d\lambda^j \wedge d\lambda^k \tag{4.9}$$

since the Q's are normal ordered with respect to $\Omega(\lambda)$.

Example: We return to the Dirac operator $h_1 = \frac{1}{i}\frac{d}{dx}$ on the interval and take $U = 1$ and $\Lambda_i(1) = \Lambda_i(0)$ to be two multiplication operators. As our model Hamiltonian we take

$$h(\underline{\lambda}) = e^{\lambda_i\Lambda_i} h_0 e^{-\lambda_k\Lambda_k}, \qquad [\Lambda_i, \Lambda_k] = 0. \tag{4.10}$$

The parameter space is here the non-compact space $\mathbf{R}^2$. Since the multiplication operators commute, $F_n = 0$ for all n and no geometric phases occur in $\mathcal{H}$. According to (4.2) a Schwinger term nevertheless shows up

$$F = \int_0^1 \frac{dx}{4\pi}\Lambda_k(x)\Lambda_\ell'(x) d\lambda^k \wedge d\lambda^\ell. \tag{4.11}$$

As a result we note that a nontrivial phase may be obtained during the transport of $\Omega(\lambda)$ even if no one-particle state requires any phase. We note that everything is well-defined. Due to periodic boundary conditions all representations are equivalent. The Hilbert-Schmidt criterion is also fulfilled, but Λ_i are *not* trace class.

Example: We start again from h_1 and do a unitary transformation on two eigenvectors $\phi_1(\lambda = 0)$ and $\phi_0(\lambda = 0)$:

$$\begin{aligned} V(\vartheta,\varphi)\phi_1(0) &= \cos\frac{\vartheta}{2}\phi_1(0) - e^{i\varphi}\sin\frac{\vartheta}{2}\phi_0(0) \\ V(\vartheta,\varphi)\phi_0(0) &= e^{i\varphi}\sin\frac{\vartheta}{2}\phi_1(0) + \cos\frac{\vartheta}{2}\phi_0(0) \\ V(\vartheta,\varphi)\phi_n(0) &= \phi_n(0) \quad \text{for } n \neq 1, 0 \end{aligned} \tag{4.12}$$

where ϑ, φ parametrizes $S^2 : 0 \leq \vartheta \leq \pi$, $0 \leq \varphi < 2\pi$. The parameter depending Hamiltonian is defined as

$$h(\vartheta,\varphi) = V(\vartheta,\varphi)h_1V^\dagger(\vartheta,\varphi). \tag{4.13}$$

Since we mix only two states the generators $\Lambda_j(\vartheta, \varphi)$ are represented by 2×2 matrices and are therefore trace class. A straightforward calculation yields the well-known curvature form

$$F_1(\vartheta,\varphi) = \frac{1}{2}\sin\vartheta d\vartheta \wedge d\varphi = -F_0(\vartheta,\varphi) \tag{4.14}$$

which is the volume form on S^2. From (4.9) we obtain the curvature form for the Dirac vacuum to be $F(\vartheta,\varphi) = -\frac{1}{2}\sin\vartheta d\vartheta \wedge d\varphi$, which is just the contribution from $\phi_0 \in P_-\mathcal{H}$.

Such a relationship is more general [8]: If Λ_j is a trace class operator we may express $A(\lambda) = i\text{Tr}\,(P_-\Lambda_j P_-)d\lambda^j$ which is the sum over the connection forms for the negative energy states in the Dirac sea. Note that Tr $P_-\Lambda_j P_-$ is the *finite* c-number shift resulting from normal ordering Q_j. From this A we immediately calculate $F = dA$ and obtain (4.9).

We may consider a line bundle for an r-particle and s-antiparticle state of the form

$$v(\lambda) = B^\dagger_{m_1}(\lambda)\ldots B^\dagger_{m_r}(\lambda)D^\dagger_{n_1}(\lambda)\ldots D^\dagger_{n_s}(\lambda)\Omega(\lambda) \tag{4.15}$$

where $B^\dagger_m$ and $D^\dagger_n$ are creation operators for particle and antiparticle modes. $v(\lambda)$ picks up a phase

$$\langle v|dv\rangle = \langle\Omega|d\Omega\rangle + \sum_{\ell=1}^{r}\langle\phi_{m_\ell}|d\phi_{m_\ell}\rangle - \sum_{k=1}^{s}\langle\phi_{n_k}|d\phi_{n_k}\rangle \tag{4.16}$$

which results from adding particle contributions to the vacuum phase but subtracting antiparticle contributions.

We summarize our findings: Transporting the vacuum yields a phase coming from the Schwinger term. The relation to one-particle phases is more subtle and depends on the type of transformation. If Λ_j fulfills only the Hilbert-Schmidt criterion no relationship exists. If Tr $P_-\Lambda_j P_-$ is finite the resulting phase comes from summing up the negative energy state contributions.

5 Parallel Transport of Density Matrices

We may ask how to transport temperature states or density matrices along loops. In part one we mentioned that phases appear when we do the step from projection operators $|\psi\rangle\langle\psi|$ to wave functions $\psi \in \mathcal{H}\setminus\{0\}$. Similarly we start from density operators ρ and take a kind of "square root" $A = \rho^{1/2}U$ with $AA^\dagger = \rho$. According to the polar decomposition A is given up to a unitary operator U. Since Tr $\rho < \infty$, $0 \le \rho \le 1$, A becomes a Hilbert-Schmidt operator and the natural scalar product $(A, B) = \text{Tr}\,(A^\dagger B + B^\dagger A)$ can be used to define a parallel transport. We consider the set $\{\rho^{1/2}U|U \text{ unitary}\}$ as a fibre over ρ. The set of trace class operators replace the parameter space. Changes inside the fibre are given by right-multiplication with unitary operators U. Let H be the antihermitean generator of $U \cong \mathbf{1} + H$. The parallel transport in the case of a pure state is determined by requiring that $\langle\psi|\dot\psi\rangle = 0$ which means that the changes in the bundle $\delta(\psi e^{i\gamma}) \cong i\gamma\psi$ are orthogonal to the changes in the fibre. Similarly we require that $\dot A$ is orthogonal to AH for all H in the sense of the scalar product introduced before, $(\dot A, AH) = \text{Tr}\,(\dot A^\dagger AH - HA^\dagger\dot A) = 0$. We deduce that [9]

$$\dot A^\dagger A = A^\dagger\dot A \qquad \text{and} \qquad A\dot A^\dagger + \dot A A^\dagger = \dot\rho. \tag{5.1}$$

The second relation follows since A is projected down to ρ. We are almost dealing with a principal fibre bundle of Hilbert-Schmidt operators over trace class operators with fibre being the unitaries over $\mathcal{H}$. (5.1) was first obtained by Uhlmann [10]. We gave an alternative derivation [9] and shall solve (5.1) by "fixing" the gauge and introducing an operator T through the equation $\dot A = TA$ with $T = T^\dagger$. From (5.1) we obtain

$$\rho T + T\rho = \dot\rho. \tag{5.2}$$

Given $\rho(t)$ we solve (5.1) for $T(t)$ and obtain from $\dot A = TA$ the time evolution for $A(t)$.

To our opinion it is difficult to imagine a setting where one can measure this anholonomy for density matrices.

Example: We take the most general two-dimensional density matrix

$$\rho(t) = \rho_+ P_+ + \rho_- P_- = \frac{\exp(\beta\vec{\sigma}\vec{B}_t)}{\mathrm{Tr}\ \exp(\beta\vec{\sigma}\vec{B}_t)}, \qquad P_\pm(t) = \frac{1+\vec{\sigma}\cdot\vec{B}_t}{2}, \qquad \rho_\pm = \frac{1\pm\tanh\beta}{2} \tag{5.3}$$

which we decomposed into pure states $P_\pm$ and $|\vec{B}_t| = 1$ is taken for simplicity reasons. Time dependence is given by a rotation around the axis $\vec{n}$

$$\vec{\sigma}\cdot\vec{B}_t = V_t\vec{\sigma}\cdot\vec{B}_0 V_t^\dagger, \qquad V_t = e^{iht}, \qquad h = \frac{\vec{n}\cdot\vec{\sigma}}{2}, \qquad \rho(t) = V_t\rho(0)V_t^\dagger. \tag{5.4}$$

The first step consists in solving (5.2) for T

$$T = (\vec{\sigma}\vec{B}_t)\frac{\tau}{2}, \qquad \tau = \tanh\beta. \tag{5.5}$$

As for the second step we have to solve

$$\dot{A}(t) = V(t)T(0)V^\dagger(t)A(t) \tag{5.6}$$

and obtain

$$A(t) = V(t)e^{t(T(0)-ih)}\rho^{1/2}(0) \tag{5.7}$$

and finally

$$U(t) = e^{iht}\rho^{1/2}(0)e^{t(T(0)-ih)}\rho^{1/2}(0). \tag{5.8}$$

We note that $U(t)$ is given by the product of two unitary groups

$$U(t) = \exp(i\frac{\vec{n}\vec{\sigma}}{2}t)\exp(-\frac{i}{2}t\vec{\sigma}\vec{u}), \qquad \vec{u} = \vec{n}\sqrt{1-\tau^2} + \vec{B}(\vec{n}\cdot\vec{B})(1-\sqrt{1-\tau^2}). \tag{5.9}$$

The analog of the geometric phase is now given by the operator

$$U(2\pi) = e^{i\pi(1-\vec{u}\cdot\vec{\sigma})}. \tag{5.10}$$

We note that the limit for pure states gives

$$\lim_{\tau\to 1} P_+(2\pi)U(2\pi) = P_+(2\pi)\exp(2\pi i\frac{1-\sin\vartheta}{2}) \tag{5.11}$$

as it should be. The angle between $\vec{B}$ and $\vec{n}$ is given by $\pi/2-\vartheta$. There are certain loops which surround no area, nevertheless a nontrivial U matrix shows up [9].

Acknowledgement: We would like to thank the organizers of this nice conference for all their efforts during this exciting time of big changes in their country and for inviting me to Liblice. I thank also Peter Falkensteiner for a very fruitful collaboration.

References

[1] M. Berry, Proc. Roy. Soc. London *A 392* (1984) 45.

[2] M. Berry "Quantum Adiabatic Anholonomy" in Anomalies, Phases, Defects, eds. M. Bregda, G. Marmo and G. Morandi, Naples 1990.

[3] W. Bulla, P. Falkensteiner and H. Grosse, Phys. Lett. *215* (1988) 359.

[4] H. Grosse and W.L. Kennedy, "The Geometric Phase in a Simple Model", to be published in Phys. Lett. A.

[5] H. Araki, in "Quantum Theories and Geometry", Kluwer Academic Publishers 1988, p. 1.

[6] P. Falkensteiner and H. Grosse, Journ. Math. Phys.*28* (1987) 850.

[7] M. Stone and W.E. Goff, Nucl. Phys. *B295* [FS1] (1988) 243.

[8] P. Falkensteiner and H. Grosse, Phys. Lett. *B* to be published.

[9] L. Dabrowski and H. Grosse, Lett. Math. Phys. *19* (1990) 205.

[10] A. Uhlmann, Rep. Math. Phys. *24* (1986) 229.

The relationship between Berry's phase and energylike first integrals of time-dependent Hamiltonians

P G L LEACH*

Department of Computational and Applied Mathematics and Centre for Non-linear Studies, University of the Witwatersrand, Johannesburg, Wits 2050, Republic of South Africa

Abstract

Time-dependent linear canonical transformations are used to derive Berry's phase for non-autonomous Hamiltonian systems which admit an energylike first intergral

1. Introduction

In this paper we derive Berry's phase[1] for one-dimensional time-dependent Hamiltonian systems which possess an energylike first integral. This work was motivated by Morales' treatment[2] of the problem in the case of a one-dimensional time-dependent quadratic Hamiltonian, but the derivation here is based on the earlier work of Leach[3,4], Lewis and Leach[5,6] and Leach et al[7] related to the application of time-dependent linear canonical transformations to time-dependent Hamiltonian systems.

2. Berry's phase for a general quadratic Hamiltonian in one dimension

The general quadratic Hamiltonian can be written as

$$H = \frac{1}{2}\mathbf{z}^T A\mathbf{z} + \mathbf{b}^T\mathbf{z} + c, \tag{1}$$

where $A(t)$ is a $2n \times 2n$ real symmetric matrix, $\mathbf{b}(t)$ a $2n$ vector, $c(t)$ a scalar and $\mathbf{z}^T = (\mathbf{q}^T, \mathbf{p}^T)$. Under the time-dependent linear canonical transformation

$$\bar{\mathbf{z}} = S\mathbf{z} + \mathbf{r} \tag{2}$$

* present address: Department of Mathematics and Applied Mathematics, University of Natal, King George V Avenue, Durban 4001, Republic of South Africa

(1) is transformed to

$$\bar{H} = \frac{1}{2}\bar{\mathbf{z}}^T\bar{A}\bar{\mathbf{z}} + \bar{\mathbf{b}}^T\bar{\mathbf{z}} + \bar{c} \tag{3}$$

and

$$\dot{S} = J\bar{A}S - SJA \tag{4}$$

$$\dot{\mathbf{r}} - 2J\bar{A}\mathbf{r} = J\bar{\mathbf{b}} - SJ\mathbf{b}, \tag{5}$$

where J is the symplectic matrix and the requirement that the transformation be canonical is expressed as

$$SJS^T = J. \tag{6}$$

In general the wave functions of the Schrödinger equations for the Hamiltonians (1) and (3) are related by an integral transform, but this collapses to a geometric transform if the transformation (2) is a point transformation[8]. In practice one wants $\bar{H}$ to be as simple as possible and takes $\bar{A} = \rho^{-2}I$, $\bar{\mathbf{b}} = 0 = c$, where the $\rho^{-2}(t)$ is introduced to compensate for the restriction to a point transformation. When

$$H = \frac{1}{2}\left\{Zp^2 + Y(qp + pq) + Xq^2\right\}, \tag{7}$$

the transformation matrix is

$$S = \begin{bmatrix} \left(\rho Z^{\frac{1}{2}}\right)^{-1} & 0 \\ Z^{-\frac{3}{2}}(\rho YZ - \dot{\rho}Z - \frac{1}{2}\dot{Z}) & \rho Z^{\frac{1}{2}} \end{bmatrix} \tag{8}$$

and $\rho(t)$ is a solution of

$$\ddot{\rho} + \left\{XZ - Y^2 + \frac{\dot{Z}Y - Z\dot{Y}}{Z} + \frac{1}{2}\frac{\ddot{Z}}{Z} - \frac{3}{4}\left(\frac{\dot{Z}}{Z}\right)^2\right\}\rho = \rho^{-3} \tag{9}$$

which has the form of Pinney's equation[9].

The solution of the Schrödinger equation for H is

$$\Phi(q,t) = |\rho Z^{\frac{1}{2}}|^{-\frac{1}{2}}\mu_n\left(\frac{q}{\rho Z^{\frac{1}{2}}}\right)\exp\left\{-i(n+\frac{1}{2})T + \frac{iq^2}{2}\left[\frac{\dot{\rho}}{\rho Z} + \frac{\dot{Z}}{2Z^2} - \frac{Y}{Z}\right]\right\}, \tag{10}$$

where $\mu_n(Q)$ is the usual wave function for the time-independent simple harmonic oscillator and $T = \int q^{-2}dt$. Berry's phase is calculated from the T-independent part of Φ and is

$$\begin{aligned}\gamma_n(c) = & -\frac{1}{2}\langle\mu_n|Q^2|\mu_n\rangle\int_0^{\tau}\left(\rho\ddot{\rho} - \dot{\rho}^2 - \frac{\rho\dot{\rho}\dot{Z}}{Z} + \frac{\rho^2\ddot{Z}}{2Z}\right. \\ & \left. -\frac{\rho^2\dot{Z}^2}{Z^2} - \rho^2\dot{Y} + \rho^2\frac{Y\dot{Z}}{Z}\right)dt,\end{aligned} \tag{11}$$

where the expectation value of Q^2 is just $n + \frac{1}{2}$.

3. Berry's phase for a Hamiltonian with an energylike first integral

Up to a time-dependent translation term $q - \alpha(t)$ the most general Hamiltonian of $T+V$ type which possesses a first integral quadratic in the momentum is

$$H = \frac{1}{2}p^2 - \frac{1}{2}\frac{\ddot{\rho}}{\rho}q^2 + \frac{1}{\rho^2}V\left(\frac{q}{\rho}\right). \tag{12}$$

Under the generalized[10] canonical transformation

$$Q = \frac{q}{\rho} \qquad P = \rho p - \dot{\rho}q \qquad T = \int \rho^{-2}dt, \tag{13}$$

(11) is transformed to

$$\bar{H} = \frac{1}{2}P^2 + V(Q). \tag{14}$$

If the solution of the Schrödinger equation for (13) is $\mu_n(Q)\exp(-i\lambda_n T)$, the wave function for (11) is

$$\begin{aligned} \phi_n(q,t) &= |\rho|^{-\frac{1}{2}}\mu_n\left(\frac{q}{\rho}\right)\exp\left[-i\lambda_n T + \frac{i}{2}\frac{\dot{\rho}}{\rho}q^2\right] \\ &= \xi_n(q,t)\exp[-i\lambda_n T]. \end{aligned} \tag{15}$$

Berry's phase is

$$\gamma_n(c) = -\frac{1}{2}\langle\xi_n|Q^2|\xi_n\rangle\int_0^T(\rho\ddot{\rho} - \rho^2)\,dt.$$

Note the product structure of $\gamma_n(c)$ which is independent of the potential.

Acknowledgements

This work was supported by the Centre National de la Recherche Scientifique (France) and the Foundation for Research Development (RSA).

References

[1] M V Berry, *Proc R Soc* **A392** (1984) 45.

[2] D A Morales *J Phys A : Math Gen* **21** (1988) L 889

[3] P G L Leach *J Math Phys* **18** (1977) 1608

[4] P G L Leach *J Math Phys* **18** (1977) 1902.

[5] H R Lewis and P G L Leach *J Math Phys* **23** (1982) 165

[6] H R Lewis and P G L Leach *J Math Phys* **23** (1982) 2371

[7] P G L Leach, H R Lewis and W Sarlet *J Math Phys* **25** (1984) 486

[8] K B Wolf *Integral Transforms in Science and Engineering* (Plenum, New York, 1979) 381

[9] E Pinney *Proc Amer Math Soc* **1** (1950) 681

[10] A Munier, J R Burgan, M R Feix and E Fijalkow *J Math Phys* **22** (1984) 1219

Quantum chaos in time dependent systems

Petr Šeba+
Institute of Mathematics, Ruhr University Bochum
Bochum, FRG

Abstract
We discuss the properties of quantum systems with Hamiltonian depending periodically on time.

Introduction

It is well known that all stationary one dimensional classical Hamiltonian systems are integrable (the equation of motion is in this case nothing but an ordinary differential equation). The only possibility to construct chaotic stationary systems is therefore to investigate models with two and more degrees of freedom. An one dimensional model display chaotic behavior only if it has time-dependent Hamiltonian. Since we are interested in the *dynamical chaos* originating from the dynamics itself and not say from the complicated structure of the Hamiltonian, we will consider only a very simple time dependence - the Hamiltonians will be time periodic. There is a large amount of examples of this kind (one usually speaks about systems with 1.5 degree of freedom) which have been thoroughly analyzed numerically as well as theoretically.

The question is what is going on when these systems are quantized. The general believe is that the quantum interference effect will stop the development of the chaos - a phenomenon which in known as the *quantum suppression of chaotic diffusion* . This phenomenon has been originally discovered by the numerical simulations on simple time periodic systems [CCHIF],[CFGV],[DGP]. Its heuristic explanation has been given in the work of the Maryland group [FGP],[FPG],[GPF],[PGF]. It has been demonstrated that the suppression of the chaotic diffusion is similar to the Anderson localization. This is, however, only a heuristic argument which does not endure a rigorous mathematical analysis [Bel1],[Bel2]. It is known that for some systems and for some distinguished frequencies of the external time-periodic force the quantum effects do not suppress the classical chaos [IS],[CaGu]. On the other hand one has been able to observe the quantum suppression effects experimentally by measuring the ionization rate of highly excited hydrogen atom in an external microwave field [BaSo],[BSPB],[BlSm2],[BlSm3],[CCHS].

The situation is therefore unclear. The numerical results, however illustrative, cannot decide whether the quantum system is unstable. The point is that one is able (even on the supercomputers) follow the time evolution only over a limited time interval. What is needed is a mathematical analysis which would be able to prove or disprove the existence of the quantum instability. The recent stay of arts in this field is nicely illustrated by the following quotation from the work of Casati and Molinari: " Mathematical rigor though desirable is a rare occurrence: only few are the landmarks in this terra incognita" [CaMo].

1. Quantum mechanics with time periodic Hamiltonians

Let us suppose that we have a quantum system which is described by a time dependent Hamiltonian H(t). The operator family H(t) is assumed to be periodic and reasonably defined, i.e.,

$$H(t) : \mathcal{H} \mapsto \mathcal{H} \tag{1.1}$$

is a self-adjoint operator defined on a separable Hilbert space $\mathcal{H}$;

$$D(H(t)) = D \tag{1.2}$$

its domain does not depend on the parameter t, and

$$H(t) = H(t+T) \tag{1.3}$$

for some $T > 0$ and all $t \in \mathbb{R}$. We will moreover assume that the family H(t) is strongly continuous (later on this assumption will be weakened). The dynamics of the corresponding quantum system is governed by the Schroedinger equation

$$i\frac{\partial|\psi(t)>}{\partial t} = H(t)|\psi(t)> \tag{1.4}$$

(we choose units with $\hbar = 1$). For our purposes it will be convenient to use the unitary evolution operator U(s,t) fulfilling

$$U(s,t) = U(s,u)U(u,t) \tag{1.5a}$$

$$U(s,t)^* = U(t,s) = U(s,t)^{-1} \tag{1.5b}$$

$$U(s+T,t+T) = U(s,t) \tag{1.5c}$$

(the relation 1.5c is a consequence of the time periodicity of the Hamiltonian) Using U(s,t) we can rewrite the solution of (1.4) as

$$|\psi(t)>= U(t,0)|\psi(0)> \tag{1.6}$$

where $|\psi(0)>$ is the initial conditions for $|\psi(t)>$.

Our aim is to describe the long-time behavior of this solutions. In particular we would like to investigate the stability of such time dependent system, i.e., we would like to know more about behaviour of the mean energy E(t)

$$E(t) =< \psi(t)|H(t)|\psi(t)> . \tag{1.7}$$

The quantum system will be unstable if

$$\lim_{t\to\infty} E(t) = \infty \tag{1.8}$$

for almost all initial conditions $|\psi(0)>$. The limit (1.8) can be, however, hardly investigated. The reason is that due to the time dependence of the Hamiltonian we have not a good expression for the evolution operator.

Let us assume that the Hamiltonian H(t) has the form

$$H(t) = H_0 + V(t) \tag{1.9}$$

where H_0 is the kinetic energy operator and V(t) is some time dependent perturbation. We are interested primarily in the time behavior of spatially bounded systems. We will therefore assume that H_0 is a self adjoint bellow bounded operator with *compact resolvent*. We will also assume that the perturbation V(t) is bounded, i.e., that

$$\sup_{0<t<T} \|V(t)\| \leq \infty \tag{1.10}$$

In this case one can build the evolution operator U(t) using the Dyson expansion [ReSi1]

$$U(s,t) = \sum_{k=0}^{\infty}(-i)^k \int_{s<s_1...\leq s_k\leq t} e^{-iH_0(t-s_1)}V(s_1)e^{-iH_0(s_1-s_2)}V(s_2).. \\ ...e^{-iH_0(s_{k-1}-s_k)}V(s_k)e^{-iH_0 s_k}ds_1 ds_2...ds_k \tag{1.11}$$

which is uniformly norm convergent.

The relations (1.5) imply

$$|\psi(nT)>= U^n|\psi(0)> \tag{1.12}$$

for all $n = 1, 2, ...$ with U being the evolution operator over one period

$$U = U(T, 0) \ . \tag{1.13}$$

In such a way the "stroboscopic picture" of the time evolution is described by a *single unitary operator* U, the so called Floquet operator. (The eigenvalues of U are called the quasi-energy levels)

For the evolution of the system the spectral nature of U is of importance. In order to illustrate this let us compute the probability $p_{n,m}(\nu)$ to excite the n-th state of H_0 after ν cycles to the state m

$$p_{n,m}(\nu) = |P_{n,m}(\nu)|^2 \tag{1.14}$$

with $P_{n,m}$ being the probability amplitude

$$P_{n,m}(\nu) = \langle n|U^\nu|m\rangle \tag{1.15}$$

and $|n>, |m>$ the corresponding eigenvectors of H_0. Decomposing the Floquet operator U formally into its eigenvectors

$$U|\omega>= e^{i\omega}|\omega> \tag{1.16}$$

we get

$$P_{n,m} = \sum_{\omega} e^{i\nu\omega} < n|\omega >< \omega|m > \tag{1.17}$$

Hence the transition from n to m is possible only if at least one quasi-energy state $|\omega>$ connects these two states, i.e., if $< n|\omega >< \omega|m > \neq 0$ for at least one ω.

Now the importance of the spectral nature of U becomes apparent: If the spectrum of U is pure point all the quasi-energy states are localized (quadratic integrable). Consequently they can connect only few states of the original Hamiltonian H_0 and a strong recurrence is expected. (this is the essence of the well known theorem of Hogg and Huberman [HoHu1],[HoHu2]). If U has on the other hand a continuous spectrum the quasi-energy states $|\omega>$ are not normalizable. They must therefore connect an infinite number of states $|n>$ leading in such a way to nonrecurrent (unstable) evolution. Summarizing this arguments we get

Theorem 1 (cf.[Bell])

Let H_0 be a positive self adjoint operator with compact resolvent. Let further $E_0(t)$ denotes the mean kinetic energy

$$E_0(t) =< \psi(t)|H_0|\psi(t) > .$$

Then

(i) if $|\psi(0)>$ belongs to the continuous subspace of U one has

$$\limsup_{t\to\infty} E_0(t) = \infty$$

(ii) if $|\psi(0)>$ belongs to the pure point subspace of U and the coordinates of $|\psi(0)>$ in the basis of the eigenvectors of U are absolutely summable the mean kinetic energy $E_0(t)$ is an almost periodic function of time.

It is really hard to say something about the spectrum of U directly. The point is that we usually do not know how this operator looks like. The only thing known is that it exist and that it can be expressed by the Dyson expansion which, however, is not very useful if dealing with the spectral properties. Therefore in order to study the spectrum of U we will use a linearization trick proposed by Howland which turns the time-dependent problem into time-independent one.

The classical equations of motion are

$$\frac{dq_i}{dt} = \frac{\partial H(t)}{\partial p_i}$$
$$\frac{dp_i}{dt} = -\frac{\partial H(t)}{\partial p_i}, \qquad i = 1, ...n \tag{1.18}$$

The energy of the system is not conserved . One can, however, turn this system formally into a time independent one introducing the time t as a new coordinate and the energy E as its conjugate momentum. The new Hamiltonian can be written as

$$K(\vec{p},\vec{q},t,E) = E + H(\vec{p},\vec{q},t) \tag{1.19}$$

Denoting a new time by σ we get equations of motion

$$\begin{aligned} \frac{dq_i}{d\sigma} &= \frac{\partial K}{\partial p_i}, & \frac{dp_i}{d\sigma} &= -\frac{\partial K}{\partial q_i} \\ \frac{dt}{d\sigma} &= \frac{\partial K}{\partial E} = 1, & \frac{dE}{d\sigma} &= \frac{\partial K}{\partial t} \end{aligned} \tag{1.20}$$

which are, of course, equivalent to (1.18).

The same strategy can be used also in the quantum case. The standard quantization leads in this case to the operator [Hol]

$$K = -i\frac{\partial}{\partial t} + H(t) \tag{1.21}$$

We are interested in time-periodic Hamiltonians. It is therefore reasonable to define the operator K on a Hilbert space

$$\mathcal{K} = L^2(0,T) \otimes \mathcal{H} \tag{1.22}$$

with periodic boundary conditions with respect to the variable t, i.e., $|f(t)> \in \mathcal{K}$ belongs to D(K) if $|f(0)>=|f(T)>$, where T is the period of the Hamiltonian , see (1.3).

The relation between the Floquet operator U and the operator K is described by the following Theorem

Theorem 1 [Ya]

There is an unitary operator J acting on the Hilbert space $\mathcal{K}$ such that

$$e^{-iKT} = J(1 \otimes U)J^* \tag{1.23}$$

The operators U and $exp(-iKT)$ are therefore isospectral and one can investigate the spectrum of K instead of U. The advantage of this approach is that the operator

K is a standard differential operator and its spectral properties can be studied more easily those of U.

At a first glance it seems that one can investigate the spectrum of K using perturbative arguments. Actually, introducing a coupling λ of the time-dependent part of the Hamiltonian defining

$$K_\lambda = K_0 + \lambda V(t) \tag{1.24a}$$

with

$$K_0 = -i\frac{\partial}{\partial t} + H_0 \tag{1.24b}$$

and having in mind that the spectrum of K_0 is pure point one can try to prove that K_λ is pure point for small λ and for smooth V(t). This will mean that the quantum system is stable for small λ. Though reasonable this statement is not valid in the general situation. The technical reason for it is that the point spectrum of the operator K_0 is *dense* and the standard perturbation theory is not applicable because of small divisors [Ka],[Bel3]. There are, however, also deep physical reasons for the fail of the perturbation theory which have its roots in the classical mechanics.

Let us assume that the classical system described by the Hamiltonian H_0 is integrable, i.e., that it posses (together with the Hamiltonian) N integrals of motion, where N is the number of its degrees of freedom. It is well known that the phase space trajectories of the system is in this case confined on N-dimensional regions of the 2N dimensional phase space. This regions are N-dimensional tori and can be parametrized by a set of N angles. If one perturb such a system with a sufficiently small and smooth perturbation the integrals of motion are lost. Nevertheless most of the invariant tori will exist also after the perturbation (they are usually called the KAM surfaces). Only the most unstable tori will be destroyed and will be replaced by thin channels in the phase-space within which the phase-space trajectory is chaotic. The "thickness" of the chaotic layer tends, however, to zero if the perturbation vanish.

The phase space of one-dimensional systems is divided by the KAM surfaces into various peaces and the stochastic channels between them remain disconnected. This means that the phase-space trajectory remains confined in one particular layer and cannot wander to infinity (Note that the trajectory cannot cross the KAM surface). This, however, implies the stability of the system since if the trajectory exposes only a finite region of the phase space its energy is bounded. The situation changes for $N \geq 2$. In this case the N-dimensional KAM surface does not divide the 2N dimensional phase space and the stochastic channels can intersect each other generating a stochastic web. The existence of the stochastic web leads to essential physics consequences. Though the thickness of the web channels may be very small the diffusion of the trajectory via this channels in unlimited (Arnold diffusion [Ar]). In such a way the Arnold diffusion imply that the system can be unstable even for infinitely small perturbations. (See [CSZ1],[CSZ2] for the stochastic webs in concrete systems)

After this discussion it become clear that to prove the stability of the quantum system in the general situation a hopeless undertaking. We will show after a while that the mathematical proof of the quantum stability brakes down exactly for those systems the classical part of which displays the Arnold diffusion.

The spectral analysis of the operator K_λ is extremely complicated. From the mathematical point of view is the basic difficulty connected with the fact that the unperturbed operator K_0 has dense point spectrum. The perturbation theory for operators with dense spectra became available only recently in connection with the study of the Anderson localization [SiWo],[Ho2]. The break through in the quantum stability of the time periodic systems has been achieved very recently by Howland [Ho3],[Ho4]. His results can be formulated as follows:

Theorem 1.2 [Ho4]

Let H_0 be a positive discrete operator with eigenvalues E_n of simple multiplicity and satisfying

$$\Delta E_n = E_n - E_{n-1} \geq cn^\alpha \tag{1.25}$$

for some $c > 0$ and $\alpha > 0$. Let further V(t) be a periodic strongly N-times differentiable family of bounded operators with period T and

$$\int_0^T V(t)dt = 0. \tag{1.26}$$

Then the spectrum of the Floquet operator K_λ

$$K_\lambda = -i\frac{\partial}{\partial t} \otimes I + I \otimes H_0 + \lambda V(t) \tag{1.27}$$

is pure point for almost every λ provided

$$N \geq [2\alpha^{-1}] + 1. \tag{1.28}$$

([z] denotes the integer part of z)

Howland has proved the theorem using adiabatic analysis and the operator version of the Kotani trick. We will not describe the details here and refer to the original papers [Ho3],[Ho4]. The Theorem 2 represents in fact the quantum version of the Kolmogorov-Arnold-Moser theorem from the classical mechanics [LL].

Let us now discuss it physical consequences. First of all we would like to mention that the fact that the Theorem 2 works only for almost every λ is of technical origin. It reflect the mathematical method used in the prove and has not a physical background. It is therefore reasonable to expect that the spectrum of the Floquet operator is pure point for all λ provided that all the assumptions of the Theorem are satisfied. The first interesting feature of the Howland theorem is that once (1.25), (1.26) and (1.28) are satisfied the spectrum of the Floquet operator is pure point for all λ. *This implies that the quantum system is stable regardless on the strength of the perturbation.* This is in a sharp contrast with the classical systems which are stable only for sufficiently weak perturbations. As the strength of the perturbation increases the KAM surfaces are destroyed and the system undergoes finally a transition to global stochasticity and becomes unstable. The Theorem 2.2 shows that there is not such a transition in the quantum case. The fact that the quantum systems do not display a transition to unstable regime is usually denoted as *quantum suppression of the classical chaotic diffusion* or shortly *quantum suppression of chaos* . There is an impressive amount of numerical results which shows that the time periodic quantum systems are stable also for perturbations which are more singular that those considered above. (The phenomenon of the quantum suppression of the classical chaotic diffusion has been discovered in the numerical experiments performed by G.Casati and collaborators in the beginning of eighties see for instance [BCPG], [CCGS], [CCS], [Sh2] and references given therein.) We will discuss this question later in the connection with quantum maps.

Let us now shortly discuss the conditions (1.25) and (1.28). The last of them concerns the smoothness of the perturbation V(t). As already mentioned the numerical results show that the system remains stable even if this condition is violated. From the mathematical point of view the smoothness is required to ensure that the higher Fourier components of the perturbation are kept under control. We will show later that in the extremely singular case of the kicked systems (the time perturbation is a sequence of delta functions) the higher Fourier components of V(t) may lead to quantum resonance effects and destroy the stability.

The condition (1.25) seems to be more substantial. The numerical results show that the quantum system becomes unstable as soon as this condition is violated [ATI]. Let us therefore briefly discuss the physics behind this phenomenon. In order to get some insight into what is going on when the condition (1.25) is not satisfied we will consider a system with the operator H_0 given by a Hamiltonian of a free particle inside n-dimensional rectangular box:

$$\mathcal{H} = L^2(\mathcal{B}_n) \tag{1.29}$$

$$H_0 = -\Delta \tag{1.30a}$$

$$D(H_0) = \{f \in \mathcal{H}^1(\mathcal{B}_n) \; ; \quad f = 0 \; on \; \delta\mathcal{B}_n\} \tag{1.30b}$$

where $\mathcal{B}_n$ denotes an n-dimensional rectangular box with incommensurate sites $l_1, l_2, ...l_n$

$$\mathcal{B}_n = [0, l_1]x[0, l_2]x...x[0, l_n]. \tag{1.30c}$$

and $\delta\mathcal{B}_n$ denotes its boundary. Using the classical result of Hermann Weyl [We1],[We2] we find for the asymptotic behavior of the eigenvalues of H_0 (the eigenvalues are supposed to be ordered with respect to the magnitude)

$$E_k = (2\pi)^2(\mathcal{S}_n)^{-2/n}\left(\frac{k}{|\mathcal{B}_n|}\right)^{(2/n)}, \quad as \;\; k \to \infty \tag{1.31}$$

where $|\mathcal{B}_n|$ denotes the volume of the box $\mathcal{B}_n$ and $\mathcal{S}_n$ is the volume of the n-dimensional sphere. The asymptotic behavior of the eigenvalues E_n can be also deduced from that of the number N(E) of the positive eigenvalues of H_0

$$N(E) = \#\{i \; ; \;\; E_i < E\} \tag{1.32}$$

for which holds [ReSi4],[BaBl]

$$N(E) \sim (2\pi)^{-n}\mathcal{S}_n|\mathcal{B}_n|E^{n/2}, \quad as \;\; E \to \infty \tag{1.33}$$

The Weyl's results follows from (1.33) taking into account that the eigenvalues E_n are not degenerate. Using (1.31) we get for ΔE_n

$$\Delta E_k := E_k - E_{k-1} \sim k^{-1+2/n} \tag{1.34}$$

This means that the condition (1.25) is fulfilled only for dimension equal to one. In this case the Theorem 2 implies the stability of the system under the time-periodic perturbation. (The stability has been tested numerically in [CoFi].) For all higher dimensions is the condition (1.25) violated and the systems are probably unstable. (The instability of the two-dimensional rotator has been numerically demonstrated in [ATI]). The physical reason becomes clear if we investigate the corresponding classical systems which display instability due to the Arnold diffusion.

2 The Fermi accelerator

One of the standard classical time periodic models exhibiting chaotic behavior is the Fermi accelerator. It has been introduced by E.Fermi in order to explain the origin of the high-energy cosmic radiation [Fe]. Fermi tried to investigate an analogue to a possible cosmic ray acceleration mechanism in which the charger particles accelerate by collisions with moving magnetic field structures. The proposed model consists of a ball bouncing between fixed and oscillating wall. The collisions between the particle and the walls are assumed to be perfectly elastic.

Introducing appropriate phase space coordinates one can rewrite the corresponding classical equations of motion in a form of a highly nonlinear mapping which connect the phase-space position of the system after the n-th and the (n+1)-th collision with the oscillating wall. (Its simplified version is the well known Ulam map.) Investigating this mapping it has been shown [ZaCh] that the orbits the particle runs in the phase space are stochastic. The energy the particle can gain from the walls remains, however, bounded [ZaCh], [Pu1], [Pu2], i.e., the unlimited particle acceleration originally proposed by Fermi does not take place. (Although the amount of the gained energy can be very high [ZaCh])

Our aim is to discuss the quantum analogy of this model. First some general remarks: The quantum particle (due to its non-locality) is continuously influenced by the oscillating wall while its classical counterpart feels the wall only during the collisions, (in discrete times). This fact makes the system particularly promising since its inherent "quantum" properties are expected to manifest themselves in a more clear way than, say, in the kicked quantum rotator.

From the mathematical point of view the wall oscillations represent a very strong perturbation of the system. The point is that the domain between the walls which is accessible to the particle is time dependent. This in turn means that the quantum system has *a time dependent Hilbert space* and cannot be investigated by the Theorem 1.2. This fact has started some discussion in the literature which has been connected with the speculations that the Fermi accelerator may have continuous quasi-energy spectrum [Kar1], [Kar2], [Kar3]. The numerical investigations showed, however, that its quasi-energy spectrum is very probably pure point. [Jo1], [JoCo], [Jo2].

We will prove that the time evolution of the quantum accelerator is recurrent for most wall oscillations (the system has pure point quasi-energy spectrum). The unbounded acceleration is possible only for some special oscillations of the wall for which the quasi-energy spectrum becomes absolutely continuous.

The model is described by the Schroedinger equation (1.4) with Hamiltonian H(t) defined on the Hilbert space $\mathcal{H}(t) = L^2(0, \ell(t))$ as

$$H(t) = -\frac{d^2}{dx^2} \tag{2.1a}$$

with Dirichlet boundary conditions describing the elastic collisions with the walls

$$\psi(0,t) = \psi(\ell(t),t) = 0. \tag{2.1b}$$

The position of the oscillating wall is given by the function $\ell(t)$ which is assumed to be smooth, positive, periodic with period T

$$\ell(t) = \ell(t+T) \tag{2.2}$$

and to fulfill

$$\begin{aligned} \dot{\ell}(0) &= \dot{\ell}(T) = 0 \\ \ell(0) &= \ell(T) = 1 \end{aligned} \tag{2.3}$$

($\dot{\ell}$ denotes the derivative with respect to time.) It is very difficult to investigate the spectrum of the Floquet operator U using directly the Hamiltonian (2.1). The difficulty comes from the fact the Hilbert space $\mathcal{H}(t)$ as well as the domain of H(t) are time dependent. To face this problem we will use a method developed in the heat equation theory for the moving and free boundary problems [Ta], [Mu], [Ra]: Starting with the equation for the evolution operator U(t)

$$i\frac{\partial}{\partial t}U(t) = H(t)U(t) \tag{2.4}$$

we fix the box introducing a unitary mapping

$$W(t): \quad L^2(0,\ell(t)) \to L^2(0,1) \tag{2.5}$$

$$(W(t)f)(x) = \ell(t)^{1/2} e^{-\frac{i}{4}\dot{\ell}(t)x^2} f(\ell(t)x) \tag{2.6}$$

and write U(t) as

$$U(t) = W(t)\tilde{U}(t) \tag{2.7}$$

where $\tilde{U}(t)$ is some unitary operator defined on the *fixed Hilbert space* $L^2(0,1)$. Inserting (2.7) into (2.4) we get

$$i\frac{\partial}{\partial t}\tilde{U}(t) = \left[W(t)^{-1}H(t)W(t) - iW(t)^{-1}\frac{\partial}{\partial t}W(t)\right]\tilde{U}(t) \tag{2.8}$$

and after a direct computation

$$i\frac{\partial}{\partial t}\tilde{U}(t) = \frac{1}{\ell(t)^2}\left[H_0 + \frac{1}{4}\ell(t)^3\ddot{\ell}(t)x^2\right]\tilde{U}(t) \tag{2.9}$$

with H_0 being the standard particle–in–a–box Hamiltonian:

$$H_0 = -\frac{\partial^2}{\partial x^2} \tag{2.10}$$

with Dirichlet boundary conditions

$$\psi(0) = \psi(1) = 0. \tag{2.11}$$

The last step consists in the appropriate time rescaling. We define

$$\tau = f(t) = \int_0^t \frac{ds}{\ell(s)^2} \tag{2.12}$$

and

$$V(\tau) = \tilde{U}(f^{-1}(\tau)) \tag{2.13}$$

Inserting $V(\tau)$ into (2.9) we get finally

$$i\frac{\partial}{\partial \tau}V(\tau) = \tilde{H}(\tau)V(\tau) \tag{2.14}$$

where

$$\tilde{H}(\tau) = H_0 + g(\tau)x^2 \tag{2.15}$$

and $g(\tau)$ is given by

$$g(\tau) = \frac{1}{4}\ell^3\left(f^{-1}(\tau)\right)\ddot{\ell}\left(f^{-1}(\tau)\right). \tag{2.16}$$

(2.14) is a Schroedinger equation with a simple time dependent potential. The coupling constant $g(\tau)$ is a periodic function of time with a rescaled period $\tilde{T}$

$$\begin{aligned} \tilde{T} &= f(T) \\ g(\tau + \tilde{T}) &= g(\tau). \end{aligned} \tag{2.17}$$

The Floquet analysis is now possible without any troubles. Using (2.14) we introduce the one cycle propagator V

$$V = V(\tilde{T}) \tag{2.18}$$

and relate it to the original Floquet operator U obtaining

$$U = V \tag{2.19}$$

(Here we use the fact that W(T) = 1 which follows from (2.3).) *The spectrum of U can be therefore obtained investigating the spectrum of V.*

To study the spectrum of V we use the Howlands method and replace V by the self-adjoint operator K

$$K = -i\frac{\partial}{\partial\tau} \otimes I + \tilde{H}(\tau) \tag{2.20}$$

acting on the Hilbert space $\mathcal{K}$

$$\mathcal{K} = L^2(0,\tilde{T}) \otimes L^2(0,1) \tag{2.21}$$

The spectrum of K can be investigated applying the Theorem 1.2. This theorem applies, however, only to perturbations which have zero mean. We define therefore

$$a = \int_0^{\tilde{T}} g(\tau)d\tau \tag{2.22}$$

and change the unperturbed operator to

$$H = H_0 + ax^2 \tag{2.23}$$

The time-periodic perturbation reads now

$$V(t) = (g(\tau) - a)x^2 \tag{2.24}$$

and satisfies the condition 1.26. The eigenvalues E_n of H fulfill

$$E_n = \pi^2 n^2 + \frac{1}{3}a + \frac{a}{4n^2} + \mathcal{O}(\frac{1}{n^3}) \tag{2.25}$$

(this formula follows from the theory of Sturm-Liouville operators, see for instance [Bo]). The condition (1.25) is therefore satisfied with $\alpha = 1$ and the Theorem 2.1.2

implies that the spectrum of K is pure point for almost every ℓ providing $g \in C^3$. Recalling the definition of $g(\tau)$ we get the sought result:

The spectrum of the Floquet operator U is pure point for almost every wall oscillation $\ell(t)$ provided the functions $\ell(t)$ are five times differentiable. Under this conditions is the quantized Fermi accelerator stable and its behaviour is recurrent.

The condition $\ell(t) \in C^5$ which is needed for the Howland's analysis is of technical origin and can be probably weakened. On the other hand to ensure the pure pointness of U some differentiability of $\ell(t)$ is necessary . The physical intuition suggests that the adiabatic regime inside the Fermi accelerator (which is responsible for the upper bound in the gained energy) breaks down if the acceleration of the wall (i.e. the second derivative of $\ell(t)$) becomes infinite at some points. Mathematically one expects that the continuous spectrum of U will be not empty in this case implying an unlimited particle acceleration. In order to illustrate the existence of the nonadiabatic resonance regime we give an explicitly solvable example of $\ell(t)$ for which the spectrum of U is purely absolutely continuous.

To construct such wall oscillations we use the fact that the Schroedinger equation for the particle inside a linearly expanding wall is directly solvable [DoRi] and choose

$$\ell(t) = \begin{cases} B + \frac{A-B}{T}(1 - 2nT) & ; \quad t \in [2nT, (2n+1)T], \; n \in \mathbb{Z} \\ B - \frac{A-B}{T}(1 - (2n+1)T) & ; \quad t \in [(2n+1)T, (2n+2)T], \; n \in \mathbb{Z} \end{cases} \tag{2.26}$$

where A,B are some positive constants. Applying the transformations (2.7) and (2.13) we get for $V(\tau)$

$$i\frac{\partial}{\partial\tau}V(\tau) = [H_0 + \Delta(\tau)x^2]V(\tau) \tag{2.27}$$

with $\Delta(\tau)$ given by a sequence of delta kicks:

$$\Delta(\tau) = f_1 \sum_{n\in\mathbb{Z}} \delta\left(\tau - \frac{2nT}{AB}\right) - f_2 \sum_{n\in\mathbb{Z}} \delta\left(\tau - \frac{(2n+1)T}{AB}\right) \tag{2.28}$$

and

$$\begin{aligned} f_1 &= \frac{B(A-B)}{2T} \\ f_2 &= \frac{A(A-B)}{2T} \end{aligned} \tag{2.29}$$

The Floquet operator V can be therefore written as a two step quantum map

$$V = exp\left(i\frac{Tf_2}{AB}x^2\right) exp\left(-i\frac{T}{AB}H_0\right) exp\left(-i\frac{Tf_1}{AB}x^2\right) exp\left(-i\frac{T}{AB}H_0\right) \quad (2.30)$$

Choose now T equal to the basic resonance, i.e.,

$$T = T_0 = \frac{2AB}{\pi} \quad (2.31)$$

which implies

$$exp\left(-i\frac{T}{AB}H_0\right) = \sum exp\left(-i\frac{n^2\pi^2 T_0}{AB}\right) |n><n| = 1 \quad (2.32)$$

($|n>$ are the eigenfunctions of H_0) and therefore

$$V = exp\left(i\frac{T(f_1 - f_2)x^2}{AB}\right) = exp\left(i\frac{(A-B)^2}{2AB}x^2\right) \quad (2.33)$$

Hence V is a simple multiplication by a continuous function and its spectrum is absolutely continuous:

$$\sigma(V) = \sigma_{a.c.}(V) = \left\{exp\left(i\frac{(A-B)^2}{2AB}\varphi^2\right); \quad \varphi \in [0,1]\right\} \quad (2.34)$$

The system is now unstable. Calculating its mean energy after N wall oscillations we get

$$\begin{aligned} E(N) &=< \psi(2NT_+), H(2NT_+)\psi(2NT_+) >=< U(2NT_+)\psi(0), H(0)U(2NT_+)\psi(0) >= \\ &=< W(2NT_+)V\left(\frac{2NT_+}{AB}\right)\psi(0), H(0)W(2NT_+)V\left(\frac{2NT_+}{AB}\right)\psi(0) >= \\ &=< W(2NT_+)V^N\psi(0), H(0)W(2NT_+)V^N\psi(0) >= \\ &= \frac{(A-B)^4}{A^2B^2}N^2\int_0^B x^2|\psi(x)|^2 dx + \mathcal{O}(N) \end{aligned} \quad (2.35)$$

The energy depends quadratically on the number of oscillations . A similar instability appears also for T equal to a rational multiple of the basic resonance period T_0. The same behaviour has been found for the kicked rotator under the quantum resonance [IS].

The concrete functional dependence of $\ell(t)$ is not substantial. The only essential point is the absence of adiabaticity in some suitably chosen periodic sequence of points which leads to the resonance excitation. Let us therefore discuss a general case assuming that the function $\ell(t)$ is periodic and continuous with a first derivative which is discontinuous at points Tn, $n \in \mathbb{Z}$, i.e.

$$\ell(t) \in C(\mathbb{R}) \cap C^1(\mathbb{R} \setminus \{T\mathbb{Z}\}) \tag{2.36a}$$

$$\dot{\ell}(T_+) \neq \dot{\ell}(T_-) \tag{2.36b}$$

After the transformations (2.7) and (2.13) we get for $V(\tau)$

$$i\frac{\partial}{\partial\tau}V(\tau) = \left[H_0 + g(\tau)x^2 + \Delta(\tau)x^2\right]V(\tau) \tag{2.37}$$

where $g(\tau)$ is a smooth periodic function with period $\tilde{T} = f(T)$ and $\Delta(\tau)$ represents a sequence of kicks

$$\Delta(\tau) = f\sum \delta(\tau - n\tilde{T}) \tag{2.38}$$

with the coupling constant f depending on the detailed shape of $\ell(t)$. The Floquet operator V is now given by

$$V = V_0 e^{-i\tilde{T}fx^2} \tag{2.39}$$

where V_0 is the Floquet operator of the regular part of (2.37):

$$V_0 = V_0(\tilde{T}) \tag{2.40}$$

with $V_0(\tau)$ solving

$$i\frac{\partial}{\partial\tau}V_0(\tau) = \left[H_0 + g(\tau)x^2\right]V(\tau) \tag{2.41}$$

To investigate the spectrum of V_0 we rewrite it as

$$V_0 = e^{-iM\tilde{T}} \tag{2.42}$$

where M is the corresponding quasi-energy operator. Applying the adiabatic hypothesis we find that the spectrum of M is for high enough states given (up to some corrections decreasing rapidly with n) by the spectrum of the steady Hamiltonian $H_0 + g(\tilde{T})x^2$. The point is that for high enough states the particle oscillates very quickly inside the box and the smooth wall oscillations lead only to adiabatic changes.

Using this heuristic argument we find for the eigenvalues μ_n of M

$$\mu_n = \pi^2 n^2 + \frac{1}{3}g(\tilde{T}) + \mathcal{O}(1/n^2), \quad n \to \infty \tag{2.43}$$

Rewriting the result into the mathematical language we obtain that the operator M can be divided into two parts

$$M = M_0 + M_1 \tag{2.44}$$

where M_0 is unitary equivalent to $H_0(\tilde{T})$ and M_1 belongs to the trace class. (This is exactly the result of Howland, see [Ho3] Theorem 5.1 for the precise formulation). Inserting this decomposition into (2.39) and using the fact that $[M_0, M_1] = 0$ we find

$$V = V_0 + V_1 \tag{2.45}$$

with

$$V_0 = e^{-iM_0\tilde{T}} e^{-if\tilde{T}x^2} \tag{2.46}$$

and V_1 being of trace class. The invariance of the absolutely continuous spectrum under the trace class perturbations [ReSi4] leads finally to

$$\sigma_{a.c.}(V) = \sigma_{a.c.}(V_0) \tag{2.47}$$

The operator V_0 is, however, similar to the quantum mapping discussed previously. Using the same arguments we can show that its absolutely continuous part is not empty for $\tilde{T}$ equal to a rational multiple of $1/\pi$.

Summarizing the discussion we find that for functions ℓ fulfilling (2.36) the absolutely continuous spectrum of U is not empty providing the function $\ell(t)$ solves

$$\int_0^T \frac{ds}{\ell(s)^2} = \frac{p}{q}\frac{1}{\pi} \tag{2.48}$$

with $p, q \in \mathbb{N}$. The behaviour of the system is in this case non-recurrent and its energy increases without bounds.

Let us note that the resonance described by the condition (2.48) is of a pure quantum origin and has not classical counterpart. The existence of the resonances predicted theoretically in [Se1] has been checked numerically by Jose [Jo3]. It has been found the energy increase is in this case quadratic in time in full agreement with the estimate (2.35).

References

[Ar] V. I. Arnold: DAN SSSR 156 (1964) 9

[ATI] S. Adachi, M. Toda, K. Ikeda: Phys. Rev. Lett. 61 (1988) 629

[BaBl] R. Balian, C. Bloch: Ann. Phys. 69 (1972) 76

[BaSo] J. E. Bayfield, D. W. Sokol: Phys. Rev. Lett. 61 (1988) 2007

[BCPG] C. P. Brivio, G. Casati, L. Perotti, I. Guarneri: Physica D33 (1988) 51

[Bel1] J. Bellissard: Stability and instability in quantum mechanics. in: Trends and Developments in Eighties; World Scientific, Singapore 1985

[Bel2] J. Bellissard: Stability and chaotic behaviour of quantum rotators, Lecture Notes in Physics 262, Springer Verlag, Berlin 1985

[Bel3] J. Bellissard: Small divisors in quantum mechanics; in: "Chaotic behavior of quantum systems"; NATO ASI Series B, vol. 120, Plenum Press, New York 1985

[BlSm2] R. Blumel, U. Smilansky: Phys. Rev. Lett. 58 (1987) 2531

[BlSm3] R. Blumel, U. Smilansky: Phys. Scripta 35 (1987) 15

[Bo] G. Borg: Acta Mathematica 78 (1946) 1

[BSPB] J. N. Bardsley, B. Sundaram, L. A. Pinnaduwage, J. E. Bayfield: Phys. Rev. Lett 56 (1986) 1007

[CaGu] G. Casati, I. Guarneri: Comm. Math. Phys. 95 (1984) 121

[CaMo] G. Casati, L. Molinari: Quantum chaos with time periodic Hamiltonians, Prog. Theor. Phys. Suppl. 98 (1989) 287

[CCGS] G. Casati, B. V. Chirikov, I. Guarneri, D. L. Shepelyanski: Phys. Rev. Lett. 56 (1986) 2437

[CCHIF] G. Casati, B. V. Chirikov, F. M. Izrailev, J. Ford: Stochastic behavior of a quantum pendulum under a periodic perturbation; in "Stochastic behavior in classical and quantum Hamiltonian systems, Lecture Notes in Physics 93; Springer Verlag, Berlin 1979

[CCHS] G. Casati, B. V. Chirikov, D. L. Shepelyanski: Phys. Rev. 154 (1987) 79

[CCS] G. Casati, B. V. Chirikov, D. L. Shepelyanski: Phys. Rev. Lett. 53 (1984) 2525

[CFGV] G. Casati, J. Ford, I. Guarneri, F. Vivaldi: Phys. Rev. A 34 (1986) 1413

[CoFi] A. Cohen, S. Fischman: Int. J. Modern Phys. B 2 (1988) 1309

[CSZ1] A. A. Chernikov, R. Z. Sagdeev, G. M. Zaslavski: Physica D 33 (1988) 65

[CSZ2] A. A. Chernikov, R. Z. Sagdeev, G. M. Zaslavski: Phys. Today 41 (1988) 27

[DGP] B. Dirizzi, B. Grammaticos, Y. Pomeau: J. Stat. Phys. 37 (1984) 93

[DoRi] S. W. Doescher, H. H. Rice: Am. J. Phys. 37 (1969) 1246

[Fe] E. Fermi: Phys. Rev. 75 (1949) 1169

[FGP] S. Fischman, D. R. Grempel, R. E. Prange: Phys. Rev. Lett. 49 (1982) 509

[FPG] S. Fischman, R. Prange, M. Griniasty: Phys. Rev. A 39 (1989) 1629

[GPF] D. R. Grempel, R. E. Prange, S. Fishman: Phys. Rev. A 29 (1984) 1639

[Ho1] J. S. Howland: Math. Ann. 207 (1974) 315

[Ho2] J. S. Howland: J. Funct. Anal. Appl. 74 (1987) 52

[Ho3] J. S. Howland: Ann. Inst. H. Poincare A 50 (1989) 309

[Ho4] J. S. Howland: Ann. Inst. H. Poincare A 50 (1989) 325

[HoHu1] T. Hogg, B. A. Huberman: Phys. Rev. Lett 48 (1982) 711

[HoHu2] T. Hogg, B. A. Huberman: Phys. Rev. A 28 (1983) 22

[IS] F. M. Izrailev, D. L. Shepelyanski: Teor. Mat. Fiz. 43 (1980) 417

[Jo1] J. V. Jose: From localized to extended states in time-dependent quantum model. in: Quantum chaos and statistical nuclear physics; Lecture Notes in Physics 263; Springer Verlag, Berlin 1986

[Jo2] J. V. Jose: Quantum manifestation of classical chaos: statistics of spectra. in: Directions in chaos; Hao Bai-Lin edit. , World Scientific, Singapore 1988

[Jo3] J. V. Jose at al: talk given on the International Conference on Quantum Chaos, Trieste, Italy , Juni 1990; World Scientific to appear

[JoCo] J. V. Jose, R. Cordery: Phys. Rev. Lett. 56 (1986) 290

[Ka] T. Kato: Petrurbation theory for linear operators, Springer Verlag, Berlin 1966

[Kar1] G. Karner: Lett. Math. Phys. 17 (1989) 329

[Kar2] G. Karner: The dynamics of periodically kicked quantum particles; Preprint Univ. Bochum, SFB 237 (1990) No 77

[Kar3] G. Karner: Dynamics of periodically kicked particles: chaos on the line versus stability on the circle, Preprint Univ. Bielefeld, BiTP 89/22 (1989)

[LL] A. J. Lichtenberg, M. A. Lieberman: Regular and Stochastic Motion. Springer Verlag, Berlin 1983

[Mu] A. Munier at al: J. Math. Phys. 22 (1981) 1211

[PGF] R. E. Prange, D. R. Grempel, S. Fishman: Quantum chaos and Anderson localization in:"Chaotic behaviour in quantum systems", NATO ASI Series B120, Plenum Press, New York 1985

[Pu1] S. D. Pustilnikov: DAN SSSR 292 (1987) 549

[Pu2] S. D. Pustilnikov: Teor. Mat. Fiz. 57 (1983) 128

[Ra] M. Razavy: Hadronic J. 8 (1985) 153

[ReSi1] M. Reed, B. Simon: Methods of modern mathematical physics, vol 1, Academic Press, New York 1972

[ReSi4] M. Reed, B. Simon: Methods of modern mathematical physics, vol 4, Academic Press, New York 1978

[ChSh2] S. J. Chang, K. J. Shi: Phys. Rev. A 34 (1986) 7

[Se1] P. Šeba: Phys. Rev. A 41 (1990) 3206

[SiWo] B. Simon, T. Wolff: Comm. Pure Appl. Math 39 (1986) 75

[Ta] R. J. Tait: Quart. Appl. Math 37 (1979) 313

[We1] H. Weyl: Ueber die asymptotische verteilung der Eigenwerte, Gott. Nach. (1911) 110

[We2] H. Weyl: Math. Ann. 71 (1912) 441

[Ya] K. Yajima: J. Math. Soc. Japan 29 (1977) 729

[ZaCh] G. M. Zaslavski, B. V. Chirikov: DAN SSSR 159 (1964) 306

+ **On leave of absence from the Nuclear Physics Institute, Řež near Praque, Czechoslovakia**

MANIPULATION OF THE SCHRODINGER'S PARTICLE BY TIME DEPENDENT EXTERNAL FIELDS.

by Bogdan Mielnik

Departamento de Fisica, Centro de Investigacion, A. P. 14-740, 07000 Mexico, D.F., Mexico. On leave of absence from the Institute of Theoretical Physics, Warsaw University, Hoza 69, Warsaw, Poland.

ABSTRACT

A question of how to stimulate an arbitrary unitary transformation of the Schrödinger wave packet is considered. A key to the solution is the method of inverting the free evolution of the particle. Possible technical applications are considered.

1. Introduction.

A widespread opinion is that the electron is difficult to manipulate because of its wave-like aspects. The electron wave packet has tendendy to "dissolve"; when meeting an obstacle, it diffracts, flows around and produces interference phenomena. By applying external fields one can control its average motion (as it happen for the electron falling into the TV screen) but not the details of the wave-like propagation (dissolution, difraction, interference). These define some insuperable limits for our control, essential, e.g., for the electronic microscopy. It will be my purpose to show, that the situation is not so bad. Due to its electric charge the electron can be manipulated and one of most effective techniques consists in applying the time-dependent external fields.

The systems driven by time-dependent external forces are recently studied for independent reasons, mostly with the aim to describe the generation of chaos [1-3]. What I shall report here belongs to a different trend which sails, in a sense, in an opposite direction, seeking to restore order rather than to generate the chaos. The problem has already its little history perhaps, worth outlining.

In 1969 W. E. Lamb Jr. [4] has raised the following question: given a particle in a pure state described by a wave function ψ and given another state φ, is there a time-dependent potential $V(x,t)$ such that the solution of the Schrödinger evolution equation brings ψ into φ after a finite time? By applying an elementary though difficult construction Lamb has shown that the answer was positive and has found the required $V(x,t)$ [4]. The motivations of Lamb were philosophical and fundamental rather than technical (to show that all the wave functions $\varphi\in L^2(\mathbb{R})$ are indeed necessary in the scheme of quantum mechanics) and no hint about practical applications was made.

In 1974 E. Lubkin [5] raised a more general problem. Assume, one has a system whose pure states are vectors in a certain Hilbert space $\mathcal{H}$. Can the system be forced to perform an arbitrary unitary operation $U:\mathcal{H}\to\mathcal{H}$ as an output of a certain physical evolution process? For a quantum system with a finite dimensional space of states Lubkin has proved that the answer was positive [5]. The work of Lubkin too was motivated by fundamental questions (as to, whether all self adjoint operators in $\mathcal{H}$ correspond indeed to some measuring devices which can be effectively constructed). The interrelation of Lubkin's work with more empirical problems of *spin echoes* [6,7], apparently, slipped unnoticed.

In 1977 a *kicked free particle* in $\mathbb{R}^1$ was considered by the present author [8]. A dynamical *scenario* was found, which permits one to invert the free evolution operation $e^{-i\tau p^2/2}$ by forcing the wave packet to perform the dynamical operation $e^{i\tau p^2/2}$ (for any given $\tau>0$). This has been, in fact, an extension of the *spin echo* effect to the non-spin variables q and p (an interpretation which, however, escaped attention at that time).

In 1979 the techniques of generating the scale transformations of the wave packet in 1 dimension were described by L.S. Brown and L.J. Carson [13] (see also L.S. Brown [14]). About the same time J. Waniewski has studied a general problem of evolution of many particle systems driven by intrincate sequences of potential kicks [9]. His results, though formal, contain a strong indication that the technology of inverting the natural evolution is not something restricted to one particle.

In 1984 Brewer and Hahn published a provocative article "Atomic Memory" [10] in which the hypothesis was raised that an analogue of the spin-echo effect can occur in the configuration space. The dynamical model used to illustrate the effect (the "rotating dye") was fairly simple though the idea was far reaching.

Shortly afterwards it has been shown [11,12] that the free evolution operations in $L^2(\mathbb{R})$ and $L^2(\mathbb{R}^3)$ can be effectively inverted in a class of almost realistic dynamical *scenarios* involving the free particle kicked by pulses of ellastic forces. This time, the fields used to invert the free evolution in $L^2(\mathbb{R}^3)$ were *harmonic*, and therefore interpretable as an *external influence*, arriving from *distant sources*.

Some practical uses of such inversion might be expected. Suppose a Schrödinger wave's packet diffracts on a microscopic obstacle. Then appart of the very act of collision a significant part of the subsequent diffraction process consists just of the free evolution described by the operator $e^{-i\tau\vec{p}^2/2}$ (the decaying influence of the obstacle left behind still felt as a boundary condition). Similarly, if the wave packet emerges from a system of slids a significant part of the subsequent interference consists just of the free propagation, described again by $e^{-i\tau\vec{p}^2/2}$ (Fig.1). The possibility of inverting that process by generating $e^{i\tau\vec{p}^2/2}$ might mean that we are able to control some phenomena, until now considered uncontrolable.

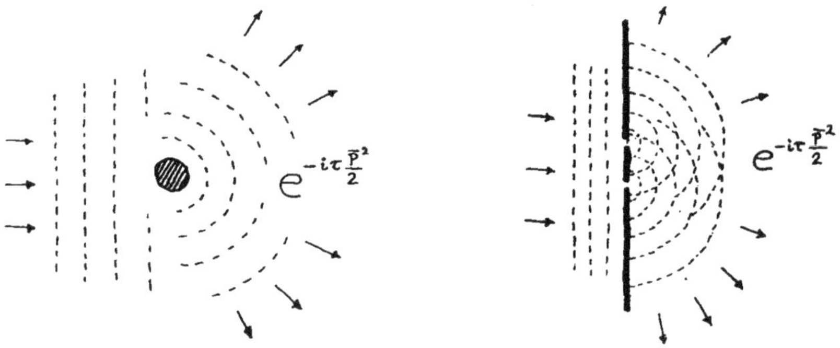

FIG.1. The free evolution element in the quantum diffraction and interference phenomena.

2. The free evolution inversion.

The prescription of how to achieve dynamically the operation $e^{i\tau p^2/2}$ inverse to the free evolution (for non-spin variables) has emerged from an unrelated study of Baker-Campbell-Hausdorff formula. Consider the product of 12 exponential operators in $L^2(\mathbb{R})$:

$$[\exp(-i\tau\frac{p^2}{2})\exp(-i\frac{1}{\tau}\frac{q^2}{2})]^6, \tag{2.1}$$

where q and p are the canonical position and momentum operators, $[q,p]=qp-pq=i$ (we put $\hbar=1$). Were q and p commuting, the 12 term product would be equal $\exp(-i6\tau\frac{p^2}{2})\exp(-i\frac{6q^2}{\tau 2})$. Since, however, $[q,p]=i$, one finds a simpler formula [8]:

$$\underbrace{e^{-i\tau\frac{p^2}{2}}\; e^{-i\frac{1}{\tau}\frac{q^2}{2}} \ldots\ldots\, e^{-i\tau\frac{p^2}{2}}\; e^{-i\frac{1}{\tau}\frac{q^2}{2}}}_{12\ \text{terms}} = 1 \tag{2.2}$$

Notice that all the signs in (2.1) are identical and all the terms have a certain dynamical sense. The operators $e^{-i\tau\frac{p^2}{2}}$ represent the free evolution incidents accomplished within the time lapse $\tau>0$. The terms $e^{-i\frac{1q^2}{\tau 2}}$ correspond to sudden pulses ("kicks") of the time-dependent oscillator shaped potential $\delta(t-t_j)\frac{1}{\tau}\frac{q^2}{2}$. The equality (2.2) means that the wave packet in one space dimension "entertained" by 12 dynamical events (6 intervals of the free evolution and 6 kicks of the oscillator force) returns exactly to its initial shape[1]. This fact can be illustrated by the following clock-wise figure, whose sides simbolize the free evolution intervals, and vertices the "pulses" of the ellastic forces (Fig.2).

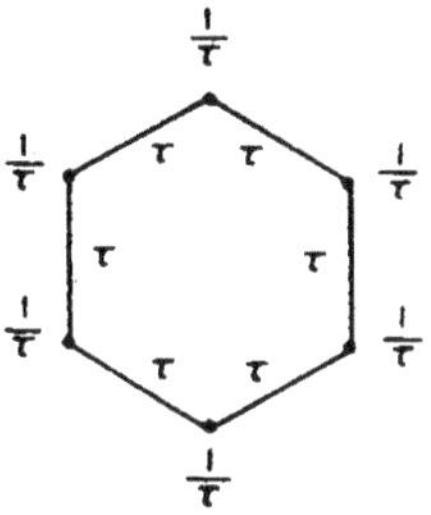

FIG.2. The simplest evolution loop. The sides of the hexagon stand for the intervals of the free evolution and the vertices represent the kicks of the oscillator potential.

An immediate consequence of (2.2) is the following formula which shows that each operation $e^{-i\tau\frac{p^2}{2}}$ can be *effectively inverted* by a sequence of 11 achievable dynamical events:

$$e^{i\tau p^2/2} = \underbrace{e^{-i\frac{1}{\tau}\frac{q^2}{2}}\, e^{-i\tau\frac{p^2}{2}} \,\ldots\ldots\, e^{-i\tau\frac{p^2}{2}}\, e^{-i\frac{1}{\tau}\frac{q^2}{2}}}_{11 \text{ terms}} \tag{2.3}$$

What curious about this prescription is that it has the form of an operator identity. Therefore, it can be applied "in blind": it permits one to restore the past state of a particle without actually knowing this state and with neither knowing the present state of the particle (it looks as if a devil different from the Maxwell's demon were at game!). The whole phenomenon is obviously an analogue of the spin-echo effect for the canonical variables q, p.

As found subsequently [11,12], the formula (2.2) is not an exception, but just a particular case in a wide family of "circular identities" involving the free evolution intervals and kicks of the ellastic potentials. Some of them are represented on Fig.3.

[1] The proof of (2.2) is obtained by checking that the left hand side commutes with both q and p; henceforth, it is a c-number. Since all terms are unitary, the number in question is just a phase factor. To show that it is actually 1 is a minor problem [8].

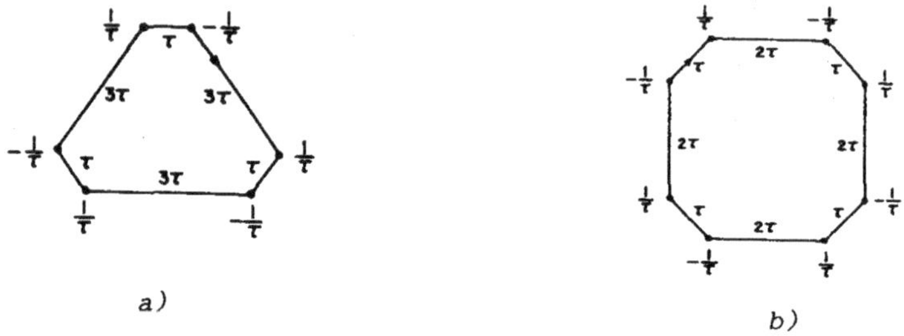

FIG.3. Cases of evolution loops including free evolution intervals.

As in case of Fig.1 they stand for "short circuits of the evolution": the kicks of the oscillator potentials (represented by vertices) and the free evolution episodes (represented by the sides) turn the evolution operation into **1**. Each of these identities when solved with respect to $e^{i\tau p^2/2}$ provides a prescription of how to invert dinamically the free evolution. Thus, e.g., Fig.3a implies:

$$e^{3i\tau p^2/2} = \left[e^{-i\frac{1}{\tau}\frac{q^2}{2}}\, e^{-i\tau p^2/2}\, e^{i\frac{1}{\tau}\frac{q^2}{2}}\, e^{-3i\tau p^2/2}\right]^2 e^{-i\frac{1}{\tau}\frac{q^2}{2}}\, e^{-i\tau p^2/2}\, e^{i\frac{1}{\tau}\frac{q^2}{2}} \quad (2.4)$$

Until now the prescriptions we refer to have a somewhat awkward "time economy". Thus, in case of (2.3) to "rejuvenate" the wave packet 1 day, one would have to work 5 days; the situation is not much better for (2.4). A more economic method of the "time-inversion" is associated with the following closed evolution diagramme (Fig.4):

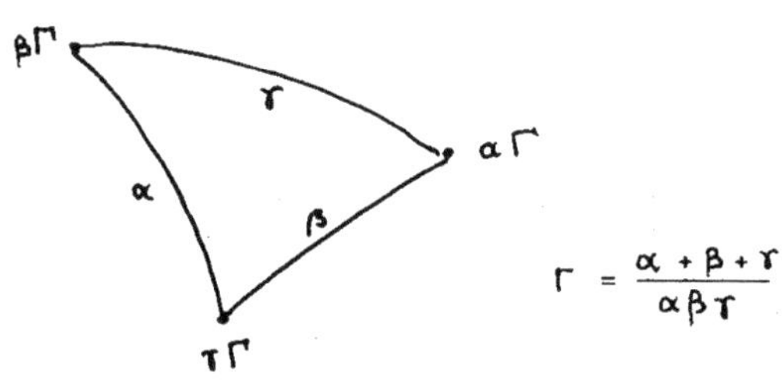

FIG.4. A general evolution loop formed by three kicks of oscillator forces and three intervals of the free evolution.

The corresponding inversion formula [12]:

$$e^{i\alpha p^2/2} = e^{-i\gamma\Gamma\frac{q^2}{2}} e^{-i\beta p^2/2} e^{-i\alpha\Gamma\frac{q^2}{2}} e^{-i\gamma\frac{p^2}{2}} e^{-i\beta\Gamma\frac{q^2}{2}} \tag{2.5}$$

where $\Gamma=(\alpha+\beta+\gamma)/\alpha\beta\gamma$, means that the free evolution operation for an arbitrary $\tau=\alpha$ can be effectively inverted at the cost of operations comprised in an arbitrarily short time interval $\tau'=\beta+\gamma$.

3. Non-singular Evolution loops.

Despite the popularity of kicked systems, the prescriptions (2.1-5) might be questioned for both practical and theoretical reasons. A sharp pulse of the electric potential is difficult to engineere. Moreover, it has to produce non-vanishing radiation effects which are neglected in our approach and might spoil the loop effect. The existence of non-singular analogs of the closed processes represented on Figs. 2-5 might be therefore of interest. The simplest one of them is obtained by considering the harmonic oscillator with a time-dependent ellastic force, its Hamiltonian defined as:

$$H(t) = \frac{1}{2} p^2 + a(\sin\frac{2\pi t}{T}) \frac{1}{2} q^2 \tag{3.1}$$

Systems with quadratic Hamiltonians have been extensively analized elsewhere [15-20], but the effect which we would like to discuss almost escaped attention in the litterature. It turns out that the general character of (either classical or quantum) motion generated by the periodic Hamiltonian (3.1) depends just on the value of one analytic function $S(\alpha)$ of one real variable $\alpha=aT^2$ [12], which represents the discriminant [19] of the periodically driven process (3.1):

$$S(\alpha) = 2 - \frac{1}{8\pi^2}\alpha^2 + \frac{8\pi^2-75}{6144\pi^6}\alpha^4 + \ldots \tag{3.2}$$

Whenever $S(\alpha)$ accepts one of special values

$$|S(\alpha)|=2\cos\frac{2\pi l}{n}, \tag{3.3}$$

(for l, n natural, $0<2l<n$) the evolution operation $U(\tau)$ generated by the Hamiltonian (3.1) turns **1** for $\tau=nT$ (i.e., after the n oscillations of the ellastic force in (3.1)). Solving numerically (3.3) one finds the following examples of the loop generating α-values [12,22]:

α	$S(\alpha)$	loop after:
15.48901	-1	3T
12.61957	0	4T
8.90445	1	6T
148.64337	-1	3T
148.97093	0	4T
149.29548	1	6T

To check our predictions a numerical integration of the classical trajectories generated by the "oscillating oscillator" (3.1) has been performed. The Figs 5,6 show the examples of "boomerang orbits" which indeed appear for the special α-values [22].

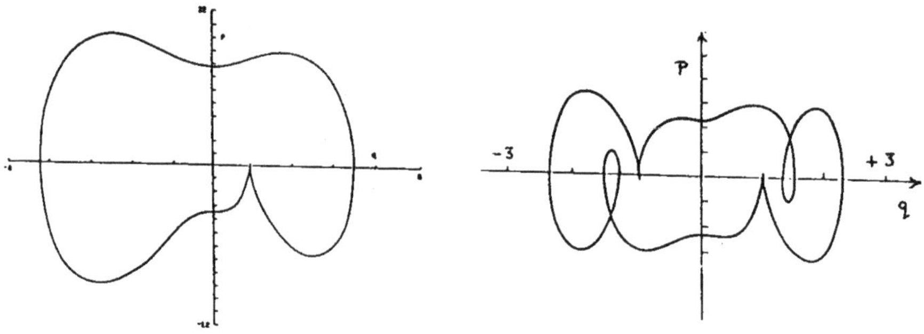

a) b)

FIG.5. The classical analogues of the quantum evolution loops. Closed phase space trajectories which occur for the time dependent Hamiltonian (3.1) when: a) $\alpha=15.48901$, b) $\alpha=8.90445$.

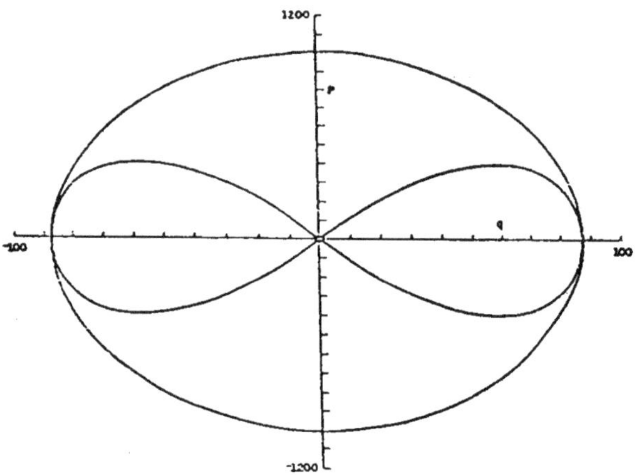

FIG.6. One of large boumerang orbits generated by the oscillator (3.1) when $\alpha=148.97093$.

An amazing property of the "sinusoidally oscillating" potentials (3.1) in $L^2(\mathbb{R})$ is that they provide quite naturally some time dependent potentials in $\mathbb{R}^3$ which, similarly as the harmonic oscillator, generate the closed evolution loops, but unlikely to the oscillator potential are source free, which in our approximation means the vanishing Laplacian $\Delta\phi=0$. Indeed, it is enough to consider the following time dependent potential in $\mathbb{R}^3$:

$$V(x,y,z,t)=\gamma(t)\frac{x^2}{2}+\gamma(t+\frac{T}{3})\frac{y^2}{2}+\gamma(t+\frac{2T}{3})\frac{z^2}{2} \tag{3.5}$$

where $\gamma(t)=a\sin\frac{2\pi t}{T}$, (curiously, the functions modulating the terms $\frac{x^2}{2}$, $\frac{y^2}{2}$, $\frac{z^2}{2}$ resemble the 3-phase electric current!.) In view of $\gamma(t)+\gamma(t+\frac{T}{3})+\gamma(t+\frac{2T}{3})\equiv0$ the potential (3.5) is harmonic (source free). Moreover, the motions in the x, y and z directions separate, and whenever the parameter $\alpha=aT^2$ accepts one of the special values (3.4), they become periodic reducing simultaneously all three partial evolution operations to **1** after $\tau=nT$. Thus, an evolution loop is produced in $\mathbb{R}^3$. One henceforth sees that a phenomenon which was thought exclusive for the time-independent harmonic oscillator (the evolution operator turning periodically 1) can be as well generated by source-free fields created by distant sources[2].

A different way of achieving the loop effects in $L^2(\mathbb{R}^3)$ involves time dependent magnetic fields. Consider a variable homogeneous magnetic field $\vec{B}(t)$. Such fields can be produced with a good accuracy in limited space regions by systems of electromagnets. If the radiation and retardation effects are neglected, the influence of $\vec{B}(t)$ on the wave packet can be described by the vector potential:

$$A(\vec{x},t) = -\frac{1}{2}\vec{r}\times\vec{B}(t)^3 \tag{3.6}$$

It was recently noticed that the potential (3.6) when introduced to the evolution equation of the Schrödinger's wave packet can provide an echo effect for the non-spin variables $\vec{x}$ and $\vec{p}$ [12]. This occurs when $\vec{B}(t)$ repeats periodically some specific "pulsation patterns". The simplest such pattern involves magnetic pulses in three orthogonal directions $\vec{n}$, $\vec{m}$, $\vec{s}$ ($|\vec{n}|=|\vec{m}|=|\vec{s}|=1$) and can be produced by a system of 3

[2] How the fields (3.5) can be created in laboratory conditions, however, remains an open question.

[3] The potential (3.6) is not exact. From the point of view of Maxwell's electrodynamics it represents a "laboratory approximation" neglecting the retardation phenomena. Yet, it works well to describe the generation of the spin-echo effects.

cylindrical electromagnets with orthogonal axes as shown on Fig.7. The pulse pattern starts when one of the electromagnets switches on creating a constant magnetic field $\vec{B}_1=a\vec{n}$ during the time interval T/6, then switches to its opposite $\vec{B}_2=-a\vec{n}$ during the next T/6-interval, then switches off. In turn, the next electromagnet awakes producing $\vec{B}_3=a\vec{m}$ and $\vec{B}_4=-a\vec{m}$ within the next time intervals T/6 and T/6 and then the last electromagnet produces $\vec{B}_5=a\vec{s}$ and $\vec{B}_6=-a\vec{s}$ in the following two time intervals T/6 and T/6. After that, the whole pattern is periodically repeated over and over again. It turns out that the general behaviour of the Schrödinger's wave packet in so constructed sequence of magnetic pulses depends on the value of one parameter $\alpha=\frac{eaT}{6mc}$. When:

$$2\cos 2\alpha-\alpha\sin 2\alpha=2\cos\frac{2\pi l}{n} \qquad (n, l\in\mathbb{N},\ n\geq 3,\ 0<2l<n) \tag{3.7}$$

the evolution operator turns **1** for $\tau=nT$ (and its multiples), producing a magnetically generated evolution loop in $L^2(\mathbb{R}^3)$.

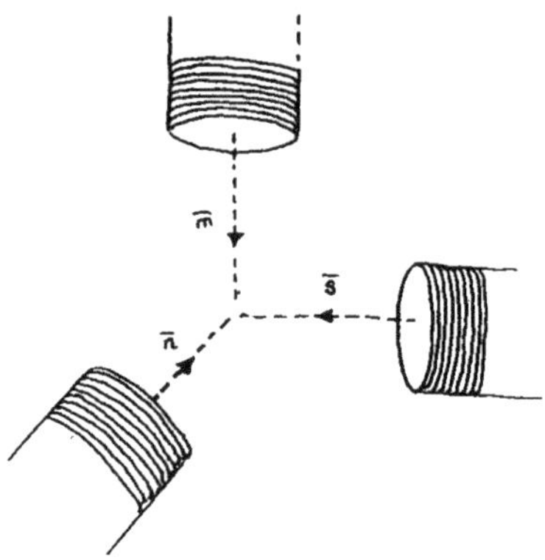

FIG.7. The three electromagnets inducing an evolution loop.

The possibility of causing such a phenomenon with the help of so simple fields might be of importance for the general manipulation technology.

4. The General Manipulation Method.

A natural method of manipulating the wave packet consists of closing the packet in an evolution loop and then applying external forces which affect the loop [12]. The perturbed loop becomes then "inexact" and the evolution operation within the former loop period no longer closes to unity. If the perturbation is small, the particle performes a "precession operation" which differs little from unity, but can have a quite general algebraic structure. In fact, indications exist that by applying to a loop an electromagnetic radiation of sufficiently fine structure, an arbitrary unitary transformation of the wave packet can be inflicted [12]. Some simple cases might be of particular interest.

A common intuition is that it would be rather difficult to "displace paralelly" a wave packet without "spilling it" because of its wave-like properties. This conviction turns out incorrect. When the evolution loop (3.5) is affected by additional pulses of an homogeneous electric field $\vec{E}(t)$, the wave packet instead of being periodically rebuild, starts to drift. If the time dependence of $\vec{E}(t)$ is adequate, the drift produces a sequence of displacements at the time moments $t=n\tau$:

$$\Psi(\vec{x},n\tau) = \Psi(\vec{x}-n\vec{a},0), \tag{4.1}$$

(where τ is the period of the unperturbed loop and the vector $\vec{a}$ depends on the detailed structure of the perturbing electric field). As can be shown, a similar sequence of parallel displacements can be also generated for the wave packet closed in one of the magnetic loops (3.7) [23]. These facts may prove essential for the recently considered masurement techniques. They permit to transform a sequence of repeated position checks into a sequence of "non-demolition measurements", useful in the metodology of measuring weak external fields [24-27]. It seems, that a part of this idea can be directly incorporated to the already existing experimental methods.

As recently noticed, the Penning trap of specially adapted parameters provides another example of an evolution loops (the "Penning loop"). The possibilities of manipulating a particle in such a loop were recently studied by D. Fernández C. [28]. As found there, a relatively unsophisticated sequence of potential shocks applied to the Penning trap surface can force the packet to perform a sequence of somewhat uncommon dynamical operations. One of them is the Fourier transformation (so that the roles of position and momentum amplitudes of the particle are reversed). Another one is the scale transformation in 3 dimensions [26]:

$$\Psi(x,y,z) = \lambda_1\lambda_2{}^2\Psi(\lambda_1x,\lambda_2y,\lambda_2z) \tag{4.2}$$

This might suggest some new experimental techniques beyond the actual problem of non-demolishing measurements [24-27]. Should the wave packet collide at some definite time moment with a microscopic dust particle or molecular object inside of a Penning loop cavity, a magnified and diffractionless image of the obstacle should appear after a definite time. Despite quite evident difficulties with controlling the time and scanning the picture a question must be asked whether the time dependent lense fields are not the predictable future of the electronic optics and other technologies. Some deeper questions concerning the "subtle manipulations" of the molecular wave packets are also imminent [12].

Acknowledgement.

The author is grateful to the Organizing Committee of the Conference "Rigorous Results in Quantum Dynamics" for the warm hospitallity and stimulating atmosphere of discussions in Liblice Castle, Czechoslovakia, June 1990. It is an honor for me to dedicate this talk to the memory of Professor Jiři Blank.

References.

1. J.Bellissard, *Trends in Eighties*, Lecture Notes in Math., Blanchard (1985).
2. G. Casati, B. Chirikov, D. Shepelyanski and I. Guarneri, Phys. Rep. **154** (2), 79 (1987).
3. W. M. Visschers, Phys. Rev. **A36**, 5031 (1987).
4. W. E. Lamb Jr., Phys. Today **22** (4), 23 (1966).
5. E. Lubkin, J. Math. Phys. **15**, 663 (1974); **15**, 673 (1974).
6. E. L. Hahn, Phys. Rev. **80**, 580 (1950).
7. D. Walgraef and P. Borkmans, Physica **59**, 47 (1972).
8. B. Mielnik, Rep. Math. Phys. **12**, 331 (1977).
9. J. Waniewski, Commun. Math. Phys. **76**, 27 (1980).
10. R. G. Brewer and E. L. Hahn, Sci. Am. **251** (12), 50 (1984).
11. B. Mielnik, in *Proceedings of 14 ICGTMP*, ed. Y M. Cho, World Scientific, Seoul, Korea (1985), p. 265.
12. B. Mielnik, J. Math. Phys. **27**, 2290 (1986); Erratum **30**, 550 (1989).
13. L.S. Brown and L.J. Carson, Phys. Rev. A **20**, 2486 (1979).
14. L.S. Brown, Phys. Rev. A **36**, 2463 (1987).
15. K. Husimi, Progr. Theor. Phys. **9**, 381 (1953).
16. H. R. Lewis and W. B. Riesenfeld, J. Math. Phys. **10**, 1458 (1969).
17. I.A. Malkin, V. Manko and D. Trifonov, Phys.Rev. **D2**, 1371, (1970).
18. K. B. Wolf, J. Math. Phys. **15**, 1295 (1974).
19. P. G. L. Leach, J. Math. Phys. **18**, 1608 (1977); **19**, **446** (1978).
20. J. F. Plebanski and D. Finley III, J. Math. Phys. **30**, 993 (1989).
21. W. Magnus and S. Winkler, *Hill's Equation*, ed. L. Bers et.al., Interscience, New York (1966).
22. G. Herrera C., Master thesis, CINVESTAV, México (1987).
23. D. Fernández C. and B. Mielnik, in preparation.
24. C. M. Caves, K. S. Thorne, R. W. P. Drewer, V. D. Sandberg and M. Zimmermann, Rev. Mod. Phys. **52** (2), 341 (1980).
25. H. P. Yuen, Phys. Rev. Lett. **51**, 719 (1983).
26. R. Lynch, Phys. Rev. Lett. **52**, 1729 (1984); **54**, 1599 (1985).
27. A. Royer, Phys. Rev. **A36**, 2460 (1987).
28. D. Fernández C., preprint CINVESTAV, 1990.

Spin Systems

ABSENCE OF FERROMAGNETISM FOR ONE-DIMENSIONAL ITINERANT ELECTRONS

Elliott H. Lieb
Department of Mathematics and Physics
Princeton University, Princeton, NJ 08544

My talk summarized results in the following two papers.

- E.H. Lieb and D.C. Mattis, Theory of ferromagnetism and the ordering of electronic energy levels, Phys. Rev. **125**, 164-172 (1962).
- M. Aizenman and E.H. Lieb, Magnetic properties of some itinerant electron systems at $T > 0$, Phys. Rev. Lett. **65**, 1470-1473 (1990).

The first paper contains a theorem about electrons (fermions) in one dimension interacting with each other via the most general possible spin-independent interaction. I.e., the Hamiltonian for N electrons is

$$H = -\frac{\hbar^2}{2m}\sum_{i=1}^{N}\Delta_i + V(x_1, \ldots, x_N),$$

and the only condition on V is that it be symmetric in the N electron coordinates. The ground state energy E_0, depends on the total spin angular momentum $j = \frac{1}{2}N, \frac{1}{2}N - 1, \ldots, \frac{1}{2}$ or 0 via the Pauli principle. The theorem is that

$$E(j+1) > E(j)$$

for each $j < \frac{1}{2}N$. In other words, the interaction V cannot reverse the ordering of the levels that is seen in the non-interacting system. As a result the system has a tendency away from ferromagnetism.

The second paper extends this result, for electrons in a finite box $[-L, L]$, to finite $\beta = 1/kT$. If $s = (N_\uparrow - N_\downarrow)/2$ is the z-component of the magnetization (with $N_\uparrow$ = number of "spin up" electrons and $N = N_\uparrow + N_\downarrow$) we can define

$$Z_S(\beta; s) = Trace\{\exp(-\beta H)\},$$

where the trace is restricted to states with the given value of s. Likewise, if j denotes the *total* spin angular momentum as before, we can define $Z_J(\beta;j)$ to be the partition function restricted to the given j value. The relation between $Z_J(\beta;\cdot)$ and $Z_S(\beta;\cdot)$ is, by the general theory of angular momentum,

$$Z_S(\beta;j) - Z_S(\beta;j+1) = \frac{1}{2j+1} Z_J(\beta;j)$$

for all j, and with $Z_S(\beta;\frac{1}{2}N+1) \equiv 0$.

In the limit $\beta \to 0$ (equivalently, the electrons have no potential energy and no kinetic energy) the dependence of $Z_S(\beta;s)$ on s is given simply by the combinatorally determined quantity

$$Y_S^{(N)}(s) = \frac{N!}{(\frac{1}{2}N+s)!\,(\frac{1}{2}N-s)!}.$$

The dependence of $Z_J(\beta;j)$ on j is, in this limit, $Y_J^{(N)}(j)$ given by

$$Y_S^{(N)}(j) - Y_S^{(N)}(j+1) = \frac{1}{2j+1} Y_J^{(N)}(j).$$

These Y's *define* the pure paramagnet. They are the relevant normalization constants for deciding the tendency of our system away from paramagnetism and toward, or away from ferromagnetism.

Our main theorem is that there are positive functions $C_N(\beta), C_{N-2}(\beta), \ldots, C_1(\beta)$ or $C_0(\beta)$ such that

$$Z_S(\beta;s) = \sum_{k=0}^{N} C_k(\beta) Y_S^{(k)}(s)$$

The monotonicity of the Y's then leads to the following two consequences:

$$Z_S(\beta;s+1)/Y_S^{(N)}(s+1) < Z_S(\beta;s)/Y_S^{(N)}(s)$$

and

$$Z_J(\beta;j+1)/Y_J^{(N)}(j+1) < Z_J(\beta;j)/Y_J^{(N)}(j)$$

for all j and s. This is the generalization to positive temperature of the earlier ground state theorem. Note that neither of the last two inequalities can be derived from the other.

A further corollary is that the magnetization $M(\beta, h)$, which can be calculated from

$$Z(\beta, h) = \sum_{s=-N/2}^{N/2} Z_S(\beta, s) \exp\{2\beta h s\}$$

satisfies

$$M(\beta, h) = \frac{d}{d\beta h} lnZ(\beta, h) < N \tanh(\beta h),$$

i.e., $M(\beta, h)$ is less than the pure paramagnetic value.

The second paper also contains a theorem stating that under some very special circumstances, the system can be *ferromagnetic* in higher dimensions. An example of this is the well known Nagaoka-Thouless theorem that the Hubbard model with infinite repulsion and one hole has a totally magnetized ground state in more than one dimension. We generalize that theorem to $T > 0$: the magnetization satisfies $M(\beta, h) > N \tanh(\beta h)$. The Hubbard model is a lattice model and, indeed, all the theorems stated here also apply to the lattice case with Δ being the discrete Laplacian.

DYNAMICS AND GROUND STATES OF A SPIN-BOSON MODEL

ANTON AMANN
Laboratory of Physical Chemistry, ETH-Zentrum
CH-8092 Zürich, Switzerland

ABSTRACT

The dynamics and the ground states of a spin-boson Hamiltonian consisting of *one* spin and infinitely many bosons interacting via a dipole-type coupling are investigated. It is argued that the associated automorphic dynamics has to be formulated on an auxiliary (C*-) "algebra of observables" since it does not leave (globally) invariant the usual one based on the canonical commutation relations. Furthermore the order parameter of the model vanishes with respect to *arbitrary* ground states if it vanishes with respect to ground states obtained as (biased) temperature to zero limits of thermic equilibrium states. Ground states of the latter special type have been investigated by H. Spohn. Spohn's respective phase diagrams are therefore valid for arbitrary ground states.

1. Introduction

The Hamiltonian

$$H = \hbar \varepsilon \sigma_1 \otimes \mathbf{1} + \mathbf{1} \otimes \sum_{n=1}^{\infty} \hbar \omega(n)\, a_n^* a_n + \hbar\, \sigma_3 \otimes \sum_{n=1}^{\infty} \lambda(n) \{a_n + a_n^*\} \qquad (1)$$

discussed here consists of *one* spin (= a two-level system with "level splitting" ε) and infinitely many harmonic oscillators (a boson field) with strictly positive frequencies $\omega(n)$, $n=1,2,\ldots$. The spin is described by 2×2-matrices, in particular the Pauli matrices σ_j, $j=1,2,3$. The boson operators a_n, $n = 1, 2, \ldots$, and their adjoints a_n^* describing the field fulfill the canonical commutation relations

$$[a_n, a_{n'}] = 0 \quad , \quad [a_n, a_{n'}^*] = \delta_{nn'} \mathbf{1}. \qquad (2)$$

The frequencies $\omega(n)$ and the coupling constants $\lambda(n)$ are supposed to fulfill the conditions

$$\sum_{n=1}^{\infty} |\lambda(n)|^2 < \infty , \qquad \sum_{n=1}^{\infty} \frac{|\lambda(n)|^2}{\omega(n)} < \infty . \qquad (3)$$

The first one is technical, whereas the second guarantees (at least on a formal level) that the Hamiltonian (1) is bounded below, i.e., the respective energies cannot be arbitrarily negative, i.e., ground states may exist.

Definition 1: A ground state of the Hamiltonian (1) is specified by an eigenfunction whose respective eigenvalue is at the lower end of the Hamiltonian's spectrum.

This definition is simple but not easy to work with: The spin-boson Hamiltonian encorporates infinitely many degrees of freedom. Hence there exist physically inequivalent representations of the canonical commutation relations (2) with respect to each of which the Hamiltonian (1) may or may not exist. Then *all* representations with properly defined Hamiltonian have to be checked for ground states. The Hamiltonian is defined in the strong resolvent limit for $N \to \infty$, where N is the number of modes of the field, or defined by the requirement that it implements the "right" (Heisenberg) dynamics of the spin-boson-model observables (represented on the underlying Hilbert space). The latter definition is more general, but for pure ground states of the spin-boson model the two definitions are equivalent[1].

A particular class of ground states has been studied by H. Spohn[2,3]. There an additional perturbation (a bias) $h\sigma_3 \otimes \mathbf{1}$ is introduced. Starting point is a *thermic equilibrium state* (KMS-state) $\omega_\beta^{\varepsilon,h}$ with respect to inverse temperature β and with respect to the Hamiltonian (1) including the additional bias $h\sigma_3 \otimes \mathbf{1}$. The strenght h of this perturbation is finally set to zero ($h \to \pm 0$) which offers the possibility to get different ground states $\phi_\pm^\varepsilon$ defined by

$$\phi^{\varepsilon,h}(A) := \lim_{\beta \to \infty} \omega_\beta^{\varepsilon,h}(A) , \tag{4a}$$

$$\phi_\pm^\varepsilon(A) := \lim_{h \to \pm 0} \phi^{\varepsilon,h}(A) . \tag{4b}$$

Here A is an arbitrary observable (a polynomial in spin and Weyl operators). The states (4b) coincide[3] if and only if $\phi_+^\varepsilon(\sigma_3 \otimes \mathbf{1}) = \phi_-^\varepsilon(\sigma_3 \otimes \mathbf{1}) = 0$. Otherwise the space-inversion symmetry ι (sending a_n, σ_2 and σ_3 into their negative but leaving σ_1 invariant) is called *broken*. Hence the additional bias offers the possibility to break the "space-inversion" symmetry of the spin-boson model. The expectation value of $\sigma_3 \otimes \mathbf{1}$ is a kind of *order parameter for the spin-boson model.* The states $\phi_\pm^\varepsilon$ are ground states in the sense of definition 1, but this is not at all obvious (cf. next section).

The spin-boson Hamiltonian is one of the few model Hamiltonians for the coupling of a "small system" (e.g., a molecule) to its environment. It arises in various contexts, as for SQUIDS, the coupling of a "small system" to phonons[4] or the coupling of a molecule to the radiation field[5].

2. Dynamics of the spin-boson system

The algebraic version of the spin-boson model is usually based on *Weyl operators*

$$W(f) := \exp\left\{ \sum_{n=1}^{\infty} a_n^* f(n) - f(n)^* a_n \right\} \tag{5}$$

with $f: \mathbb{N} \to \mathbb{C}$ being a complex-valued square summable ("test") function defined on the natural numbers. As "algebra of observables" one may choose the C*-algebra

$\mathscr{A}$ generated by the 2×2–matrices and the Weyl operators with respect to a suitable class of test functions[6–9]. This C*-algebra of observables can be defined abstractly, i.e., without any reference to a Hilbert space representation (see theorem 5.2.8 in ref. 10). It is globally invariant under the (Heisenberg-type) dynamics α^0 implemented by the Hamiltonian (1) *with the level splitting ε set to zero*[8]. For nonzero level-splitting the additional term $\hbar\, \varepsilon\, \sigma_1 \otimes \mathbf{1}$ and the bias $h\, \sigma_3 \otimes \mathbf{1}$ are considered as a *perturbation*. Due to missing continuity properties of the dynamics α^0 (it is not strongly norm-continuous[11]) one has to take into account[12] that the spin-boson dynamics with *nonzero* level splitting ε does *not leave globally invariant the C*-algebra $\mathscr{A}$*. All calculations have to be made with respect to suitable *representations* of $\mathscr{A}$ on some Hilbert space (i.e., suitable representations of the spin-boson observables). Hence this full dynamics does *not* give rise to a C*-system in the ordinary sense. This is particulary inconvenient for the discussion of limits like in Eq. (4). If the dynamics α^0 would be strongly norm-continuous, it would be no problem at all to show that the ground states (4) are ground states in the sense of definition 1 (by use of propositions 5.3.19 and 5.3.23 in ref. 10). In the present situation this is not so clear.

These problems can be circumvented when the original algebra of observables $\mathscr{A}$ is replaced by an *auxiliary C*-algebra* in such a way that the dynamics now is well-defined (on the C*-algebra and not just in some representations) and strongly norm-continuous[13,14]. Essentially, the auxiliary C*-algebra is generated by "time-smoothed" operators of the type

$$\int_s^u \tilde{\alpha}_t^{\varepsilon}(Q)\, dt \tag{6}$$

where the perturbed dynamics $\tilde{\alpha}^{\varepsilon}$ is formulated on the weak closure of the C*-algebra $\mathscr{A}$ represented on an appropriate Hilbert space $\mathscr{H}$ with dt being the ordinary Lebesgue measure on the reals ($s,u,t \in \mathbb{R}$). The operators Q are taken from $\mathscr{A}$ which is not globally invariant under the perturbed dynamics $\tilde{\alpha}^{\varepsilon}$ (this is indicated by the tilda "~"). The auxiliary C*-system helps to transfer mathematical results to the special situation of boson and spin-boson systems. *In particular, it follows that the ground states (4) are ground states in the sense of definition 1.*

3. Ground states of the spin-boson system

Spohn's ground states (4) are of very special type. It might well be that there exist ground states ϕ with nonvanishing order parameter $\phi(\sigma_3 \otimes \mathbf{1}) \neq 0$, breaking the space-inversion symmetry ι even when Spohn's analysis[3] predicts that it is *not* broken. Fortunately, such a phenomenon does not happen:

> *Theorem*[1]*: Consider the ground states ϕ_+^{ε} and ϕ_-^{ε} (4) of the discrete spin-boson model (ε arbitrary, $h = 0$) introduced by Spohn. Then the order parameter $|\phi_+^{\varepsilon}(\sigma_3 \otimes \mathbf{1})| = |\phi_-^{\varepsilon}(\sigma_3 \otimes \mathbf{1})|$ coincides with the maximal expectation value of $\sigma_3 \otimes \mathbf{1}$ with respect to an arbitrary ground state in the sense of definition 1.*

This result is a consequence of the following *variational principle* for infinite quantum systems: Let the spin-boson C*-algebra $\mathcal{A}$ be represented on some Hilbert space $\mathcal{H}$ such that the Hamiltonian (1) – with arbitrary but fixed level splitting ε – exists and is bounded below. By abuse of notation it is denoted by H again. Define the real-valued function

$$Q: [-1, +1] \rightarrow \mathbb{R}$$
$$Q(u) := \inf \{ \operatorname{Tr}(DH) \mid D \in \mathcal{B}^1_+(\mathcal{H}), \operatorname{Tr}(D) = 1, \operatorname{Tr}(D \ \sigma_3 \otimes \mathbf{1}) = u \}. \qquad (7)$$

Here Tr denotes the trace and $\mathcal{B}^1_+(\mathcal{H})$ denotes the positive trace-class operators acting on the Hilbert space $\mathcal{H}$. It follows[1] that Q is an even, convex function which does *not depend (apart from a constant function) on the particular representation chosen* and that

$$|\phi^{\varepsilon}_+(\sigma_3 \otimes \mathbf{1})| = |\phi^{\varepsilon}_-(\sigma_3 \otimes \mathbf{1})| =$$
$$= \sup \left\{ |u| \,\middle|\, u \in [-1, +1], Q(u) \textit{ takes its minimal value} \right\}. \qquad (8)$$

Hence phase transitions of the spin-boson model with respect to ground states can in principle be investigated by looking at a particular version of the Hamiltonian (1), e.g., in a *Fock space* setting.

4. References

1. A. Amann, *Ground states of a spin-boson model,* to appear in *Annals of Physics.*
2. H. Spohn and Dümcke R., *J. Stat. Phys.* **41** (1985) 389.
3. H. Spohn, *Commun. Math. Phys.* **123** (1989) 277.
4. A. J. Leggett, Chakravarty S., Dorsey A. T., Fisher M. P. A., Garg A. and Zwerger W., *Rev. Mod. Phys.* **59** (1987) 1.
5. P. Pfeifer, *Chiral Molecules - a Superselection Rule Induced by the Radiation Field,* (Thesis ETH-Zürich No. 6551, ok Gotthard S+D AG, Zürich, 1980).
6. M. Fannes, Nachtergaele B. and Verbeure A., *Europhys. Lett.* **4** (1987) 963.
7. M. Fannes, Nachtergaele B. and Verbeure A., *J. Phys. A* **21** (1988) 1759.
8. M. Fannes, Nachtergaele B. and Verbeure A., *Commun. Math. Phys.* **114** (1988) 537.
9. M. Fannes, *Temperature states of spin-boson models,* in: *Quantum Probability and Applications IV. Lecture Notes in Mathematics Volume 1396,* ed. by L. Accardi and W. von Waldenfels, (Springer, Berlin, 1989).
10. O. Bratteli and Robinson D. W., *Operator Algebras and Quantum Statistical Mechanics Vol.* 2, (Springer, New York, 1981).
11. A. Amann, *J. Phys. A* **23** (1990) L783.
12. M. Fannes and Verbeure A., *Commun. Math. Phys.* **35** (1974) 257.
13. A. Amann, *Perturbation theory of C*-systems without norm-continuity properties,* preprint (1990).
14. A. Amann, *C*-systems without norm-continuity properties,* to appear in: *Proceedings of "Current Topics in Operator Algebra", Nara (Japan), August 16 - 20, 1990,* ed. by Y. Nakagami, (World Scientific, Singapore, 1991).

SLOW DECAY IN A SPIN MODEL

RAF DEKEYSER
Instituut voor Theoretische Fysica, Katholieke Universiteit Leuven, B–3001 Heverlee, Belgium

and

M. HOWARD LEE
Nuclear Physics Institute, Czechoslovak Academy of Sciences 250 68 Rez, Czechoslovakia and
Department of Physics, University of Georgia Athens, GA 30602, USA

ABSTRACT

The time evolution of a spin in the spin—1/2 van der Waals model has been obtained by solving the Heisenberg equation of motion exactly. Using this solution, we have obtained the long time behavior of the autocorrelation function of a spin. The origin and mechanism of slow decay are described.

1. Introduction

For a microscopic spin interaction model known as the spin van der Waals model,[1] it is possible to show analytically that the autocorrelation function of a spin can decay algebraically. In this model one can thus understand the origin and mechanism of slow decay. In this paper we shall describe qualitatively a few salient features of our recent work.[2] Let us consider a model of N lattice spins defined by the following energy H:

$$H = -N^{-1}\Sigma\, \mathbf{J}\, \underset{\sim}{s}\cdot\underset{\sim}{s} \tag{1}$$

where $\mathbf{J} = (J_x, J_y, J_z)$, $\underset{\sim}{s}$ is a spin–1/2 operator at a given lattice site and the sum over all pairs of spins is implied. The model is a generalized Heisenberg model. We shall simplify it by taking $\mathbf{J}$ to be the same for every pair of spins. We shall further confine ourselves to a cylindrical form of the interaction: $\mathbf{J} = (J, J, J_z)$. Let $g = J/N$ and $g_z = J_z/N$. Then there is only one energy parameter, say, $R = J_z/J$.

We are interested in the time evolution behavior of one spin which may be any one of N spins. The total spin may be decomposed as follows:

$$\Sigma\, \underset{\sim}{s} = \underset{\sim}{s} + \underset{\sim}{S} \tag{2}$$

where $\underset{\sim}{s}$ and $\underset{\sim}{S}$ will be referred to as a small spin and a large spin, respectively. The two spins commute. The time evolution of the small spin $\underset{\sim}{s}$ is to be obtained by solving the equation of motion, $\dot{\underset{\sim}{s}} = i[H,\underset{\sim}{s}]$. We shall consider the case where $N \to \infty$, that is, the large spin $\underset{\sim}{S}$ is very "large," behaving very nearly like a macroscopic classical vector. With this solution, the autocorrelation function $<\underset{\sim}{s}(t)\cdot\underset{\sim}{s}(0)>$ is to be evaluated when $t \to \infty$. Henceforth, we shall denote $\underset{\sim}{s}(0) = \underset{\sim}{s}$ and $\underset{\sim}{S}(0) = \underset{\sim}{S}$.

There is another operator, denoted $\underset{\sim}{\tau}$, which turns out to simplify the time evolution problem enormously. This operator, more precisely its zero eigenvalue, is found to be connected to the existence of slow decay. Let this operator be defined:

$$\underset{\sim}{\tau} = 2g(S_x,S_y,RS_z) = 2g(\underset{\sim}{S} - (1-R)S_z\hat{z}). \tag{3}$$

2. Time Evolution

In terms of this operator, our energy may be expressed as

$$H = -(\underset{\sim}{S}/2 + \underset{\sim}{s})\cdot\underset{\sim}{\tau} \equiv H_o + V. \tag{4}$$

Observe that $\underset{\sim}{\tau}$ plays the role of an effective field, albeit time dependent. As a result, the equations of motion for the small and large spins are those of an effective field model:

$$\dot{\underset{\sim}{s}} = \underset{\sim}{s} \times \underset{\sim}{\tau} \text{ and } \dot{\underset{\sim}{S}} = \underset{\sim}{S} \times \underset{\sim}{\tau}. \tag{5a,b}$$

Now observe from (5b) that $\dot{S}_z = 0$ (strictly valid if $N \to \infty$). Hence, from (3), we have $\dot{\underset{\sim}{\tau}} = 2g\dot{\underset{\sim}{S}}$. Both $\underset{\sim}{S}$ and $\underset{\sim}{\tau}$ have the same time evolution behavior.

Also, using (3) in (5b), we obtain

$$\dot{\underset{\sim}{S}} = -2\omega\,\underset{\sim}{S} \times S_z\hat{z} \tag{6}$$

where $\omega = g - g_z = g(1-R)$. According to (6), the large spin $\underset{\sim}{S}$ rotates simply about the z axis in spin space with a frequency $\Omega_z = 2\omega\, S_z$. This frequency vanishes at $R = 1$. Hence, when $R = 1$, $\dot{\underset{\sim}{S}} = \dot{\underset{\sim}{\tau}} = 0$. Equation (6) implies that the time evolution of the large spin $\underset{\sim}{S}$ has a solution of the form: $\underset{\sim}{S}(t) = \mathbf{M}(t)\,\underset{\sim}{S}$, where $\mathbf{M}(t)$ is a 3×3 unitary matrix, *i.e.*, det $\mathbf{M}(t) = 1$. The time evolution may be viewed as a rotation in spin space where the length of the spin vector is an invariant of time. The "effective field" vector $\underset{\sim}{\tau}$ is time independent when $R = 1$ and ∞. At any other values of R, it has a time evolution. It has the same equation of motion as the large spin: $\dot{\tau} = -2g\, \tau \times S = -2\omega\, \tau \times S_z\hat{z}$. Just as the large spin, τ rotates about the z axis with the same frequency Ω_z.

Let us now turn to the time evolution of the small spin s. Equation (5a) implies that $\dot{s} \cdot \tau = 0$. Hence, the small spin rotates about the vector τ which itself rotates about the z axis. Hence, the small spin precesses except when R = 1 and ∞. Given (5a and b), one can establish at once that $s(t) \cdot S(t) = s \cdot S$. That is, the "angle" between the two spin vectors is a constant of time. We shall see that a slow decay of the autocorrelation function is possible when the small spin undergoes a precession.

Now in parallel to the time evolution of the large spin, one may write $s(t) = \mathbf{L}(t)\, s$ where $\mathbf{L}(t)$ is also a 3×3 unitary matrix, *i.e.*, det $\mathbf{L}(t) = 1$. The matrix $\mathbf{L}(t)$ is evidently more complicated than the matrix $\mathbf{M}(t)$. Using (3) in (5a), we obtain

$$\dot{s} = 2g\, s \times S - 2\omega\, s \times S_z\hat{z}. \tag{7}$$

Hence, if g = 0, (*i.e.*, R = ∞), the equation of motion for the small spin reduces to that of the large spin. That is, $\mathbf{L}(t) = \mathbf{M}(t)$. The small spin rotates about the z axis with the frequency Ω_z. If $\omega = 0$ (*i.e.*, R = 1), $S(t) = S$ and $\dot{s} \cdot S = 0$. Hence, the small spin now rotates about the large spin vector with a frequency $\Omega = 2gS$. For R ≠ 1 and ∞, the time evolution of the small spin involves rotations of the frequencies Ω and Ω_z. We have obtained elements of $\mathbf{L}(t)$. Our solution for $s(t)$ is exact. It satisfies, among others, $(s(-t))^* = s^*(t)$, where * means complex conjugation, *i.e.*, invariant under the time reversal operation. It implies that the autocorrelation function obtained from this solution can never be of simple exponential decay.[3] Our exact solution is in fact quite different from those usually obtained under stochastic theoretic methods, which are a priori time irreversible.

3. Autocorrelation Function and Slow Decay

Let us define the autocorrelation function of the small spin as follows: $G(t) = 4\langle s(t) \cdot s \rangle$, where the brackets mean an ensemble average. We shall look for the time dependent part of $\mathbf{G}(t)$ in the form of $t^{-\kappa}$, $t \to \infty$, where $\kappa = \kappa(R) > 0$. Owing to the xy symmetry, $G_{xx} = G_{yy}$. A slow decay is found only in this component and only when $T > T_c$, where T_c is the critical temperature of the system. Using our solution, we obtain

$$2\,G_{xx}(t) = \langle(1-(S_z/S)^2)\cos\Omega_z t\rangle + \langle(1+S_z/S)^2\cos\phi t\rangle \tag{8}$$

where $\phi = \Omega - \Omega_z$, which is an eigenvalue of $\underset{\sim}{T}$. If $t \to \infty$, the phases in (8) become very large. They lead to rapid oscillations and cancellations upon ensemble averaging, except if the frequencies themselves vanish. Thus, the condition for the existence of a slow decay in our system is: $\Omega_z = 0$ or $\phi = 0$ or both. The condition $\Omega_z = 2\omega S_z = 0$ can always be satisfied, independently of R, since $0 \leq |S_z| \leq S$. The condition $\phi = 0$, or zero eigenvalue of $\underset{\sim}{T}$, means $S_z = S/(1-R)$. Hence, if $R \geq 2$ and $R = 0$, it can be attained. That is, for $0 < R < 2$, no slow decay can arise from the second term of eq. (8).

Now observe that if $R \to \infty$, $\phi \to \Omega_z$, *i.e.*, $\underset{\sim}{T} \to -2\omega S_z\hat{z}$. Hence, at this limit

$$G_{xx}(t) = \langle\cos\Omega_z t\rangle = \exp(-ct^2),\ c > 0. \tag{9}$$

There is no slow decay if $R = \infty$ (*i.e.*, $g = 0$, $g_z \neq 0$). Our analysis further shows that this behavior persists until $R > 2$. Thus, there is no slow decay for $2 < R < \infty$, *i.e.*, $\kappa = \infty$. For the long time behavior, where zero eignevalue of $\underset{\sim}{T}$ is dominant, $\underset{\sim}{T}$ remains coincident with the z axis as long as $R > 2$. For $0 \leq R \leq 2$ (but $R \neq 1$), there is a slow decay, $\kappa = 3$, given by the first term of (8). If $R = 0$, there is still a slower decay, $\kappa = 2$, given by the second term of (8). Thus, the small spin can have a slow decay if it precesses in spin space. A simple rotation leads to decoupling of modes, resulting in a fast decay only.

4. Acknowledgments

One of us (MHL) wishes to thank colleagues at the Czechoslovak Academy of Sciences in Rez for their hospitality during the month of June 1990 while he was a visiting member. He is especially grateful to Dr. M. Znojil. This work was supported in part by the National Science Foundation and the U.S. Army Research Office.

5. References

1. R. Dekeyser and M.H. Lee, Phys. Rev. **B19**, 265 (1979).
2. R. Dekeyser and M.H. Lee, Phys. Rev. B (to be published).
3. M.H. Lee, Phys. Rev. Lett. **51**, 1227 (1983).

Stochastic Analysis

A model of four space–time dimensional gauge fields: reflection positivity for associated random currents

S. Albeverio
Fakultät für Mathematik, Ruhr–Universität, D–4630 Bochum 1,
SFB237 Essen–Bochum–Düsseldorf,
BiBoS Research Centre, D–4800 Bielefeld 1 (Germany),
and CERFIM, Locarno (Switzerland)

K. Iwata
Fakultät für Mathematik, Ruhr–Universität, D–4630 Bochum 1,
SFB237 Essen–Bochum–Düsseldorf,
and BiBoS Research Centre, D–4800 Bielefeld 1 (Germany)

and T. Kolsrud
Department of Mathematics, Royal Institute of Technology,
S–10044 Stockholm (Sweden)

Abstract

In this lecture, we briefly expose our construction of certain generalized vector random fields over 4–dimensional space–time which are of gauge type, are invariant with respect to the (proper) Euclidean group and possess the sharp Markov property with respect to 3–dimensional hyperplanes. These fields yield relativistic functions and Wilson loops variables for which confinement bounds have been proven by Tamura. We show that associated random 2-currents satisfy the reflection positivity condition.

Introduction

The problem of the construction of local relativistic fields, satisfying some basic postulates, has been outstanding since many years. Since the basic work of Nelson this has been closely related to the construction of suitable Markov Euclidean generalized random fields, the basic correlation functions, Wightman

functions, of the relativistic fields being obtained by analytic continuation "in the time variables" of the moments, Schwinger functions, of the corresponding Euclidean fields.
In this procedure the existence of a Hamiltonian H in a Hilbert space $\mathcal{H}$ is needed, H and $\mathcal{H}$ being intrinsically tied to the Markov character of the Euclidean field. Nelson's approach has been successful essentially in 2 space–time dimensions, where models with polynomial, exponential or trigonometric interactions have been constructed. The basic construction itself was essentially obtained in the seventies with a more abstract realization of H and $\mathcal{H}$ compared to the one given by Nelson, via Euclidean axioms; the proof of "Nelson's axioms" was achieved only in the late seventies for the trigonometric interactions by Albeverio and Høegh-Krohn, in the mid–eighties for exponential interactions by Gielerak and Zegarlinski, and only two years ago in the case of polynomial interactions by Albeverio, Høegh-Krohn and Zegarlinski [AHZ] (see also [AZ]).
The constructed Markov Euclidean fields are generalized random fields but allow as test functions in particular measures of finite energy in the sense of the Yukawa kernel $(-\Delta+m^2)^{-1}$, m being a positive constant, i.e. positive Radon measures ρ s.t.

$$\int_{\mathbb{R}^2} \rho(dx) \int_{\mathbb{R}^2} \rho(dy) G(x,y) < \infty ,$$

with $G(x,y)$ the fundamental solution of $(-\Delta+m^2)$ (such measures are thus in the Sobolev space $H^2_{-1}(\mathbb{R}^2)$). Writing $x=(x^0,x)\in \mathbb{R}\times\mathbb{R}$, one has that measures of the form $\rho(dx)=\delta_t(dx^0)f(x)$, with $f\in\mathcal{S}(\mathbb{R})$, $t\in\mathbb{R}$, are of finite energy and calling $<\rho,X>$ the field tested with such ρ ($<\ ,\ >$ being in the sense of generalized functions) one has $<\rho,X>=<f,X_t>$, with X_t a $\mathcal{S}'(\mathbb{R})$-valued Markov process, indexed by $t\in\mathbb{R}$. X_t is symmetric, in the sense that the associated Markov semigroup of transition probabilities e^{-tH}, $t\geq 0$ is symmetric in $\mathcal{H}\equiv L^2(d\mu)$, where μ is the invariant measure of X_t and coincides with the restriction of the law of $<\rho,X>$ to the σ–algebra Σ_0 generated by $<\delta_0\otimes f,X>$.
The Markov property of X_t is with respect to the sharp σ–algebras Σ_0, $\Sigma_{\geq 0}$, $\Sigma_{\leq 0}$ of the present resp. future resp. past of X_t. It can be identified with the Markov property of the field X with respect to the hyperplane $x^0=0$, in the sense that $\Sigma_{\{x^0\geq 0\}}$ is conditionally independent of $\Sigma_{\{x^0\leq 0\}}$ given $\Sigma_{\{x^0=0\}}$, with Σ_Λ the σ–algebra generated by the $<\rho,X>$, ρ of finite energy with supp $\rho\subset\Lambda$. For a description of H in terms of infinite dimensional Dirichlet forms see e.g. [AHK1, 2]-[ARö1]. H coincides on a dense set of cylinder functions with the generator of the classical Dirichlet form given by μ, in the sense of [ARö1]–[ARö3]. H is, in the case of exponential interactions, the generator associated with the maximal classical Dirichlet form [AK]. In all cases e^{-tH} is the limit of uniquely defined "space–cut–off" semigroups and these are given by classical Dirichlet forms [RöZ], [ARöZ].
The fact that these results justify the strong causality principle embodied in canonical quantum field theory is underlined in [AHK1,2], see also [AHKRö].

Is such a picture possible in 4 space–time dimensions?

The discussion of fields of this type turns out to be extremely difficult, in fact an unsolved problem.

However we have been able to show that a modification of the above ideas to models of vector fields with interactions with non necessarily quadratic "kinetic energy" yields at least a partial positive answer. In fact we have been able to construct proper Euclidean Markov fields over $\mathbb{R}^4$. The analytic continuation of the moments of these fields (Schwinger functions) has yielded relativistic functions.

Our starting point is the construction of (generalized) vector–valued Markov fields A obtained as solutions of a certain system of coupled first order partial differential equations with constant coefficients. The simplest way to describe these fields is to look at them as solutions of the inhomogeneous quaternionic Cauchy–Riemann equation

$$\partial A = F \ , \tag{1.1}$$

where the given source F is a quaternionic valued Poisson and Gaussian white noise over $\mathbb{R}^4$, with a suitable invariance property under the Euclidean group in $\mathbb{R}^4$, such as to make (1.1) a covariant equation (in law). ∂ is the quaternionic Cauchy–Riemann operator. The A fields are neither Gaussian nor independent at every point. Using the moments of these fields (Schwinger functions) we constructed relativistic invariant functions (time–ordered Wightman functions). Although their physical interpretation is still open in a number of points, as we discussed in [AHKI], [AIK1], [AIK4], these functions satisfy besides the invariance properties also a "spectrum property". As relativistic functions they are neither of the free nor of the ultralocal type. The spirit of the model is somewhat similar to the one of the Born–Infeld model [BoI] see also [AHK], [AHKI]. Our basic constructions are contained in a series of papers [AHK], [AHKI], [AIK1], [AIK3]. In [T] H. Tamura proved the existence of Wilson's loop observables $\exp(i \int_C A)$ for loops C in $\mathbb{R}^4$, in the case where F is a suitable invariant Poisson field, as non trivial random variables satisfying "confinement bounds". In the present paper, after recalling the basic known properties of the model, we shall mention two recent developments in its study, one concerning the fact that dA, looked upon as a Euclidean random field indexed by smooth antiself–dual 2–forms or antiself–dual 2 simplices, is reflection positive, the other concerned with a lattice approximation of the field A.

In Section 2 we recall the basic results concerning the model of classical fields. Section 3 is dedicated to a study of the associated quantum model in terms of Euclidean random fields. 3.1 presents the model, in section 3.2 the Markov property is stated. Sections 3.3 and 3.4 present new results concerning random 1- and 2-currents associated with the model. We close with 2 remarks, one concerning extentions of the model to 2 resp. 8 dimensions, and one concerning a lattice approximation of the model.

2. The model for classical fields

2.1 The basic equations

As a real vector space, $\mathbb{R}^4$ is isomorphic to the space $\mathbb{H}$ of quaternions , the isomorphism being given by $x \equiv \sum_{\mu=0}^{3} x_\mu \tau_\mu \leftrightarrow x \equiv (x_0, \ldots, x_3) \in \mathbb{R}^4$, with $\tau_0 = 1$ and τ_k, $k = 1,2,3$ the quaternionic units (which can be related to Pauli matrices). $\mathbb{R}$ is embedded in $\mathbb{H}$ by identifying $t \in \mathbb{R}$ with $t1 \in \mathbb{H}$. We shall thus denote by x a point in $\mathbb{H}$ or equivalently in $\mathbb{R}^4$. The isomorphism is a metric one, when we give the quaternions the norm $|x|$ of the corresponding vector $x \in \mathbb{R}^4$. Let $\partial \equiv \partial_0 \tau_0 - \sum_{k=1}^{3} \partial_k \tau_k$ be the quaternionic Cauchy–Riemann operator, (with $\partial_\mu \equiv \frac{\partial}{\partial x_\mu}$) and let $\bar{\partial} \equiv \partial_0 \tau_0 + \sum_{k=1}^{3} \partial_k \tau_k$ its conjugate. Then $\partial \bar{\partial} = \bar{\partial} \partial = \Delta \tau_0$, with Δ the Laplacian in $\mathbb{R}^4$. We observe that $SO(4)$ is realized by quaternions as $(Sp(1) \times Sp(1))/\mathbb{Z}_2$ with $Sp(1) \simeq SU(2)$. Let (Λ, y) be a rotation (Λ) and a translation (by $y \in \mathbb{R}^4$) and (a, b) be a corresponding element in $Sp(1) \times Sp(1)$. If $A(\cdot)$ goes over under (Λ, y) into $aA\left(a^{-1}(\cdot - y)b\right) b^{-1}$, then it is called 4-vector action. On the other hand if $F(\cdot)$ goes over, under the same transformation, into $bF\left(a^{-1}(\cdot - y)b\right) b^{-1}$, then it is called scalar-3-vector action. The following classical equation

$$\partial A = F \tag{2.1}$$

is then covariant (in the sense that ∂ intertwines the two actions explained as above). As has been remarked in [AIK1], the equation (2.1) can be looked upon as a inhomogeneous Weyl equation and constitutes (in a suitable representation) one half of the inhomogeneous Dirac equation on $\mathbb{H}^2 \simeq \mathbb{C}^4$ (the Dirac operator can be written as $\begin{pmatrix} 0 & \bar{\partial} \\ -\partial & 0 \end{pmatrix}$, realizing the Clifford algebra over $\mathbb{R}^4$ as the algebra of 2×2 matrices with quaternionic elements).

In [AHKI] it was shown that if we introduce the space

$$\mathcal{T} \equiv \left\{ \varphi \in C^\infty(\mathbb{R}^4; \mathbb{H});\ \lim_{|x| \to \infty} \varphi(x) = 0,\ \bar{\partial}\varphi \in \mathcal{S} \equiv \mathcal{S}(\mathbb{R}^4; \mathbb{H}) \right\}$$

then the mapping from $\mathcal{T}$ into $\mathcal{S}$ given by $\varphi \in \mathcal{T} \longrightarrow \bar{\partial}\varphi \in \mathcal{S}$ is bijective, with inverse $\xi \in \mathcal{S} \longrightarrow \bar{S} * \xi \in \mathcal{T}$, with $\bar{S}(x) \equiv \frac{\bar{x}}{2\pi^2 |x|^4}$ the kernel of $\bar{\partial}^{-1}$. (2.1) is also covariant in the sense that in (2.1) $A \in \mathcal{S}'$ implies $F \in \mathcal{T}'$ and vice versa. (2.1) is solved by

$$A = S * F \ ,$$

with $S(x) \equiv \frac{x}{2\pi^2|x|^4}$ the kernel of ∂^{-1}. Covariance in (2.1) means also that if F transforms as a scalar–3–vector so A transforms as a 4-vector.

The natural action of $SO(4)$ on 2-forms has two irreducible components. The best description of this decomposition is obtained by introducing the Hodge dual $*$. The two components are labeled by the eigen value with respect to the $*$-action. The $+1$ (resp.-1)-eigen space is called self-dual (resp. anti-self-dual). For the convenience we consider the following basis:

$$\omega^i_\pm \equiv dx^0 \wedge dx^i \pm *(dx^0 \wedge dx^i), \quad i = 1,2,3.$$

We have

$$*\omega^i_\alpha = \mathrm{sgn}(\alpha)\omega^i_\alpha, \quad i = 1,2,3, \; \alpha = \pm,$$

and

$$\omega^i_\alpha \wedge \omega^j_\beta = 2\mathrm{sgn}(\alpha)\delta_{\alpha\beta}\delta^{ij}\omega,$$

where $\omega = dx^0 \wedge dx^1 \wedge dx^2 \wedge dx^3$ denotes the volume form. The correspondence

$$\vec{F} = \sum_{i=1}^{3} F_i\tau_i \longmapsto \sum_{i=1}^{3} F_i\omega^i_-$$

gives rise to an isomorphism as $SO(4)$-modules. If we interpret A as a 1–form, then we have

$$\delta A = -F_0 \; , (d - *d)A = \vec{F} \; . \tag{2.2}$$

2.2 Reflection

The reflection θ defined by $\theta(x) = (-x^0, \vec{x})$, $x \in I\!R^4$, induces a map on differential forms (by pull back). We see

$$\theta f = f \circ \theta, \; f \in \Omega^0,$$
$$\theta(dx^0) = -dx^0, \; \theta(dx^i) = dx^i.$$

By multilinearity this is carried over to the space of all differential forms. On Ω^2 one finds

$$\theta(\omega^i_\pm) = -\omega^i_\mp.$$

Hence, if we decompose $\Omega^2 = \Omega^2_+ \oplus \Omega^2_+$, the action of θ on Ω^2 can be written

$$\theta F = \theta(F_+, F_-) = -(F_-, F_+) \circ \theta = -(\rho F) \circ \theta, \tag{2.3}$$

where ρ is the flip $\rho : (\xi_+, \xi_-) \to (\xi_-, \xi_+)$. θ maps $\Omega^2_\pm$ into $\Omega^2_\mp$, and the reason for this is that θ changes the orientation, and therefore the Hodge $*$–operator. (2.3) implies

$$< \theta F, \xi > = - < F, (\rho\xi) \circ \theta > . \tag{2.4}$$

Finally let $A = A_\mu dx^\mu \in \Omega^1$. Then we have $d\theta A = \theta dA$.

3. The model for Euclidean random fields

3.1 Quaternionic random fields

A four component random field F indexed by $\mathcal{S}(\mathbb{R}^4) \otimes \mathbb{H}$ may be defined by its characteristic functional:

$$\mathbb{E}\{e^{i<F,\chi>}\} = e^{-\int_{\mathbb{R}^4} \psi(\chi(x))dx}, \quad \chi \in \mathcal{S}(\mathbb{R}^4) \otimes \mathbb{H}, \tag{3.1}$$

where

$$\begin{aligned} \psi(\lambda) &= \psi_G(\lambda) + \psi_P(\lambda), \\ \psi_G(\lambda) &= \frac{\sigma_0}{2}\lambda_0^2 + \frac{\sigma}{2}|\vec{\lambda}|^2, \\ \psi_P(\lambda) &= i\lambda_0\beta + \int_{\mathbb{H}\setminus 0} (1 - e^{i(\lambda,\alpha)} + i(\lambda,\alpha)\cdot 1_{|\alpha|\le 1})\nu(d\alpha). \end{aligned} \tag{3.2}$$

Here β is real, σ_0, $\sigma \geq 0$, and the *Lévy measure* ν is invariant under conjugation, i.e. $\nu(d\alpha) = \nu(b(d\alpha)b^{-1})$, with quaternions b of modulus one. If one consider the scalar-3-vector action, then F and $g \cdot F$ have the same law for each Euclidean transformation g. F is a sum of two independent random fields, one Gaussian, and one (generalized) Poisson distributed, corresponding to ψ_G and ψ_P respectively.

For technical reasons we also assume that for some $\varepsilon > 0$

$$\psi(\lambda) = O(|\lambda|^{4/3+\varepsilon}), \quad \lambda \to 0. \tag{3.3}$$

Condition (3.3) implies that one can solve the random field equation

$$\partial A = F \tag{3.4}$$

so that the resulting random field A is given by a measure on $\mathcal{S}'(\mathbb{R}^4) \otimes \mathbb{H}$ which also is invariant under the (proper) Euclidean group. This also involves enlarging the index set for F. We further remark that in general A is not invariant under reflection, a fact which is implicit in sect.2.

Writing $F = (F_0, \vec{F})$, F_0 can be considered as a 0–form (function), $\vec{F}$ as an anti–self–dual 2–form, and A as a 1–form, as before. This depends of course on the two representations of the Euclidean group used for F and A. Choosing the data (i.e. the function ψ) for F in (3.1) such that $F_0 = 0$ we may look upon F, identifying it with $\vec{F}$, as an anti–self–dual 2–form. Then A is a divergence free 1–form. The condition means that ψ is only a function of the imaginary quaternions, i.e. $\psi(\lambda) = \psi(\vec{\lambda})$.

3.2 The global Markov property

As discussed in [AIK4] the generalized random field $\{< \varphi, A >\}$ has the global sharp Markov property. This is actually a particular case of general results [I], but we shall here limit ourselves to the description of the special case relevant to our model. Let C be the "time zero" hyperplane in $\mathbb{R}^4$: $C = \{x \in \mathbb{R}^4;\ x^0 = 0\} \cong \mathbb{R}^3$. By (3.3) the index set for F naturally extends without destroying the stochastic continuiy. Therefore the sharp field is well defined by $A(\delta_0 \otimes \eta) = F(S * (\delta_0 \otimes \eta))$ and moreover, if $\{j_\varepsilon\}$ is a sequence of mollifiers tending to δ_0, then $< j_\varepsilon \otimes \eta, A >$ converges in probability to $A(\delta_0 \otimes \eta)$. This can be looked upon as the time zero electromagnetic potential field, tested with the space–test function η. Let B any open subset of $\mathbb{R}^4$ and let

$$\mathcal{F}_B \equiv \sigma\{< \varphi, A >,\ \varphi \in C_0^\infty(\mathbb{R}^4; \mathbb{H}),\ \text{supp}\ \varphi \subset B\} \vee \mathcal{N},$$

with $\mathcal{N}$ the trivial sub σ–field. Let $\Sigma_C \equiv \sigma\{A(\delta_0 \otimes \eta),\ \eta \in C_0^\infty(\mathbb{R}^3; \mathbb{H})\} \vee \mathcal{N}$ the sharp σ–algebra associated with the time zero field. For any open $B \supset C$ we have $\Sigma_C \subset \mathcal{F}_B$. The global sharp Markov property (with respect to hyperplanes) is then expressed by the following

Theorem

Let C_+, C_- be the connected components of $\mathbb{R}^4 \backslash C$. Then $\mathcal{F}_{C_+}$ is conditionally independent of $\mathcal{F}_{C_-}$ given Σ_C.

Remark

Above we have chosen a particular hyperplane C. However by the Euclidean invariance (homogeneity) of the field A we see that the above Markov property holds with respect to arbitrary hyperplanes. In fact it holds also for arbitrary separating regions in $\mathbb{R}^4$ cfr. [I]. For the proof of above theorem we refer to [AIK4](and [I]).

Remark

a) In the Poisson case $\sigma_0 = \sigma = 0$ with ν of bounded variation one can show, cfr.[I], that Σ_C coincides with the whole σ–algebra $\mathcal{F}_{\mathbb{R}^4}$, while in the Gaussian case Σ_C is strictly smaller than the whole σ–algebra.

b) In the Gaussian case a Markov property with respect to hyperplanes can be formulated for the so called transversal fields (similar to the one discussed for free scalar fields by using the Fock space structure in terms of the Segal isomorphism of the associated L^2 space. See e.g. [GJ], [Ne] and [Si]). For the transversal fields the reflection with respect to C acts trivially on the time zero σ-field and therefore one has reflection positivity for them (cfr. [AIK4] for details, related results were discussed before e.g. in [Lö], [GeP], [Gu]).

3.3 Random 1–currents and Wilson loops

Under the assumption (3.3), the equations

$$(d - *d)A = \vec{F}, \tag{3.5}$$

and

$$\delta A = -\partial_\mu A^\mu = 0. \tag{3.6}$$

can be solved and we obtain a system of random variables $< dA, \xi >$ where ξ runs through the 2–forms on $\mathbb{R}^4$ with coefficients in $\mathcal{S}(\mathbb{R}^4)$. This way dA becomes a random 2–current (and similarly A a 1–current): the realizations are (with probability one) 2–forms the coefficients of which are tempered distributions, i.e. in $\mathcal{S}'(\mathbb{R}^4)$.

It turns out that under further restrictions on the data for F, i.e. the function ψ in §3.1, we can actually evaluate the integral of dA over a 2–simplex. This corresponds to letting the ξ's above become singular. Assume that for some constants $c > 0$, $4/3 < p < 3/2$ and $0 \leq q < 1$

$$\frac{1}{c}|\vec{\lambda}|^p \leq \psi(\vec{\lambda}) \leq c|\vec{\lambda}|^p \;,\; |\vec{\lambda}| \leq 1 \tag{3.7}$$

$$\psi(\vec{\lambda}) \leq c|\vec{\lambda}|^q, \quad \vec{\lambda} \in \mathbb{R}^3. \tag{3.8}$$

In [T], Tamura showed that under these assumptions (stronger than (3.3)) the random variable (Wilson loop) $\exp i \int_C A$ exists, as an L^p–limit of a regularized version, for any smooth loop C in $\mathbb{R}^4$, and for any $p \geq 1$. In particular, if C is the boundary of a 2-simplex σ, then $\int_C A = \int_\sigma dA$. Incidentally, this shows that $\int_\sigma F$ makes sense.

Since $\mathbb{R}^4$ is topological trivial, all information about $\int_C A$ for loops C is in $\int_\sigma dA$, $\sigma \in S^2$, the 2–simplices in $\mathbb{R}^4$. We can give S^2 a topology, so that it can be decomposed like Ω^2:

$$S^2 = S^2_+ \oplus S^2_- \tag{3.9}$$

into self–dual and anti–self–dual 2–simplices. The elements of $S^2_\pm$ satisfy

$$\int_{\sigma\pm} \xi = \int_{\sigma\pm} \xi_\pm, \quad \xi \in \Omega^2, \tag{3.10}$$

so that, with obvious notation,

$$\int_\sigma \xi = \int_{\sigma -} \xi_- + \int_{\sigma +} \xi_+, \quad \xi \in \Omega^2, \sigma \in S^2. \tag{3.11}$$

This way A really becomes a gauge field, and the group involved is $U(1)$. Tensoring copies of A we may obtain Lie algebra valued random fields, thus also non–abelian groups could be handled (cfr. for a similar situation [AHK3], [AHKH], [AHKHK1,2], [D], [F], [GKS]).

Remark:

To visualize some anti-self-dual 2-simplices, it is simpler to go to a lattice, $\mathbb{Z}^4$, say. Consider the positively oriented 2-surface (square) τ_1 lying in the x^0x^1-plane with corners at $(x^0,x^1,0,0)$, $x^0,x^1=0,1$. We denote by τ_1' the positively oriented 2-surface obtained by 'reflecting' τ_1 to the square in the x^2x^3-plane with corners at $(0,0,x^2,x^3)$, $x^2,x^3=0,1$. Then $\sigma_1 \equiv \tau_1 - \tau_1'$ is in S^2_-. Similarly for other cyclic permutations of $(1,2,3)$. We may use the σ_i, $i=1,2,3$ thus obtained as elementary building blocks to produce other anti-self-dual 2-simplices. For instance any finite sum $\sigma = \sum_{ig} a_{ig} g\cdot\sigma_i$ where the g are Euclidean transformations or dilations and the coefficients a_{ig} are real numbers, defines an element of S^2_-.

3.4 Reflection positivity of the random 2-currents

Consider now the collection of random variables $< dA,\xi >$, with $\xi \in \Omega^2_-$. Denote by $\Phi = \Phi_\xi$ the random variable $\exp(i < dA,\xi >)$. We define the *reflection* $R\Phi$ of Φ by the formula

$$R\Phi = R\Big(e^{i<dA,\xi>}\Big) \equiv e^{i<d(\theta A),\rho\xi>}, \quad \xi \in \Omega^2_-, \tag{3.12}$$

where we recall that ρ was defined in (2.3). Then $R(R\Phi)=\Phi$. For $\xi\in\Omega^2_-$ we get

$$\begin{aligned} < d\theta A,\rho\xi > &=< \theta dA,\rho\xi >=< dA,\theta\rho\xi > \\ &=< dA,\xi\circ\theta >= -\frac{1}{2} < F,\xi\circ\theta >, \end{aligned} \tag{3.13}$$

so that

$$R\Phi_\xi = e^{-\frac{i}{2}<F,\xi\circ\theta>} \stackrel{d}{=} e^{-\frac{i}{2}<F,\xi>} = \bar{\Phi}_\xi, \tag{3.14}$$

since F is reflection invariant, which is obvious from (3.1). It follows from the independence of F that the random field $\{< dA,\cdot >\}$ restricted on anti-self-dual 2-forms is reflection positive when the reflection R is extended to finite linear combinations of exponentials in the obvious manner:

$$R(\sum z_n\Phi_{\xi_n}) = \sum \bar{z}_n R\Phi_{\xi_n}.$$

One proves easily that reflection positivity holds also in the singular case, i.e. for anti-self-dual 2-simplices. (The associated Hilbert space, after analytic continuation, is one-dimensional, similarly as in 2-dimensional gauge theories, cfr. [AIK3], [Be].)

Remark

a) The study of models of the type of those analyzed in this paper, initiated in [AHK], has been extended to group-valued fields in [AHK3], to octavic fields in [Os2,3] and to complex fields in [AHK3], [Os1], [Be], [AIK2,3]. In all cases the equation is of the type $\partial A = F$, with ∂ the relevant Cauchy-Riemann operator. In the case of the octavic space a weaker form of Markov property is proven in [Os2] (the methods of [I] should however extend also to this case, yielding the result on the Markov property corresponding to the one mentioned above in the quaternionic case). In the case of complex valued or $\mathbb{R}^2$-valued fields in [Os1], [Be] the basic function space is taken to be $S_0(\mathbb{R}^2, \mathbb{R}^2) \equiv \{\varphi \in S(\mathbb{R}^2; \mathbb{R}^2) \int \varphi dx = 0\}$ (to give meaning to Δ^{-1}). In this 2-dimensional case reflection symmetry holds for A also in the non Gaussian case (provided the Lévy measure ν is even) and the transversal fields are reflection positive. The analytic continuation is discussed in [Os1], [AIK], [Be]. In [Be] also Wilson's loops variables are shown to exist, by adapting the method of [T] under suitable assumptions on the Lévy function ψ. In [AIK2,3] invariant fields of the above type on a 2-dimensional Riemann manifold, where Euclidean invariance is replaced by conformal invariance are investigated, and an interesting connection with the theory of automorphic forms has been found.

b) We briefly mention that a lattice version of the model has also been studied, cfr. [AHK], [Sch]. In this version $\mathbb{R}^4 \equiv H$ is replaced by $\Lambda_\delta \equiv \delta\mathbb{Z}^4 \cap \Lambda, \Lambda$ a bounded regular subset of $\mathbb{R}^4$, for some $\delta > 0$. The source field $F(x), x \in \mathbb{R}^4$ is replaced by a source field $F_\delta(\delta n)$, $\delta n \in \Lambda_\delta$ the coordinate process with underlying probability measure of the form

$$dP_{\Lambda_\delta}(q) = (Z_\delta^{-1})^{|\Lambda_\delta|} \cdot e^{-W_\delta(q)} \Pi dq(\delta n),$$

with $q : \Lambda_\delta \to \mathbb{R}$,

$$W_\delta(q) \equiv \sum_{\delta u \in \Lambda_\delta} \delta^4 f_\delta(q(\delta n)),$$

for a suitable choice of functions $f_\delta : \mathbb{R}^4 \to \mathbb{R}^+$, called 'admissible density functions'. Examples of such functions are given in [Sch] (and in [Be] for the corresponding 2-dimensional analogue). Defining suitably a discrete analogue ∂_δ of the Cauchy-Riemann operator ∂ and a discrete analogue $A_\delta(\delta n)$ of $A(x)$ one can show that

$$E[e^{i<\varphi, A_\delta>}] \to E[e^{i<\varphi, A>}]$$

for all $\varphi \in C_0^\infty(\mathbb{R}^4; H)$, in the Gaussian case or in the non Gaussian case with $\psi(\lambda) = -|\lambda|^p$ for $\frac{4}{3} < p \leq 2$. This then gives a certain control on the limit continuum field A through its lattice counterpart A_δ, exhibiting also rather explicitly the probability law of A_δ. Let us also recall that a Poisson approximation of the field A in the Gaussian case has also been briefly pointed out in [AHKI], together with its possible physical applications.

Acknowledgements:

The first named author would like to thank Professors Jaroslav Dittrich and Pavel Exner for a very kind invitation to a most stimulations workshop. We are very grateful to Regine Kirchhoff, Carsten Welge and Bozena Zegarlinski for their skilfull help with the setting of the manuscript. The editors must be also thanked for their great understanding and in particular Pavel Exner for preparing the manuscript for publication. Financial support of the DFG (SFB 237 Bochum-Essen-Düsseldorf and BiBoS Projects) as well as of the Swedish Natural Science Research Council, the Swedish National Board for Technical Development, and the Royal Swedish Academy of Sciences are gratefully acknowledged.

References

[AHK] S.Albeverio, R.Høegh-Krohn: Construction of interacting local relativistic quantum fields in four space-time dimensions, Phys. Lett. B200, 108–114 (1988), with erratum in *ibid.* 202, p. 621 (1988)

[AHK1] S. Albeverio, R. Høegh-Krohn: Dirichlet forms and diffusion processes on rigged Hilbert spaces, Z. Wahrsch. verw. Geb. 40, 1–57 (1977)

[AHK2] S. Albeverio, R. Høegh-Krohn: Hunt processes and analytic potential theory on rigged Hilbert spaces, Ann. Inst. H. Poincaré (Probability Theory) B13, 269–291 (1977)

[AHK3] S. Albeverio, R. Høegh-Krohn: Quaternionic non abelian relativistic quantum fields in four space-time dimensions; Phys. Letts. B189, 329–336 (1987)

[AHKH] S.Albeverio, R.Høegh-Krohn, H.Holden: Markov cosurfaces and gauge fields. Acta Phys. Austr. Suppl. XXVI, 211-231 (1984)

[AHKHK1] S. Albeverio, R. Høegh-Krohn, H. Holden, T. Kolsrud: Construction of quantised Higgs like fields in two dimensions , Phys. Lett. B. 222 (1989), 263-268

[AHKHK2] S. Albeverio, R. Høegh-Krohn, H. Holden, T. Kolsrud: Representation and construction of multiplicative noise; J. Funct. Anal. 87, 250-272 (1989)

[AHKI] S. Albeverio, R. Høegh-Krohn, K. Iwata: Covariant Markovian random fields in four space-time dimensions with nonlinear electromagnetic interaction; pp.69–83 in Proc. Dubna Conf. 1987, Edts. P. Exner, P. Seba, Lec. Notes in Physics 324, Springer, Berlin (1989)

[AHKRö] S. Albeverio, R. Høegh-Krohn, M. Röckner: Canonical quantum fields and infinite dimensional Dirichlet processes, in preparation

[AHKZ] S. Albeverio, R. Høegh-Krohn, B. Zegarlinski: Uniqueness and global Markov property for Euclidean fields: The case of general polynomial interactions; Commun. Math. Phys. 123, 377–424 (1989)

[AIK1] S. Albeverio, K. Iwata, T. Kolsrud: Random fields as Solutions of the Inhomogenous Quaternionic Cauchy-Riemann Equation. I. Invariance and Analytic Continuation . Commun. Math. Phys. 132, 555-580 (1990)

[AIK2] S. Albeverio, K. Iwata, T. Kolsrud: Conformally invariant random fields and processes – Old and New; Stockholm Preprint 1990, to appear in Proc. Lisboa Conf. "Stochastic Analysis", Ed. A.B. Cruzeiro, J.C. Zambrini, Birkhäuser, New York (1991)

[AIK3] S. Albeverio, K. Iwata, T. Kolsrud: Conformally invariant and reflection positive random fields in two dimensions; Stockholm Preprint 1990, to appear in Volume dedicated to M.Zakai, Academic Press (1991)

[AIK4] S. Albeverio, K. Iwata, T. Kolsrud: Homogeneous Markov generalized vector fields and quantum fields over 4-dimensional space-time; SFB 237-Preprint, to appear in Proc. Trento Conf., Ed. Da Prato, Tubaro (1991)

[AK] S. Albeverio, S. Kusuoka: Maximality of infinite dimensional Dirichlet forms and R. Høegh-Krohn's model of quantum fields; to appear in Memorial Volume for Raphael Høegh-Krohn, Cambridge Univ. Press (1991)

[ARö1] S. Albeverio, M. Röckner: New developments in theory and applications of Dirichlet forms; pp. 27–76 in Proc. 2nd Int. Conf. Ascona - Locarno - Como 1988, Ed. S. Albeverio, G. Casati, U. Cattaneo, D. Merlini, R. Moresi, World Scient. (1990)

[ARö2] S. Albeverio, M. Röckner: Stochastic differential equations in infinite dimension: solutions via Dirichlet forms; SFB 237 - Preprint, to appear in Prob. Th. Rel. Fields (1991)

[ARöZ] S. Albeverio, M. Röckner, T. S. Zhang: Girsanov transform of symmetric diffusion processes in infinite dimensional space; SFB Preprint (1991)

[AZ] S. Albeverio, B. Zegarlinski: A survey on the global Markov property in quantum field theory and statistical mechanics, to appear in Memorial Volume for Raphael Høegh-Krohn, Cambridge Univ. Press (1991)

[Be] C. Becker, Diplomarbeit, in preparation, Bochum (1991)

[BoI] M.Born, L.Infeld: Foundations of the new field theory; Proc. Roy. Soc. A 144, 425-451 (1934)

[D] B. Driver: YM_2: Continuum expectation, lattice convergence and lassos; Comm. Math. Phys. 123, 576-616 (1989)

[F] D.S.Fine, Quantum Yang–Mills on a Riemann surface; submitted to

Commun. Math. Phys. (1991)

[GeP] I. Georgescu, R. Purice: On the Markov property for the free Euclidean electromagnetic field; Letts. Math. Phys. 4, 465–467 (1980)

[GJ] J. Glimm, A. Jaffe: Quantum physics, a functional integral point of view; 2nd edition. Springer, Berlin (1987)

[GKS] L.Gross, C.King, A.Sengupta: Two dimensional Yang-Mills theory via stochastic differential equations; Ann. Phys. 194, 65–112 (1989)

[Gu] F. Guerra: Local algebras in Euclidean quantum field theory; pp 13–26, Symp. Math. XX, Academic Press, New York (1976)

[I1] K. Iwata: On linear maps preserving Markov properties and applications to multicomponent generalized random fields; Ph D Thesis, Bochum (1990)

[I2] K. Iwata: The inverse of a local operator preserves the Markov property, Bochum, SFB 237-Preprint (1991)

[Lö] J. Löffelholz: The Markov property of the free electromagnetic field; Karl Marx Universität (1982)

[Ne] E. Nelson: The free Markov field; J. Funct. Anal. 12, 211–227 (1973)

[Os1] E. P. Osipov: Two–dimensional random fields as solution of stochastic differential equation; Bochum Preprint (1990)

[Os2] E. P. Osipov: Markov properties of solutions of stochastic partial differential equations, Novosibirsk Preprint (1989)

[Os3] E. P. Osipov: Euclidean Markov fields from stochastic partial differential equations in eight–dimensional space, Novosibirsk Preprint (1987)

[RöZ] M. Röckner, T.S.Zhang: On uniqueness of generalized Schrödinger operators and applications, Preprint (1991)

[Sch] M. Schmitt: Die Gitterapproximation nicht–linearer elektromagnetischen Felder in vier Dimensionen, Diploma Thesis, Bochum (1991)

[Sei] E. Seiler: Gauge theories as a problem of constructive quantum field theory and statistical mechanics; Lect. Notes Phys. 159, Springer 1982

[Si] B. Simon: The $P(\varphi)_2$ Euclidean (quantum) field theory; Princeton Univ. Press 1974

[T] H.Tamura: On the possibility of confinement caused by nonlinear electromagnetic interaction; Kanazawa preprint 1990, to appear in J. Math. Phys.

ON POTENTIAL CONTINUITY IN STOCHASTIC SPECTRAL ANALYSIS

MICHAEL DEMUTH
Karl-Weierstraß-Inst. für Mathematik
Mohrenstr. 39
Berlin 1086, Germany

and

JAN VAN CASTEREN
Dept. of. Math., Univ. Inst. Antwerpen
Universitetsplein 1
Wilrijk 2610; Belgium

This article is dedicated to the late Dr. Jiri Blank. The first author mourns his death at both the personal and professional levels of their relationship. We will miss his support for the development of mathematical physics.

ABSTRACT

The spectral behaviour of perturbed Feller generators is considered if the potential varies in Kato-Feller-norm.

1. Introduction

Stochastic spectral analysis means the spectral theory for Feller generators K_0 using essentially the stochastic process theory. Its main feature is, that the form of K_0 is not fixed. One starts with the properties of a transition density function, i.e. one starts with introducing certain semigroups. That allows a large variety of applications to very general K_0.

Here we consider only the behaviour of some spectral data for $K_0 + V$, where V is a Kato-Feller-potential, if V varies in the Kato-Feller-norm. The operator norm convergence of resolvent differences implies the continuity of wave operators, scattering operators, and scattering matrices.

The present article is based on [1] and [2]. More detailed information and all the proofs are given there. Because the number of pages is limited here, we will give only some necessary definitions and explain some results.

2. Feller generators and Kato-Feller-potentials

We start with a function p_F having all the properties of a transition density function of a Feller process.

Assumption 1: Assume a function $p_F : [0,\infty) \times \mathbb{R}^n \times \mathbb{R}^n \to [0,\infty)$, satisfying the following properties:

Existence: Let $0 \leq p_F(t,x,y) < ct^{-m}e^{bt}$

with an arbitrary $m \in \mathbb{N}$ and positive constants b, c.

Chapman-Kolmogorov-equation: Let

$$\int_{\mathbb{R}^n} p_F(t,x,u)p_F(s,u,y)du = p_F(t+s,x,y).$$

Symmetry: Let $p_F(t,x,y) = p_F(t,y,x)$.

Feller property: Let

$$x \to \int_{\mathbb{R}^n} f(y)p_F(t,x,y)dy \quad \text{be continuous}$$

for $f \in L^\infty(\mathbb{R}^n)$.

C_0-property: Let

$$\lim_{t\to 0} \int f(y)p_F(t,x,y)dy = f(y)$$

for all continuous and bounded f. □

Then the corresponding semigroup of p_F in $L^2(\mathbb{R}^n)$ has a selfadjoint generator K_0. Assumption 1 allows transition density functions of many processes. That means restrictions of K_0 are given by very general differential expressions. The most prominent example is the Laplacian where p_F is the Wiener density function. For canonical diffusion processes one has generators of the form $\sum_{i,j} \frac{\partial}{\partial x_i}(a_{ij}(x)\frac{\partial}{\partial x_j})$ with bounded diffusion matrix elements. But also unbounded diffusion coefficients can be allowed. For n = 1 the solution theory of stochastic differential equations allows $a(x) = 1 + x^2$. Possible further K_0 (e.g. pseudo-differential operators) are treated in [2].

Kato-Feller-potentials $V: \mathbb{R}^n \to \mathbb{R}$ are introduced by the condition

$$\lim_{t\to 0} \sup_x \int_0^t ds \int_{\mathbb{R}^n} p_F(s,x,y)|V(y)| \, dy = 0.$$

This class is natural from the stochastic point of view. In particular, it follows that there is a closed selfadjoint operator extending $K_0 + V$ which generates a semigroup represented by a generalized Feynman-Kac-formula.

Having Kato-Feller-potentials one is able to introduce the Kato-Feller-norm:

$$\|V\|_{KF} = \sup_x \int_0^1 ds \int_{\mathbb{R}^n} p_F(s,x,y)|V(y)|dy.$$

In the final section we will announce spectral theoretical results if a sequence V_α tends to a V_{α_0} with respect to the Kato-Feller-norm as $\alpha \to \alpha_0$. (In the Wiener case $\|V\|_{KF}$ contains all L^p-convergences with $p \geq \frac{n}{2}$.).

3. Continuity of scattering data

Assume again a function p_F given in Assumption 1, with the corresponding generator K_0. Let V_α, V_{α_0} be Kato-Feller-potentials. For this set we suppose a stability condition:

$$\sup_x E_x\{\exp(-\int_0^t V_\alpha(\omega(s))ds)\} \leq Be^{At}$$

where the positive constants A, B are assumed to be independent of α.

Theorem 2: Let $|V_\alpha| \leq |W|$ and $|V_{\alpha_0}| \leq |W|$ such that the Cook-criteria are satisfied for $\|We^{-isK_0}f\|$, $f \in P_{ac}(K_0)\mathcal{H}$, $\mathcal{H} = L^2(\mathbb{R}^n)$. As usual we define the wave operators

$$\Omega_\pm(K_\alpha,K_0) := \underset{t\to\pm\infty}{\text{s-lim}}\ e^{itK_\alpha}e^{-itK_0}P_{ac}(K_0)$$

where $K_\alpha := K_0 + V_\alpha$ and in the same way $\Omega_\pm(K_{\alpha_0},K_0)$. Assume now $\lim_{\alpha\to\alpha_0} \|V_\alpha - V_{\alpha_0}\|_{KF} = 0$, then follows

$$\underset{\alpha\to\alpha_0}{\text{s-lim}}\ \Omega_\pm(K_\alpha,K_0) = \Omega_\pm(K_{\alpha_0},K_0),$$

which implies also the weak convergence of the corresponding scattering operators. □

Theorem 3: Assume the existence of wave operators for the scattering systems $\{K_0,K_\alpha\}$, $\{K_0,K_{\alpha_0}\}$. All scattering operators commute with K_0. Denote the scattering matrices with $S_\alpha(\lambda)$, $S_{\alpha_0}(\lambda)$.

Let φ be a positive function, such that

$$[(K_0+a)^{-1}\varphi^4](x) \leq c\,\varphi^4(x)$$

(pointwise where a is assumed to be sufficiently large),

$$\|\varphi^{-1}(K_0+V_{\alpha_0} - \lambda\pm i0)^{-1}\varphi^{-1}\| < \infty.$$

Let V_α, V_{α_0} be bounded operators in $L^2(\mathbb{R}^n)$ such that

$$\lim_{\alpha\to\alpha_0} \|V_\alpha - V_{\alpha_0}\|_{L^2} = 0,$$

$$\lim_{\alpha\to\alpha_0} \|(V_\alpha - V_{\alpha_0})\varphi^4\|_{KF} = 0.$$

Then we have for the operatornorm convergence of the scattering matrices

$$\lim_{\alpha \to \alpha_0} \| S_\alpha(\lambda) - S_{\alpha_0}(\lambda) \| = 0. \qquad \square$$

The proofs of these two theorems are based on detailed estimates of the resolved differences

$$(K_0+V_\alpha-z)^{-1} - (K_0+V_{\alpha_0}-z)^{-1}$$

as $V_\alpha \to V_{\alpha_0}$ in the Kato-Feller-norm. For Theorem 2 regular z with large Rez are sufficient. For Theorem 3 one has to study the limit absorption case. The detailed proofs are given in [2].

References

1. M. Demuth and J. van Casteren, On spectral theory for selfadjoint Feller generators, Reviews in Math. Phys. 1 (1989), 325-414
2. M. Demuth and J. van Casteren, On spectral theory for selfadjoint Feller generators, Part II, (to appear)

Miscellaneous

EXACT RESULTS IN THE STATISTICAL MECHANICS OF INTEGRABLE MODELS IN (1+1) DIMENSIONS

R. K. BULLOUGH
LAMF, The Technical University of Denmark,
DK-2800 Lyngby, Denmark, and
Department of Mathematics, UMIST, P.O.Box 88,
Manchester M60 1QD, U.K.

D.J. PILLING, YU-ZHONG CHEN
Department of Mathematics, UMIST, P.O. Box 88,
Manchester M60 1QD, U.K

and

J. TIMONEN
Department of Physics, University of Jyväskylä
P.O. Box 35, 40351 Jyväskylä, Finland

ABSTRACT

A method of functional integration for the partition function Z of integrable models in 1+1 dimensions is given. The method exploits the integrability of the models, classical or quantum, and makes a canonical transform on the classical action S to an expression of this action in terms of action–angle variables. Additional constraints are found which allow the functional integral to be evaluated in thermodynamic limit at finite density. The quantum cases prove to be distinguished from the classical cases. Breather solutions contribute as phonons in the classical free energies but as a finite number of distinct fermion like quantum breathers in the quantum free energy.

1. Introduction

This paper, written (we hope and believe) very much in the spirit of this meeting is a restatement of previous work [1–12] concerning exact results found for the *quantum and classical* statistical mechanics of two model dynamical systems – the sine–Gordon (s–G) model

$$\phi_{xx} - \phi_{tt} = m^2 \sin \phi, \tag{1}$$

where ϕ is a real field, m is a "mass", $\phi_{xx} \equiv \partial^2\phi/\partial x^2$ etc., and the Nonlinear Schrödinger (NLS) model

$$-i\phi_t = \phi_{xx} - 2c\phi^*\phi^2, \tag{2}$$

where ϕ is a complex field and c is a real "coupling constant".

This meeting is concerned with 'Rigorous results in quantum dynamics'. The work we describe in its quantum theories concerns exact results in the quantum dynamics of model systems like (1) and (2) in one space and one time (1+1) dimensions in equilibrium with a heat bath with fixed (finite) temperature.

We note too that we are concerned with *exact* results. The level of rigour we adopt is below much of that as exemplified by other papers at this meeting. However, and in a very complicated theoretical situation where rigour would be almost impenetrable to readers beyond the authorship, the results we achieve are certainly *exact*. And in this lies their interest.

After the discovery of the 'inverse scattering method' for solving the classical Korteweg–de Vries (KdV) equation

$$u_t + 6uu_x + u_{xxx} = 0 \tag{3}$$

by Kruskal and co-workers in 1967 [13,14], large numbers (infinitely large numbers) of 'integrable' nonlinear wave equations have been solved by this method in 1+1 dimensions, and there are extensions of the method to 2+1 dimension [15]. We say something about 'integrable' (classical) models and 'integrability' later. We remind the reader that amongst the classical integrable models in (1+1) are the s–G model eqn.(1) and the sinh–Gordon (sinh–G) model

$$\phi_{xx} - \phi_{tt} = m^2 \sinh \phi; \tag{4}$$

the *attractive* (c<0) and *repulsive* (c>0) NLS models eqn.(2); the Landau–Lifshitz $(L - L')$ model

$$\vec{S}_{,t} = \vec{S} \times \vec{S}_{,xx} + \vec{S} \times (\vec{J} \cdot \vec{S}); \tag{5}$$

the Heisenberg ferromagnet

$$\vec{S}_{,t} = \vec{S} \times \vec{S}_{,xx} + \vec{B} \times \vec{S} \tag{6}$$

(where $\vec{B}$ is a longitudinal field); the Toda lattice

$$\ddot{Q}_n = \frac{ab}{M}\left[e^{-(Q_{n+1}-Q_n)} - e^{-(Q_n-Q_{n-1})}\right] \tag{7}$$

and the classical 'massive Thirring' model (MTM)

$$(-\partial_\mu\gamma^\mu + m_0)\psi + g^2\gamma^\mu\psi(\bar{\psi}^\gamma{}_\mu\psi) = 0. \tag{8}$$

For (8) $\gamma^\mu(\mu = 0,1)$ are Pauli matrices; $\psi = \binom{\psi_1}{\psi_2}$; $\bar{\psi} = \psi^*\gamma^0$; and $\{\psi_i(x), \bar{\psi}_j(x')\}_\pm$ $= \delta_{ij}\delta(x - x')$, $\{\psi^*(x), \bar{\psi}(x')\}_\pm = \delta(x - x')$. Here $\{\cdot,\cdot\}_\pm$ means anticommuting and commuting Poisson brackets, respectively. To these models we certainly add KdV, eqn.(3), if only because this model started everything off! However, all of the models (1), (2), (4)–(8) have obvious interest to physicists. The KdV is concerned with water wave theory and a very interesting history [16] but we know of no place where the *quantised* KdV is manifestly relevant so we shall not treat it further in this paper.

In fact although all the other models have been solved for their free energies by the methods sketched in this paper we restrict attention to s–G, sinh–G, and the NLS models. The point is that the *methods* sketched for the free energy F (partition function Z) of these four models in this paper can, and have [10,11,17,18], been applied successfully to the other models and exact results obtained.

2. Integrability

All the models cited, and there are many many more, are classical *integrable* models: they are *completely integrable* in the Liouville–Arnold sense [19]. However the theorem [19] that a Hamiltonian system of N degrees of freedom is completely integrable when there exist N independent first integrals I(k) (say) commuting under the Poisson bracket, i.e. $\{I(k),\ I(\ell)\} = 0$, is proved for the case N finite. A new feature is that this theorem for $N<\infty$ degrees of freedom must now be extended to a continuously infinite number of degrees of freedom because the *functions* $\phi(x,t)$ are fields with co-ordinates $\phi(x,t)$ labelled by $x \in R$. To our knowledge no rigorous result is proved in this case. For repulsive lattices a "fairly rigorous" result can be found in limit $N \rightarrow \infty$ for N lattice points in L. For s–G etc. (attractive cases) there is a problem of keeping the count as the phonons go out (see below) and the breathers with two degrees of freedom replace them. However, the classical models we consider are all Hamiltonian: they have Poisson brackets $\{\cdot,\cdot\}$; *and* they have enough independent first integrals $I(k)$ in 'involution':

$$\{I(k),\ I(\ell)\} = 0, \tag{9}$$

where $k,\ \ell\ \in\ R$ – a result which is usually proved by explicit integration of the equations of motion.

Amongst the $I(k)$ is the classical Hamiltonian $H(k)$. The mapping to this k–space description is achieved through the spectral transform of the inverse method [20], and $H(k)$ can be expressed in action–angle form as a functional of action variables alone. Examples of these "diagonal" Hamiltonians will be given later in this Section. If the manifold of level lines $\{I(k) = \text{constant}|\ k \in R\}$ is compact and connected, motions lie on tori or *generalised* tori: for the former a continously infinite number of canonical pairs of constant action variables $P(k)$ $(0 \leq P(k) < \infty)$, and angle variables $Q(k)$ $(0 \leq Q(k) < 2\pi)$, can be found, and the $Q(k)$ evolving in time define the tori. If there are soliton solutions one finds in addition p_k, q_ℓ; $k, \ell = 1, 2, ..., N_s$, such that $\{p_k, q_\ell\} = \delta_{k\ell}$ and $-\infty < p_k, q_\ell < \infty$: the generalised tori defined by q_ℓ have infinite period.

The spectral transform and its inverse spectral transform are canonical transformations which preserve the symplectic structure (the Poisson brackets). If the original field variable $\phi(x,t)$ is real, i.e. $\phi(x,t) \in R$, there is a canonical momentum $\Pi(x,t)$ and

$$\{\Pi(x,t),\ \phi(x',t)\} = \delta(x - x'). \tag{10}$$

The spectral transform then provides "enough" first integrals – namely the action variables $P(k)$, $k \in R$, together with the $p_i; i = 1, ..., N_s$, if there are also soliton solutions. For field variables $A[\Pi, \phi]$ and $B[\Pi, \phi]$ (say) the equal time Poisson bracket can be defined as

$$\{A, B\} = \int dx \left(\frac{\delta A}{\delta \Pi}\frac{\delta B}{\delta \phi} - \frac{\delta A}{\delta \phi}\frac{\delta B}{\delta \Pi}\right), \tag{11}$$

then $\{\Pi, \Phi\} = \delta(x - x')$. Both the s–G and sinh–G models have this bracket. However in the NLS models $\phi \in C$ and it is convenient to take as canonical variables $\phi(x,t)$ and $\phi^*(x,t)$. Then

$$\{\phi(x,t),\ \phi^*(x',t)\} = i\delta(x - x'). \tag{12}$$

A slightly more complicated situation arises in the case of the $L - L'$ model or the Heisenberg ferromagnet for which the Poisson bracket structure reflects the algebra of spin operators.

There is an obvious extension of these classical results to quantum theory by following Dirac and replacing the Poisson brackets by commutators: $\{\cdot,\cdot\} \to i[\cdot,\cdot]$ $(\hbar = 1)$. For a well defined ground state of the quantum model the operators of the model must be placed in normal order. For the two NLS models the Heisenberg's equations of motion become

$$-i\phi_t = \phi_{xx} - 2c\phi^+\phi^2 \tag{13}$$

with commutation relation $[\phi,\ \phi^+] = \delta(x - x')$ and $c > 0$ or $c < 0$. The structure of the s–G model is more complicated, and the Heisenberg's equation of motion for it is

$$:\phi_{xx} - \phi_{tt} = m^2 \sin\phi: \tag{14}$$

with commutation relation $[\Pi, \phi] = -i\delta(x - x')$. Here colons around the equation means normal order. The corresponding equation of motion for the sinh–G model can be found from (14) by replacing $\sin\phi \to \sinh\phi$.

The classical integrable models can also be quantised algebraically via the so–called *quantum groups* [17,21]. This method of quantisation is most easily introduced by working first of all with the "little" r–matrix formulation [22] of the classical *integrable lattices* (and see references in[10]). There are also other reasons to introduce integrable lattice versions of the classical integrable models. The partition function of a field either classical or quantum means calculating a functional integral, and we need to discretize the functional integral both to define it and to evaluate it. We may also need to negotiate an ultra-violet divergence (classical or quantum) and we may have an infra-red divergence: the finite mass $m>0$ eliminates the latter for the s-G model [1-12]. For the ultra-violet divergence it is obviously convenient to work with an integrable lattice, spacing a, and let $a \to 0$ after renormalisation if necessary. Furthermore, there is a problem of a proper thermodynamic limit: this is taken at finite density and it is convenient to impose periodic boundary conditions (b.c.s.) period L (say) and for N particles in L, $\lim_{L\to\infty} L^{-1}N = \bar{n}>0$. This thermodynamic limit is most easily carried out on a lattice.

Consider now a lattice with $2N$ points in a period L, $L = 2Na$. A "transfer matrix" $M_n \equiv M(n+1,n)$, a local monodromy matrix, acts on a lattice variable v_n so that

$$v_{n+1} = M_n v_n. \tag{15}$$

The elements of M_n are functions of the discretized field variables ϕ_n, Π_n (or ϕ_n, $\phi_n{}^*$) of the lattice model to be solved: M_n plays the role of the Lax operator for a continuous field $\phi(x,t)$, and for lattice forms of the integrable models of $AKNS$ type [20] like s-G, sinh-G and the NLS models, M_n is a 2×2 matrix. The global monodromy matrix T is then

$$T = \overleftarrow{\prod_n} M_n = M_{N-1}...M_{-N+1}M_{-N} \equiv \begin{pmatrix} A(\lambda) & B(\lambda) \\ C(\lambda) & D(\lambda) \end{pmatrix}, \tag{16}$$

where $\lambda \in C$ is the spectral parameter. Notice that the elements of T are also functionals of $\{\phi_n | -N \le n \le N-1\}$. It proves important that, following from the symmetry properties of the models, one finds $D = A^*$, $B = \pm C^*$, and therefore $detT = |A|^2 \mp |B|^2 (=1)$.

Discrete versions of the Poisson bracket relations (11) on the ϕ_n induce sixteen bracket relations on the four elements of $T = T(\lambda)$. One can introduce the classical r-matrix $r(\lambda,\mu)$ such that these bracket relations can be expressed in a compact form as [22]

$$\{T(\lambda) \overset{\otimes}{,} T(\mu)\} = [T(\lambda)\otimes T(\mu), r(\lambda,\mu)]. \tag{17}$$

As usual $\otimes$ is Kronecker product: $T(\lambda)\otimes T(\mu)$ is a 4×4 matrix and the right side of (17) is the usual matrix commutator. The left side is simply the corresponding matrix of Poisson brackets. If we define

$$\Delta(\lambda) \equiv trT(\lambda) \tag{18}$$

then one finds from (17) that

$$\{\Delta(\lambda),\ \Delta(\mu)\} = 0. \tag{19}$$

Since $\lambda,\mu \in C$ (simply), (19) is a very large number of commuting constants. A lattice with $2N$ lattice points in a period $L = 2Na$ has $2N<\infty$ degrees of freedom. However, the lattice is constructed for the continuous field $\phi(x,t)$ and the r-matrix is really associated with this integrable field. Thus (19) is the integrability condition for the continously infinite number of degrees of freedom of this field $\phi(x,t)$. The lattice which becomes the integrable field $\phi(x,t)$ as the lattice spacing $a \to 0$ is in general not integrable. However it is possible to construct an integrable lattice which becomes the integrable field $\phi(x,t)$ as $a \to 0$ and both lattice and field have the same r–matrix [22] and the brackets (17).

The result (19) means also that both $\Delta(\lambda)$ and $\ell n\Delta(\lambda)$ are generators of classical Hamiltonians. It may be helpful to the reader to take a specific example, and so we give here some of the details for a sinh–G lattice given as an integrable lattice with the r–matrix of the integrable sinh–G field $\phi(x,t)$. An integrable s–G lattice was reported in [23] and a sinh–G lattice is obtained by replacing ϕ_n by $i\phi_n$: $\phi_n \to \phi(x)$, $x = na$, in continuum limit. In the subsequent development certain changes of normalisation are convenient[24] as compared with[23].

The chosen integrable sinh–G lattice has the monodromy matrix M_n,

$$M_n = (C^+C^-)^{-\frac{1}{2}}\bar{M}_n;\ \bar{M}_n = \begin{pmatrix} p_n^-\phi_n & S^{\frac{1}{2}}(\lambda\chi_n^- - \frac{1}{\lambda}\chi_n^+) \\ S^{\frac{1}{2}}(\frac{1}{\lambda}\chi_n^- - \lambda\chi_n^+) & p_n^+\phi_n \end{pmatrix}, \tag{20}$$

where $S = (\frac{1}{4}ma)^2$, $S^{\pm} = 1 + S \perp iS^{\frac{1}{2}}(\lambda - \lambda^{-1})$; $\chi_n^{\pm} = exp(\pm\frac{1}{2}u_n)$ and $p_n^{\pm} = \exp(\pm\gamma_0 p_n)$ in which $\gamma_0>0$ is the "coupling constant". It is a coupling constant of the classical sinh–G model and will appear in the Hamiltonian. It may be removed in the classical case but not in the quantum case: in practice it pays to keep it in both cases.

The canonical variables of this integrable sinh–G lattice are u_n and the conjugate momenta p_n : $\{p_n, u_{n'}\} = \delta_{nn'}$. Because of the complicated form of the actual lattice Hamiltonian it has been convenient in M_n to introduce also the functions $p_n^{\pm}$, $\chi_n^{\pm}$ and ϕ_n of the fundamental canonical variables p_n and u_n:

$$\phi_n = (1 + S^2 + S[(\chi_n^+)^2 + (\chi_n^-)^2])^{1/2}. \tag{21}$$

Because of the normalisation we have chosen we shall in fact use $\ell n(z^{2N+1}\Delta(\lambda))$ as the generator of the classical Hamiltonian. Here $z = C^+/C^-$ plays a role as complex variable (generalised spectral parameter) in the analysis. For fixed N the Hamiltonian is $H_N = \bar{H}_N - \bar{P}_N$ where

$$\bar{H}_N = \frac{2m}{i\gamma_0}\lim_{\lambda\to iS^{\frac{1}{2}}}\frac{\partial}{\partial\lambda}\ell n(z^{2N+1}\Delta(\lambda)) \tag{22a}$$

and

$$\bar{P}_N = \frac{2m}{i\gamma_0}\lim_{\lambda\to iS^{-\frac{1}{2}}}\frac{\partial}{\partial\lambda^{-1}}\ell n(z^{2N+1}\Delta(\lambda)). \tag{22b}$$

Without going into a detailed account of the actual calculation we merely notice that, when $a \to 0$ with $x \equiv na$, $u_n \to u(x)$ and $p_n \to ap(x)$, then $H_N \to H[u]$ where

$$H[u] = \gamma_0^{-1} \int dx [\frac{1}{2}\gamma_0^2 p^2 + \frac{1}{2} u_x^2 + m^2(\cosh u - 1)]. \tag{23}$$

If ϕ replaces u then this is the Hamiltonian $H[\phi]$ of the classical integrable sinh–G model. Hamilton's equations are now exactly eqn. (4).

We can find action–angle variables under *periodic* b.c.s. for the *lattice* with Hamiltonian (22). These action–angle variables are complicated. But in the thermodynamic limit $N \to \infty$ matters simplify considerably. In this limit we find [24] in fact that H_N of (22) can be expressed in action–angle form: $H_N = H[p]$ where

$$H[p] = \sum_{n=-N}^{N} \omega(\tilde{k}_n) P_n + \mathcal{O}(L^{-1}) \tag{24}$$

with $\omega(k) = (m^2 + k^2)^{\frac{1}{2}} + \mathcal{O}(a)$. When $L \to \infty$ (see below) $H[p] \to \int \omega(k) P(k) dk$ and the action variables $P(k)$ are formally the same as those more familiar ones found for b.c.s. decaying fast enough at $\pm\infty$. The modes $\tilde{k}_n$ are suitable phonon modes defined in terms of the generalised spectral parameter z essentially as $z = exp(i\tilde{k})$ which go dense as $L \to \infty$.

The first surprise is that Hamiltonian (24) seems to describe, in action–angle form, a collection of $2N$ (actually and conveniently $2N + 1$ but this change has no consequences) independent harmonic oscillators. The appearance of this Hamiltonian is however deceptive because the modes $\tilde{k}_n$ are not independent: they *depend on the action variables* P_n, so that the oscillators are in fact nonlinear (as they should be). We can find the conditions which the allowed modes $\tilde{k}_n$ have to satisfy by imposing periodic b.c.s. on the sinh–G lattice. Periodic b.c.s. mean that the auxiliary "scattering problem" (15) is that for a periodic potential. We can therefore apply Floquet theory to this problem with the result in the $L \to \infty$ limit [1–3,18,24] that

$$\tilde{k}_n = k_n - L^{-1} \sum_{n' \neq n} \Delta_c(\tilde{k}_n, \tilde{k}_{n'}) P_n. \tag{25}$$

Here the k_n result from a "standing wave" condition and are $k_n = 2\pi n L^{-1}, n$ an integer. The same analysis also explicitly determines the phase shifts Δ_c and for the classical sinh–G model this proves to be [1,3,18,24]

$$\Delta_c(k, k') = -\frac{1}{4} \frac{m^2 \gamma_0}{k\omega(k') - k'\omega(k)}. \tag{26}$$

We note in passing that a similar analysis can be made for the *repulsive* NLS model with analogous results: the $\omega(k)$ of sinh–G is replaced by $\omega(k) = k^2$ and (26) is replaced by[25]

$$\Delta_c(k, k') = -2c(k - k')^{-1}. \tag{27}$$

We still have to work out with proper care the expression for the Hamiltonian (24) in the actual thermodynamic limit $L \to \infty$. In this limit the *allowed* modes $\tilde{k}_n$

become dense and $P_n \to P(\tilde{k})d\tilde{k}$. Now $P_n = \mathcal{O}(1)$ for the sinh–G lattice, and since $d\tilde{k}_n = \mathcal{O}(L^{-1})$, it follows that $P(\tilde{k}) = \mathcal{O}(L)$ as $L \to \infty$. It is at this point that a significant difference from the usual theory of action–angle variables for decaying b.c.s. on R appears. Because $P(\tilde{k}) = \mathcal{O}(L)$ we can define a finite density $\rho(k)>0$ for phonons as

$$\rho(k) \equiv \lim_{L\to\infty} L^{-1}P(k). \tag{28}$$

We find therefore in the thermodynamic limit [6] that

$$\lim_{L\to\infty} L^{-1}H[p] = \int_{-\infty}^{\infty} dk\rho(k)\omega(k). \tag{29}$$

There is also an equation which involves the density of states in the $\tilde{k}$–space$f(\tilde{k})$: $f(\tilde{k}) = (2\pi)^{-1}dk/d\tilde{k}$ so from (25)

$$f(k) = (2\pi)^{-1} + (2\pi)^{-1} \int \frac{d}{dk}\Delta(k,k')\rho(k')dk'. \tag{30}$$

The phonon density can be easily recognised as the density of occupied phonon states amongst the possible states $f(\tilde{k})$. These two densities are connected via eqn. (30), an expression which also determines the zero temperature properties of the model. In a fermion description (see below) $f(\tilde{k})$ and $\rho(\tilde{k})$ can be identified at $T = 0$. On the other hand, as we shall see in the next Section, at any finite temperature the entropy of the model is finally determined by the relation (30), and it is in this way that the thermal properties of the model become non–trivial i.e. non–harmonic.

Evidently we need to extend the previous arguments for models like s–G which have soliton solutions. The Hamiltonian of the s–G model is (23) with coshu-1 replaced by 1-cos$u$. The kink (antikink) soliton of the classical s–G model have energy $E = (M^2+p^2)^{\frac{1}{2}}$, $M = 8m\gamma_0^{-1}$ and p is the momentum. A similar lattice analysis as that of sinh–G shows that in thermodynamic limit the allowed momenta p_n are packed as $p_n = 2\pi n L^{-1}$ for integers n, and $L \to \infty$ makes the p_n dense. It is easy to show that in this thermodynamic limit the kink (antikink) contribution to the energy density becomes

$$\lim_{L\to\infty} L^{-1}E_{kinks} = \int_{-\infty}^{\infty} (M^2+p^2)^{\frac{1}{2}}\rho_s(p)dp, \tag{31}$$

where we have introduced a finite soliton (kink or antikink) density $\rho_s(p)$.

The phonons of s–G seem to behave very much like those of sinh–G, and the asymptotic lattice contribution of them to the energy density is (24). But there is one crucial difference. The normalisation condition on the monodromy matrix T of sinh–G is $detT = |A|^2 - |B|^2 = 1$ when for s–G it is $detT = |A|^2 + |B|^2 = 1$. This means that for s–G $|A|$ is bounded above and therefore the phonon action variables $P(k)$ are bounded above. Hence we cannot define a *finite* phonon density in thermodynamic limit for the phonons of s–G, and in this limit $H_{ph}[p]$ makes no contribution.

We still have to treat the breathers of classical s–G (and the similar soliton solutions of other integrable models such as attractive NLS eqn. (2)) in thermodynamic limit. Breather solutions have *two* degrees of freedom: for s–G in continuum limit the breather energies are $(4M^2 \sin^2\theta + p^2)^{\frac{1}{2}}$, where $4\gamma_0^{-1}\theta$, $0 \le \theta < \frac{\pi}{2}$, is the action variable for an internal oscillation mode. Canonical to this is the angle variable Φ, $0 \le \Phi < 8\pi$ [20]. Evidently we can write the breather energies in the discretized form

$$E_{breather} = \sum_n \sum_\ell \frac{2M \sin\theta_{n\ell}}{\sqrt{1 - v_n^2}}; \tag{32}$$

by using $p = 2Mv(1 - v^2)^{-\frac{1}{2}} \sin\theta$. The translational momenta p_n of breathers are packed in the limit $L \to \infty$ in the same way as the translational momenta of kinks, namely $p_n = 2\pi n L^{-1}$, and $\sum_n \to (2\pi)^{-1} L \int dp$ in (32). However there is the *double* sum in (32) consequent on the point that for every classical breather with translational momentum p_n there is a range $[0, \frac{\pi}{2})$ of values of its internal momentum $\theta_{n\ell}$. The *labels* ℓ must also go dense as $L \to \infty$ so we choose to pack this label in the same way as the label n so that $\sum_\ell \sim L \int d\theta$. This choice means that $\lim_{L\to\infty} L^{-1} E_{breather}$ diverges as $\mathcal{O}(L)$!

There is a way out of this nonphysical result – namely to let the θ's, as values, lie in a narrow angle $\mathcal{O}(L^{-1})$ densely. If we now define new variables $\bar{P}_n$, $\bar{Q}_n$ such that

$$\begin{aligned} \bar{P}_n &= 16\gamma_0^{-1} \sum_\ell \theta_{n\ell} = \mathcal{O}(1) \\ \bar{Q}_n &= (4L)^{-1} \sum_\ell \Phi_{n\ell};\ 0 \le \bar{Q}_n < 2\pi, \end{aligned} \tag{33}$$

the brackets $\{4\gamma_0^{-1}\theta_{n\ell}, \Phi_{n'\ell'}\} = \delta_{nn'}\delta_{\ell\ell'}$ mean that $\{\bar{P}_n, \bar{Q}_{n'}\} = \delta_{nn'}$ while (32) becomes equivalent to

$$H_{breather} = \sum_n \omega(k_n) \bar{P}_n, \tag{34}$$

since $\sin\theta_{n\ell} \sim \theta_{n\ell} = \mathcal{O}(L^{-1})$ as $L \to \infty$. Here $\omega(k) = (k^2 + m^2)^{\frac{1}{2}}$ and $k = mv(1 - v^2)^{-\frac{1}{2}}$. The somewhat ashtonishing conclusion is that for the s–G model the (bounded) phonon contributions to the Hamiltonian, which are eliminated in thermodynamic limit, are *replaced* by phonons of arbitrarily large amplitude (i.e. *nonlinear* phonons with $\bar{P}_n = \mathcal{O}(1)$) derived from the classical breathers. In other words, in thermodynamic limit the classical s–G breathers break up into interacting phonons. Because $\bar{P}_n = \mathcal{O}(1)$ and $\bar{P}_n \to \bar{P}(k)dk$, $\bar{P}(k) = \mathcal{O}(L)$, and we can define a finite phonon density $\rho(k) = \lim_{L\to\infty} L^{-1} \bar{P}(k)$ as for sinh–G. In thermodynamic limit therefore

$$\lim_{L\to\infty} L^{-1} E_{breather} = \int_{-\infty}^{\infty} \omega(k)\rho(k)dk, \tag{35}$$

and the total energy density of s–G becomes

$$\lim_{L\to\infty} L^{-1} E[p] = 2\int_{-\infty}^{\infty} (M^2 + p^2)^{\frac{1}{2}} \rho_s(p)dp + \int_{-\infty}^{\infty} \omega(k)\rho(k)dk. \tag{36}$$

The factor 2 in the first term on the right side takes note of the fact that kinks and antikinks each give an identical contribution. When evaluating the functional integral for the free energy of the s–G model we must use the discretized version of (36). To $\mathcal{O}(L^{-1})$ this is the Hamiltonian

$$H[p] = 2\sum_{n=1}^{N_s}(M^2 + \tilde{p}_n^2)^{\frac{1}{2}} + \sum_{n=-\frac{1}{2}N_{ph}}^{\frac{1}{2}N_{ph}} \omega(\tilde{k}_n)P_n \tag{37}$$

and $2N_s + N_{ph} = N$, the total number of degrees of freedom. Note again that neither (36) nor (37) contains any breather contribution given as such: the phonon contribution replaces these. This feature and these results for sinh–G and s–G are apparently generic, they apply to the repulsive and attractive NLS models and the results listed in the next Section for the classical free energies all follow from them.

Before evaluating the functional integral for the classical free energy let us first return briefly to the problem of how the classical theory sketched above can be extended to the quantum case. The classical bracket (17) has a natural extension to the quantum case which is [26]

$$R(\lambda,\mu)\hat{T}(\lambda) \otimes \hat{T}(\mu) = \hat{T}(\mu) \otimes \hat{T}(\lambda)R(\lambda,\mu) : \tag{38}$$

$R(\lambda,\mu)$ is the quantum R-matrix[26] and $\hat{T}(\lambda)$ is the quantum monodromy matrix. Much as in the classical case, the matrix trace of (38) means that

$$[\hat{\Delta}(\lambda), \hat{\Delta}(\mu)] = 0, \tag{39}$$

where $\hat{\Delta}(\lambda) \equiv tr\hat{T}(\lambda)$. For a quantum integrable system with N quantum degrees of freedom one can find N commuting operator constants $\hat{I}_k$: $[\hat{I}_k, \hat{I}_\ell] = 0$; $k, \ell = 1, ..., N$, one of which can be the quantum Hamiltonian of the system. Motions are no longer on well defined tori: the canonical commutation relations $[\hat{P}_i, \hat{Q}_j] = -i\delta_{ij}$ ($\hbar = 1$) induce an N–dimensional uncertainty relation on the *states* of the system. These states span an N–fold infinite dimensional Hilbert space, one each for each quantum degree of freedom. Evidently (39), where $\lambda, \mu \in C$, forms quantum integrability conditions for a field (with infinite number of degrees of freedom).

Although the quantisation (38) derives from the quantum groups [21], the Heisenberg equations of motion derived from the Hamiltonian $\hat{H}$ for these systems are the same as those found from simple Dirac quantisation. Thus for s–G, for example, Heisenberg's equations prove to be (14).

Just as in the classical case the periodic b.c.s. introduce constraints in the quantum case which appear as "phase shift equations". These are like (25) given for classical sinh–G in the classical case. Note however that these phase shift equations are different in the two cases and so are the measures to be used in the functional integrals for the free energies. The phase shift equations derive[18] from the commutation relations (38) and determine the quantum measures. Thus all ordering problems in the functional integrals are transferred to the measures which now contain the deep quantum mechanics.

There is a crucial difference in the quantum and classical thermodynamic limits for "attractive" models like s–G and attractive NLS which have non–topological breather solutions. First of all the ordinary phonons of these models are absorbed in the quantum renormalisation so that they do not contribute to the Hamiltonians. We found this was also true for the classical models in thermodynamic limit, but the real difference lies in the thermodynamic limits for the breather contributions. We find[12,17,27] that the quantum breathers contribute *a pure phase*

$$i\int_0^\beta dt\sum_\ell 4(\gamma_0'')^{-1}\theta_\ell\frac{\partial\Phi_\ell}{\partial t} \tag{40}$$

to the classical action S: the classical action S is the exponent in contributions $expS$ in the functional integral which is the partition function Z (compare e.g. (53), (54) below). The number γ_0'' is a renormalised coupling constant: $\gamma_0''=\gamma_0(1-\gamma_0/8\pi)^{-1}$. Our strategy is to *remove* all such phases in the theory by replacing them by an equivalent set of constraints. It will become plain later that it is only in such phase terms that $\hbar$ appears. Thus all of the quantum features are in this phase. The equivalent constraints imposed are therefore constraints we imposed on the classical action by quantum mechanics i.e. the constraints are *quantum* constraints.

Evidently the phase (40) vanishes if each term in the sum over ℓ satisfies

$$\int_0^\beta dt\;4(\gamma_0'')^{-1}\theta_\ell\frac{\partial\Phi_\ell}{\partial t}=2\pi m_\ell \tag{41}$$

and m_ℓ is a positive integer. Since t must be periodic period β (see later) we then have that

$$\oint\frac{4}{\gamma_0''}\theta_\ell d\Phi_\ell=2\pi m_\ell \tag{42}$$

and for $0\le\Phi_\ell<8\pi$, noting that θ_ℓ is a constant, we find

$$\theta_\ell=\frac{1}{16}m_\ell\gamma_0''. \tag{43}$$

Since $0\le\theta_\ell<\frac{\pi}{2}$ there is a bound on the integers m_ℓ for each lable ℓ : $m_\ell=1,...,N_{b-1}$ where $N_b-1=[8\pi/\gamma_0'']$ and $[\cdot]$ means integral part. (Notice that $[8\pi/\gamma_0'']=[8\pi/\gamma_0]-1$ so $N_b=[8\pi/\gamma_0]$). It turns out that to the highest value of m_ℓ, $m_\ell=N_b-1$, corresponds a quantum kink antikink pair of mass $2M$. To the N_b-2 remaining integers m_ℓ there are N_b-2 *distinct* quantum breathers with mass $2M\sin(m_\ell\gamma_0''/16)$. Each of these quantum breathers is a distinct quantum particle which preserves its identity in thermodynamic limit: each has a well–defined finite density limit. In diagonal form the Hamiltonian of the quantum s–G model proves to be [20]

$$\hat{H}[p]=\sum_j\sum_\ell 2M\sin(\ell\gamma_0''/16)\cosh\alpha_j+\sum_n M\cosh\alpha_n, \tag{44}$$

where we have used rapidities α : $p_\nu=M_\nu\sinh\alpha_\nu$.

In the *semiclassical* limit $\gamma_0'' \to \gamma_0 \to 0$ the discrete breather levels become $2M\sin\theta_\ell = 2M\sin(\ell\gamma_0''/16) \to M\ell\gamma_0/8 = \ell m;\ \ell = 1, ..., N_b - 2;\ N_b \to \infty$. In this limit the set of breather levels is thus replaced by the levels of one quantum oscillator[4,10] with level spacing m. In a further *classical* limit these oscillator levels become dense because the soliton mass $M = 8m\gamma_0^{-1}$ is kept fixed as $\gamma_0 \to 0$ and therefore the level spacing $m \to 0$ in this limit. The dense packing of breather levels only at the bottom of the energy band, $\theta = \mathcal{O}(L^{-1})$, can then be understood as the minimisation of the total breather energy in the classical thermodynamic limit with finite density.

3. Free Energy

For a finite number N of degrees of freedom the classical partition function Z is given by a phase space integral

$$Z = \mathcal{N}^{-1}\int\prod_{i=1}^{N} dp_i dq_i\ exp(-\beta H[p,q]), \tag{45}$$

where $\{p_i, q_j\} = \delta_{ij}$ and $\mathcal{N}$ is a normalisation. All properties of thermal *equilibrium* can be derived from the partition function, e.g. the (Helmholtz) free energy F is $F = -\beta^{-1}\ell n Z,\ \ \beta = T^{-1}$ $(k_B = 1)$ and T is the temperature. For fields as in the s–G, sinh–G, etc., models classical Z becomes a functional integral

$$Z = \int \mathcal{D}\Pi\mathcal{D}\phi\ \exp(-\beta H[\phi]), \tag{46}$$

where, for real Π, ϕ, $\{\Pi, \phi\} = \delta(x - x')$: $H[\phi]$ is a notation for $H[\Pi, \phi]$, i.e. H depends on $\Pi(x,t)$ and $\phi(x,t)$. The functional integral (46), taken with the measure $\mathcal{D}\Pi\mathcal{D}\phi$ on an infinite dimensional symplectic manifold co–ordinatised by Π, ϕ, has a natural extension to the quantum case which, unlike the more usual Feynman description[28], retains the symplectic manifold. One can show[29] that the quantum partition function $Z = tr\ \exp(-\beta\hat{H})$, in which $\hat{H}$ is the Hamiltonian operator, is the functional integral

$$Z = tr\int \mathcal{D}\Pi\mathcal{D}\phi\ \exp S[\phi], \tag{47}$$

where S is formally the Wick rotated *classical* action in Hamiltonian form

$$S[\phi] = \int_0^\beta d\tau\{i\int \Pi(x,\tau)\phi_\tau(x,\tau)dx - H[\phi]\} \tag{48}$$

and $H[\phi]$ is the classical Hamiltonian. The tr means periodic b.c.s. in τ of period β, and the thermodynamic limit means periodic b.c.s. in x of period L, i.e. the functional integral (47) is taken over a space–time torus.

It is important that, despite the formal Wick rotation, τ is real time, i.e. the classical orbits of $H[\phi]$ determine the quantum statistical mechanics. Thus in particular in classical limit, where $\int_0^\beta d\tau$ is $\hbar^{-1}\int_0^{\beta\hbar} d\tau$ and $\hbar \to 0$ ($\hbar = 1$ in (48)) the $\lim \hbar \to 0$ eliminates the $i\Pi(x,\tau)\phi_\tau(x,t)$ terms – trajectories are vanishingly small loops – and (47) reduces to the classical form (46) as it should. In (46) $H[\phi]$ is the usual Hamiltonian in real time. We can already see that the term $\hbar^{-1}\int_0^{\beta\hbar} d\tau(i\int \Pi\phi_\tau dx)$ is a pure phase and that $\hbar$ appears only in this phase ($\hbar^{-1}\int_0^{\beta\hbar} H[\phi]d\tau = \beta H[\phi]$ simply). In effect we remove this whole phase by imposing appropriate equivalent quantum constraints.

Because the classical Hamiltonians of the integrable models can be transformed into action–angle form, and this can be done also in thermodynamic limit with finite density as discussed in the previous section, we wish of course to transform the functional integrals (46) and (47) to ones expressed in terms of these new canonical variables. Thus we want for example to transform the classical partition function (46) to

$$Z = \int \mathcal{D}\mu \; exp(-\beta H[p]). \tag{49}$$

where e.g. $H[p]$ for s–G is (37). In other circumstances, since the Jacobian of a canonical transformation is unity, there would be no problem. But for the continuous infinity of degrees of freedom involved in (49) we must introduce the measure $\mathcal{D}\mu$. It is a pragmatic fact that one can work in thermodynamic limit with action–angle variables on the real line if the measure $\mathcal{D}\mu$ looks after the problem of the thermodynamic limit. Notice that action–angle variables on the real line $x \in R$ with vanishing b.c.s. at $x = \pm\infty$, otherwise correspond to a zero–density limit.

For the classical sinh–G model we find ($\hbar = 1 \Rightarrow h = 2\pi$)[3]

$$\mathcal{D}\mu = \lim_{N\to\infty} \prod_{n=-\frac{1}{2}N}^{\frac{1}{2}N} (2\pi)^{-1} dP_n dQ_n \tag{50}$$

which is very much like the classical Liouville measure. The $P_n(Q_n)$ are the discretized action (angle) variables and the *labels* n are found from the 'phase shift equation' (25). Notice that in Hamiltonian (24), which is to be used in the functional integral (46) together with the measure (50), the labels n are determined by the modes $\tilde{k}_n$ given by (25) so that the apparent simplicity of (24) is deceptive. For the classical s–G model $\mathcal{D}\mu$ is more complicated because there are kinks and antikinks as well as phonons (but no breathers). We find [2] that

$$\mathcal{D}\mu = Z_{KG} \sum_{N_K} \sum_{N_{\bar{K}}} (N_K! N_{\bar{K}}!)^{-1} \prod_j \frac{dp_j dq_j}{2\pi} \prod_j \frac{d\bar{p}_j d\bar{q}_j}{2\pi} \mathcal{D}\mu_{KG}. \tag{51}$$

We here count N_K kinks and $N_{\bar{K}}$ antikinks separately, and $0 \le N_K,\ N_{\bar{K}} < \infty$ while

$$\mathcal{D}\mu_{KG} \equiv \lim_{N\to\infty} \prod_n \frac{dP_n dQ_n}{2\pi Z_{KG}(N)} \tag{52}$$

and $Z_{KG}(N)$ is the partition function for N *linear* Klein–Gordon modes: $Z_{KG} = \lim_{N\to\infty} Z_{KG}(N)$. The labels n and the 'shifted' momenta $\tilde{p}_j$, which appear in the Hamiltonian (37), derive from 'phase shift equations' which generalise (25) to include also kinks ans antikinks[2,18,24].

The canonical transformation which transforms the classical Hamiltonian to action–angle form, transforms the quantum Z (47) to

$$Z = \int \mathcal{D}\mu \ \exp S[p], \tag{53}$$

where $\mathcal{D}\mu$ is the quantum version of the classical measure in (49). If, as for sinh–G, there are phonon variables P, Q but no soliton variables,

$$S[p] = \int_0^\beta d\tau \{ i \int P(k) Q_\tau(k) dk - H[p] \} \tag{54}$$

in which $H[p]$ in the classical Hamiltonian in action–angle form found in thermodynamic limit with finite density. In the quantum propagator, from which (53) with (54) derives by formal Wick rotation through $t \to -i\tau$[1,28,29], quantum contributions arise as fluctuations about the classical trajectories which are determined from $\delta S = 0$. Since the imaginary unit i multiplies the (real) PQ_τ terms in (54) but not $H[p]$, a similar consideration does not apply to (53).

In fact we notice that since $P(k)$, $Q(k)$ are real the quantity $\int_0^\beta d\tau i \int P(k) Q_\tau(k) dk$ is a pure phase (which alone contains $\hbar : \hbar = 1$). We remove *this* phase by first discretizing to

$$i \int_0^\beta d\tau \sum_n P_n \frac{d}{d\tau} Q_n. \tag{55}$$

Then we impose the equivalent constraints

$$\int_0^\beta d\tau P_n dQ_n = \oint P_n dQ_n = 2\pi m_n \tag{56}$$

so that $P_n = m_n$ (for one turn of dQ_n round the classical torus). These constraints reduce quantum Z, eqn. (53), to the exact form (49) for classical Z. But the *quantum constraints* (56) now distinguish the quantum from the classical case. These constraints are an exact set of constraints compatible with the deeper quantisation which is taken care of by the measure $\mathcal{D}\mu$. Nevertheless (56) is formally Bohr semiclassical quantisation.

Since the $P_n \in R$ are constants $P_n = m_n$ and $P_n \geq 0$ means $m_n \geq 0$. Two obvious *alternative* choices for these quantum constraints are then

$$P_n = 0, 1, 2, \ldots \tag{57a}$$

for *bosons*, and

$$P_n = 0, 1 \tag{57b}$$

for *fermions*. There are also other choices which apparently lead to 'fractional statistics'. Indeed we have preliminary results for an anyon statistical mechanics of models of sinh–G and repulsive NLS types[12,17,27], but we do not consider these more exotic cases here. It is clear from (57) that the quantum $\mathcal{D}\mu$ includes a discrete summation over the phonon action variables labelled by n. It is also clear that (56) could provide *equivalent* descriptions in terms of fermions or in terms of bosons – bose–fermi equivalence [2,3,5–7,10–12] – and it does so as we show below. Note that the 'phase shift equations' which also appear as constraints for the functional integrals have different phase shifts for bosons, fermions and classical particles[2–12].

Similar arguments for solitons yield $p_n = 2\pi m_n L^{-1}$ for kinks and antikinks consistent with the packing of soliton momenta in classical thermodynamic limit. So far we find that the kinks and antikinks act like fermions only, i.e. we do not yet have any equivalent boson description of these. For breathers the translational momenta behave similarly as for kinks and antikinks i.e. these are fermions-like. The internal momenta on the other hand lie at discrete 'angles' $\theta_\ell = \ell\gamma_0''/16$; $\ell = 1, ..., N_b - 2$, as discussed in the previous section so that there are $N_b - 2$ distinct fermion-like breathers and $\mathcal{D}\mu$ must include a summation over these.

4. Explicit Results

We finally quote results for the free energies for some of the integrable models. The classical sinh–G and repulsive NLS models have Hamiltonian (24) with $\omega(\tilde{k}) = (m^2 + \tilde{k}^2)^{\frac{1}{2}}$ and $\omega(\tilde{k}) = \tilde{k}^2$ respectively. The shifted modes $\tilde{k}$ are determined from eqn. (25) with classical phase shifts (26) (sinh–G) and (27) (repulsive NLS). The partition function is given by functional integral (49) with $\mathcal{D}\mu$ given by (50). Actual evaluation of the functional integral has thus far been possible only by iterating (25). This produces an iterative solution for the free energy (in the limit $L \to \infty$)[2–12],

$$\begin{aligned} FL^{-1} = & (2\pi\beta)^{-1}\int_{-\infty}^{\infty} dk\ell n(\beta\omega(k)) \\ & - (2\pi\beta)^{-2}\int_{-\infty}^{\infty}\frac{dq}{\omega(q)}\int_{-\infty}^{\infty} dk\Delta_c(k,q)\frac{d}{dk}\ell n\omega(k) \\ & + (2\pi\beta)^{-3}\int_{-\infty}^{\infty}\frac{dp}{\omega(p)}\int_{-\infty}^{\infty} dq\Delta_c(q,p)\frac{d}{dq}\Big[\frac{1}{\omega(q)}\int_{-\infty}^{\infty} dk\Delta_c(k,q)\frac{d}{dk}\ell n\omega(k)\Big] \\ & - \frac{1}{2}(2\pi\beta)^{-3}\int_{-\infty}^{\infty}\frac{dq}{\omega^2(q)}\Big[\int_{-\infty}^{\infty} dk\Delta_c(k,q)\frac{d}{dk}\ell n\omega(k)\Big]^2 + ... \quad . \end{aligned} \tag{58}$$

For sinh–G we find from $(58)^3$ that

$$FL^{-1} = m\beta^{-1}\Big[\frac{1}{4}t - \frac{1}{8}t^2 + \frac{3}{16}t^3 - \frac{53}{128}t^4 + ...\Big] + F_{KG}, \tag{59}$$

where $t \equiv (M\beta)^{-1}$ and F_{KG} is the classical free energy of the free Klein–Gordon field, $F_{KG} = (\beta a)^{-1}[\ell n(\beta a^{-1}) + \frac{1}{2}ma]$. We also find (59) from (46) by the transfer integral method[3]. Notice that the classical ultra-violet divergence is contained only in this linear contribution ($a \to 0$ cannot be taken in F_{KG}) to the free energy. It is given by the first term on the right side of (58), and the integral is only defined if a proper cut–off procedure is specified: we use the lattice form for $\omega(k)$, $\omega(k) = [m^2 + (\frac{2}{a})^2 \sin^2(\frac{ka}{2})]^{1/2}$ and $|k| \leq \pi/a$. All the other terms in (58) are well defined.

We find further that the iterative series (58) is in fact summable: it is exactly the formal iteration of the coupled integral equations [2–12]

$$\begin{aligned} FL^{-1} &= (2\pi\beta)^{-1} \int_{-\infty}^{\infty} \ell n(\beta\epsilon(k))dk \\ \epsilon(k) &= \omega(k) + (2\pi\beta)^{-1} \int_{-\infty}^{\infty} dk' \frac{d}{dk} \Delta_c(k,k') \ell n(\beta\epsilon(k')) \end{aligned} \tag{60}$$

(so the low temperature strictly asymptotic series (59) is also summable).

Notice that (60) is a formal expression in the sense that its ultra-violet divergence must be handled appropriately. It is also important to notice that in (60) the expression for the free energy is formally that for a classical ideal gas. All the interactions go into the definition of the "dressed" (temperature dependent) excitation energy $\epsilon(k)$. This result is generic for the integrable models, and their free energies can always be expressed in a form like (60). This form for the free energies can also be derived from a variational approach to the statistical mechanics of the integrable model, the 'generalised Bethe ansatz method' [2–12,18], which does not involve functional integration: it is based on minimisation of a formal expression for the free energy in terms of densities of excitations and densities of allowed states [2–12,18].

The classical repulsive NLS model differs from the (relativistic) sinh–G only in that $\omega(k) = k^2$ and Δ_c is given by (27). Because of the *infra-red* divergence inherent in this model the individual terms in (58) apparently are not well defined. We find from (60) for this model however that

$$FL^{-1} = (\beta a)^{-1} \ell n(\beta a^{-2}) + Kc^{1/3}T^{4/3}, \tag{61}$$

where $L \to \infty$ is understood. This is a kind of scaling result (the generic form of the second term is $K\beta^{-1}(cT)^{1/3}$), and we were able to fix the constant K by appeal to the corresponding calculation by the transfer integral method[25]: $K = 0.59$. The non-analytic behaviour in T is consistent with failure of the system to iterate. The infra-red divergence is avoided in the quantum versions of the model and the quantum analogue of (60) does have a solution[30]. Because the classical thermodynamic limit on the attractive NLS replaces the breather like solitons by large amplitude phonons we also find (60) for this case (with $c \to -c$) and a form like (61) is expected for this attractive case.

For the quantum sinh–G model the measure (50), and the phase shift equations (25), must be modified according to the quantum constraints (56). Also, in the phase shift equations (25) the classical phase shift Δ_c must be replaced by

$$\Delta_f(k,k') = -2\tan^{-1}\left[\frac{m^2\sin(\frac{1}{8}\gamma_0'')}{k\omega(k') - k'\omega(k)}\right] \tag{62}$$

for fermions, $\gamma_0'' \equiv \gamma_0(1+\gamma_0/8\pi)^{-1}$, and by $\Delta_b(k,k') = \Delta_f(k,k') + 2\pi\theta(k'-k)$ for bosons; $\theta(k)$ is the unit step. It is important that the smooth branch $-2\pi<\Delta_f<0$ is taken for Δ_f : $\Delta_f(k=k') = -\pi$ and there is no singularity. Note that the bose form Δ_b has a singularity at $k = k'$, and that Δ_c is the classical limit of Δ_b. We then find[3] that for bosons

$$\begin{aligned} FL^{-1} &= (2\pi\beta)^{-1}\int_{-\infty}^{\infty} \ell n(1-e^{-\beta\epsilon(k)})dk \\ \epsilon(k) &= \omega(k) + (2\pi\beta)^{-1}\int_{-\infty}^{\infty}\frac{d}{dk}\Delta_b(k,k')\ell n(1-e^{-\beta\epsilon(k')})dk' \ , \end{aligned} \tag{63}$$

and for fermions

$$\begin{aligned} FL^{-1} &= \mu\bar{n} - (2\pi\beta)^{-1}\int_{-\infty}^{\infty} \ell n(1+e^{-\beta\tilde{\epsilon}(k)})dk \\ \tilde{\epsilon}(k) &= \omega(k) - \mu - (2\pi\beta)^{-1}\int_{-\infty}^{\infty}\frac{d}{dk}\Delta_f(k,k')\ell n(1+e^{-\beta\tilde{\epsilon}(k')})dk' \ , \end{aligned} \tag{64}$$

in which μ is a chemical potential and $\bar{n} = \lim_{L\to\infty} NL^{-1}$. These two quantum forms are bose–fermi equivalent: put $\ell n(1+e^{-\beta\tilde{\epsilon}}) = -\ell n(1-e^{-\beta\epsilon})$ (for $\mu = 0$) and use the jump on Δ_b.

In the classical s–G model there are kinks and antikinks as well as phonons (but no breathers!) and therefore this system is more complicated than either of the sinh–G or the repulsive NLS models. The partition function of the classical s–G model can be calculated from (49) with measure (51) and Hamiltonian (37). In the discrete Hamiltonian (37) the phonon modes labelled by $\tilde{k}_n$ and kink (and antikink) momenta $\tilde{p}_n$ are not independent: they are coupled through appropriate phase shift equations. We find [2] these phase shift equations for the classical s–G model in thermodynamic limit at finite density are

$$\begin{aligned} \tilde{k}_n &= k_n - L^{-1}\sum_{n'\neq n}\Delta(\tilde{k}_n,\tilde{k}_{n'})P_{n'} - L^{-1}\sum_j \Delta_K(\tilde{k}_n,\tilde{p}_j) \\ \tilde{p}_j &= p_j - L^{-1}\sum_n \Delta_K(\tilde{k}_n,\tilde{p}_j)P_n - L^{-1}\sum_{j'\neq j}\Delta_{KK}(\tilde{p}_j,\tilde{p}_{j'}) \end{aligned} \tag{65}$$

in which j, j' label both kinks and antikinks. The phase shifts in (65) are classical s–G phase shifts which are [2,18,24]

$$\begin{aligned} \Delta(k,k') &= \frac{1}{4}\gamma_0 m^2[k\omega(k') - k'\omega(k)]^{-1} \\ \Delta_K(k,p) &= -2\tan^{-1}\left\{m\gamma^{-1}[k - v\omega(k)]^{-1}\right\} \\ \frac{\partial}{\partial\alpha}\Delta_{KK}(\alpha,\alpha') &= 8\gamma_0^{-1}\ell n\left(\frac{\cosh(\alpha-\alpha')-1}{\cosh(\alpha-\alpha')+1}\right) \end{aligned} \tag{66}$$

for phonon–phonon, phonon–kink (=kink–phonon) and kink–kink (=kink–antikink, etc.) interactions, respectively: Δ_K is written in terms of the velocity v with $\gamma = (1-v^2)^{\frac{1}{2}}$; Δ_{KK} is written for simplicity in derivative form and in terms of rapidities, so that $\tilde{p} = M \sinh \tilde{\alpha}$.

The system (49), (37) and (65) iterates, the integrals can still be done in this more complicated case, and the result is a strictly asymptotic series for the classical free energy of s–G. In the limit $L \to \infty$ this free energy is [2] $FL^{-1} = F_{KG} + F^{(1)} + F^{(2)} + \cdots$, where F_{KG} is the free energy of the classical Klein–Gordon field (divergent in the limit $a \to 0$) as in the sinh–G case above, and

$$\begin{aligned} F^{(1)} = &-\beta^{-1} m (8/t\pi)^{\frac{1}{2}} e^{-1/t} \left[1 - \frac{7}{8}t - \frac{59}{128}t^2 - \frac{897}{1024}t^3 - \frac{75005}{32768}t^4 - \cdots\right] \\ &-\beta^{-1} m \left[\frac{1}{4}t + \frac{1}{8}t^2 + \frac{3}{16}t^3 + \frac{594}{2^9}t^5 + \frac{7922}{2^{11}}t^6 + \cdots\right] \end{aligned} \tag{67a}$$

$$F^{(2)} = \frac{8m}{\pi} M e^{-2/t} \left\{ \ell n \frac{4C}{t} - \frac{5}{4}t \left[\ell n \frac{4C}{t} + 1\right] - \frac{t^2}{32}\left[13 \ell n \frac{4C}{t} + 2\right] + \cdots \right\} \tag{67b}$$

$$F^{(\nu)} = \mathcal{O}(e^{-\nu/t}); \; \nu = 3, 4, \cdots \; . \tag{67c}$$

Here $t \equiv (M\beta)^{-1}$ and C is the Euler constant: $\ell n C = 0.5771\cdots$. The results (67) coincide exactly with the ones found from the transfer integral method[2]. With the problem associated with the classical ultra–violet divergence of KG understood, the asymptotic expansion (67) can be summed up to coupled integral equations

$$\begin{aligned} FL^{-1} &= (2\pi\beta)^{-1} \int_{-\infty}^{\infty} \omega(\alpha) \ell n(\beta \epsilon(\alpha)) d\alpha - 2(2\pi\beta)^{-1} \int_{-\infty}^{\infty} E(\alpha) e^{-\beta \tilde{E}(\alpha)} d\alpha \\ \epsilon(\alpha) &= \omega(\alpha) + (2\pi\beta)^{-1} \int_{-\infty}^{\infty} \frac{d\Delta}{d\alpha} \ell n(\beta \epsilon(\alpha')) d\alpha' + (\pi\beta)^{-1} \int_{-\infty}^{\infty} \frac{d\Delta_K}{d\alpha} e^{-\beta \tilde{E}(\alpha')} d\alpha' \\ \tilde{E}(\alpha) &= E(\alpha) - (\pi\beta)^{-1} \int_{-\infty}^{\infty} \frac{d\Delta_{KK}}{d\alpha} e^{-\beta \tilde{E}(\alpha')} d\alpha' - (2\pi\beta)^{-1} \int_{-\infty}^{\infty} \frac{d\Delta_K}{d\alpha} \ell n(\beta \epsilon(\alpha')) d\alpha' \end{aligned} \tag{68}$$

which are here expressed in terms of rapidities: $\omega(\alpha) = m \cosh \alpha$, $E(\alpha) = M \cosh \alpha$.

The structure of the quantum free energy of s–G is similar to that of the classical result (68) except that the phonon contribution is replaced by contributions of $N_b - 2$ distinct quantum breathers. In fermion form the quantum free energy of s–G proves to be[4,10,31,32]

$$FL^{-1} = \sum_{\ell=1}^{N_b-2} \{\mu_\ell \bar{n}_\ell - (2\pi\beta)^{-1} \int_{-\infty}^{\infty} d\alpha_\ell \, M_\ell \cosh\alpha_\ell \, \ell n(1+e^{-\beta\epsilon_\ell})\}$$
$$+ \mu_S \bar{n}_S - 2(2\pi\beta)^{-1} \int_{-\infty}^{\infty} d\alpha_S \, M \cosh\alpha_S \, \ell n(1+e^{-\beta\epsilon_S})$$
$$\epsilon_\ell = M_\ell \cosh\alpha_\ell - \mu_\ell - (2\pi\beta)^{-1} \sum_{\ell'=1}^{N_b-2} \int_{-\infty}^{\infty} d\alpha_{\ell'} \frac{d\Delta_{\ell\ell'}}{d\alpha_\ell} \ell n(1+e^{-\beta\epsilon_{\ell'}})$$
$$- 2(2\pi\beta)^{-1} \int_{-\infty}^{\infty} d\alpha_S \frac{d\Delta_{\ell S}}{d\alpha_\ell} \ell n(1+e^{-\beta\epsilon_S}) \tag{69}$$
$$\epsilon_S = M \cosh\alpha_S - \mu_S - (2\pi\beta)^{-1} \sum_{\ell=1}^{N_b-2} \int_{-\infty}^{\infty} d\alpha_\ell \frac{d\Delta_{\ell S}}{d\alpha_S} \ell n(1+e^{-\beta\epsilon_\ell})$$
$$- 2(2\pi\beta)^{-1} \int_{-\infty}^{\infty} d\alpha_{S'} \frac{d\Delta_{SS'}}{d\alpha_S} \ell n(1+e^{-\beta\epsilon_{S'}})$$

in which $M_\ell = 2M\sin(\ell\gamma_0''/16)$, $M = 8m{\gamma_0''}^{-1}$, m is a renormalised mass and μ_ℓ, μ_S are chemical potentials for each particle. We do not give here the phase shifts which appear in (69) but refer to the literature[4,10,18,31–33]. The quantum free energy of s–G we give in fermion form only: we have not worked out all the details of the bose form. For the details of how the $N_b - 2$ distinct quantum breathers in (69) become one quantum oscillator in the *semiclassical* limit we refer to Ref. 4.

We conclude this Section by noting that we hope it has now become evident that this functional integral method extends to quantum and classical models other than the ones mentioned in this paper, namely to all of the integrable models. This method can treat on equal footing fields which describe classical particles, bose particles, fermi particles and (apparently) even particles which obey fractional statistics. Bose and fermi descriptions are completely equivalent, and this equivalence plays on important role even in the classical cases for e.g. the classical free energy of sinh–G (repulsive NLS) is the classical limit of the bose form. Attractive models which have non-topological solitons of breather type have markedly different thermodynamic limits in the quantum and classical cases. The discrete breather levels which describe distinct quantum particles are replaced by one quantum oscillator in the semiclassical limit, and by interacting (nonlinear) phonons in a further classical limit. This difference in the behaviour of quantum vs. classical models has interesting connections to quantum vs. classical chaos.

References

1. R. K. Bullough, D. J. Pilling and J. Timonen, in *Nonlinear Phenomena in Physics*, ed. F.Claro (ed.), (Springer–Verlag, Berlin, 1985).

2 J. Timonen, M. Stirland, D. J. Pilling, Yi Cheng and R. K. Bullough, *Phys. Rev. Lett.* **56** (1986) 2233.
3. R. K. Bullough, D. J. Pilling and J. Timonen, *J. Phys.* **A 19** (1986) L955.
4. J. Timonen, R. K. Bullough and D. J. Pilling, *Phys. Rev.* **B 34**, (1986) 6525.
5. R. K. Bullough, D. J. Pilling and J. Timonen, in *Solitons*, ed. M. Lakshmanan, (Springer–Verlag, Berlin, 1988).
6. J. Timonen, D. J. Pilling, Y. Z. Chen and R. K. Bullough, in *Plasma Theory and Nonlinear and Turbulent Processes in Physics*, eds. V. G. Bar'yakhtar et al., (World Scientific, Singapore, 1988).
7. J. Timonen, Y. Z. Chen and R. K. Bullough, *Nucl. Phys. B*, **5A** (1988) 58.
8. R. K. Bullough, D. J. Pilling and J. Timonen, in *Integrable Systems and Applications*, eds. M. Balabane, P. Lochak and C. Sulem, (Springer–Verlag, Berlin, 1989).
9. R. K. Bullough, D. J. Pilling, Y. Cheng, Y. Z. Chen and J. Timonen, in *Nonlinear Evolution Equations: Integrability and Spectral Methods*, eds. A. P. Fordy and A. Degasperis, (Manchester University Press, Manchester, 1990).
10. R. K. Bullough, Y. Z. Chen and J. Timonen, in *Nonlinear World*, eds. V. G. Bar'yakhtar, V. M. Chernousenko, N. S. Erokhin, A. G. Sitenko and V. E. Zakharov, (World Scientific, Singapore, 1990).
11. R. K. Bullough, Y. Z. Chen, J. Timonen, V. Tognetti and R. Vaia, *Phys. Lett. A*, **145** (1990) 154.
12. R. K. Bullough and J. Timonen, to appear in *Physica D* (1991).
13. C. S. Gardner, J. M. Greene, M. D. Kruskal and R. M. Miura, *Phys. Rev. Lett.* **19** (1967) 1095.
14. R. K. Bullough and P. J. Caudrey (eds.), *Solitons*, Chap.1, (Springer–Verlag, Heidelberg, 1980).
15. P. J. Caudrey, in *Soliton Theory: A Survey of Results*, ed. A. P. Fordy, (Manchester University Press, Manchester, 1989) and references.
16. R. K. Bullough, in *Solitons*, ed. M. Lakshmanan, (Springer–Verlag, Berlin, 1988).
17. R. K. Bullough and J. Timonen, in *Proc. 19th Intl. Conf. on Diff. Geometric Methods in Theor. Phys.*, eds. U. Bruzzo and R. Cianci, (Springer–Verlag, to appear 1991).
18. Yu–zhong Chen, Ph. D. Thesis, University of Manchester (1989).
19. V. I. Arnold, *Mathematical Methods of Classical Mechanics*, (Springer–Verlag, Heidelberg, 1978).
20. See e.g. L. D. Faddeev and L. A. Takhtadjan, *Hamiltonian Methods in the Theory of Solitons*, (Springer–Verlag, Berlin, 1987); and the book cited in Ref. 14.
21. R. K. Bullough, S. Olafsson, Yu–zhong Chen and J. Timonen, in *Differential Geometric Methods in Theoretical Physics*, eds. Ling–Lie Chau and Werner Nahm, (Plenum, New York, 1990) and references.
22. E. K. Sklyanin, LOMI Preprint E–3–1979 (1979).
23. A. G. Izergin and V. E. Korepin, *Lett. Math. Phys.* **5** (1981) 166.
24. Yi Cheng, Ph. D. Thesis, University of Manchester (1987).

25. R. K. Bullough, D. J. Pilling and J. Timonen, in *Dynamical Problems in Soliton Systems*, ed. S. Takeno, (Springer–Verlag, Heidelberg, 1985).
26. E. K. Sklyanin, L. A. Takhtadjan and L. D. Faddeev, Teor. Mat. Fiz. **40** (1979) 194.
27. R. K. Bullough and J. Timonen, in *Microscopic Aspects of Nonlinearity in Condensed Matter*, eds. A. R. Bishop and V. Tognetti, (Plenum, New York, to appear 1991).
28. R. P. Feynman and A. R. Hibbs, *Quantum Mechanics and Path Integrals*, (McGraw–Hill, New York, 1965)
29. See e.g. Ref. 27 and R. K. Bullough, in *Nonlinear Phenomena in Physics*, ed. F. Claro, (Springer–Verlag, Berlin, 1985).
30. C. N. Yang and C. P. Yang, *J. Math. Phys.* **10** (1969) 1115.
31. M. Fowler and X. Zotos, *Phys. Rev.* **B 24** (1981) 2634; **B 25** (1982) 5806.
32. M. Imada, K. Hida and M. Ishikawa, *Phys. Lett.* **90A** (1982) 79; *J. Phys.* **C 16** (1983) 35.
33. V. E. Korepin, *Teor. Mat. Fiz.* **41** (1979) 169.

ON EXTENSION THEORY IN L^2–SPACES

J.F. Brasche
Institut für Mathematik
Ruhr–Universität Bochum
4630 Bochum, FRG

Abstract: Modifying a previous work by Kochubei[1] we give a method to construct self–adjoint extensions of

$$H \mid \{f \in D(H) \mid \operatorname{supp}(|f|m) \subset N^c := X \setminus N\}.$$

Here X, m, H and N denote a topological space, a Borel measure on X, a self–adjoint operator on $L^2(X;m)$ and a closed m–null set respectively, satisfying some fairly weak assumptions specified below. Our method is both more general and more simple than those given previously.[2–9]

Notation and Assumptions

X locally compact second countable space
$\mathcal{B}_b(X)$ set of bounded Borel measurable functions $f : X \to \mathbb{C}$
m positive Radon measure on X with $\operatorname{supp}(m) = X$
N closed subset of X with $m(N) = 0$
H self–adjoint operator on $L^2(X;m)$ with domain $D(H)$
$\varrho(H)$ resolvent set of H
$< f, g >_{H,z} := < (H - z)f, (H - z)g >$,
$< \, , \, >$ the scalar product on $L^2(X;m)$
$C_{N,H}$ space of $f \in D(H)$ with a version f_N continuous on some neighborhood of N. Henceforth f_N denotes any such version.

We assume that $C_{N,H}$ is a core for H, i.e. $C_{N,H}$ is dense in the Hilbert space $(D(H), < \, , \, >_{H,z})$ for one and therefore all $z \in \varrho(H)$.

Definitions and Preliminary Steps.

A positive Radon measure μ on N is admissible iff for one (and therefore all) $z \in \varrho(H)$ there exists a real number $a(z)$ such that

$$\int |f_N| d\mu \leq a(z)\sqrt{< f, f >_{H,z}} \quad \forall f \in C_{N,H} .$$

We denote by $M(N, H)$ the set of all admissible positive Radon measures on N.

Let $\mu \in M(N, H)$, $z \in \varrho(H)$. By the definition of admissibility and the assumption that $C_{N,H}$ is dense in $(D(H), < \,,\, >_{H,z})$ there exists a unique linear bounded transformation

$$I_\mu : (D(H), < \,,\, >_{H,z}) \to L^1(X; m)$$

satisfying

$$I_\mu f = f_N \quad \mu - \text{a.e.} \quad \forall f \in C_{N,H} .$$

We denote by H'_N the restriction of H on

$$D(H'_N) := \{f \in D(H) \mid I_\mu f = 0 \quad \mu - \text{a.e.} \quad \forall \mu \in M(N, H)\} .$$

Obviously

$$D(H'_N) \supset \{f \in D(H) \mid \operatorname{supp}(|f|m) \subset N^c\} .$$

Let $g \in \mathcal{B}_b(X)$. By Riesz respresentation theorem (for the Hilbert space $(D(H), < \,,\, >_{H,z})$), there exists a unique $p_z^{g\mu} \in L^2(X; m)$ satisfying

$$< p_z^{g\mu}, (H - z)f >= \int \bar{g} I_\mu f d\mu \quad \forall f \in D(H) .$$

The Results.

By von Neumann's extension theory and the following theorem we know all self–adjoint extensions of H'_N.

Theorem: (i) H'_N is closed

(ii) Let $z \in \varrho(H)$. Then

$$\operatorname{ran}(H'_N - z)^\perp = [p_z^\mu \mid \mu \in M(N,H)]$$

([...] denotes the closed linear hull of the set $\{\ldots\}$, p_z^μ is, of course, shorthand for $p_z^{1\mu}$.)

Proof: (i) Obvious.
(ii) Let $f \in D(H)$. Let $\mu \in M(N,H)$.

$$I_\mu f = 0 \quad \mu - \text{a.e.} \Leftrightarrow \int \bar{g} I_\mu f d\mu = 0 \ \forall g \in \mathcal{B}_b(X)$$
$$\Leftrightarrow < p_z^{g\mu}, (H-z)f >= 0 \ \forall g \in \mathcal{B}_b(X).$$

Thus

$$D(H'_N) = \{f \in D(H) \mid (H-z)f \in [p_z^\mu \mid \mu \in M(N,H)]^\perp\}.$$

Since $\operatorname{ran}(H-z) = L^2(X;m)$ this implies

$$\operatorname{ran}(H'_N - z) = [p_z^\mu \mid \mu \in M(N,H)]^\perp. \qquad \blacksquare$$

Often one can obtain a fairly explicit description of the deficiency subspaces of H'_N. We illustrate this fact by the following simple but useful example.[10]

Example: Let H be the Laplacian $-\Delta$ on $L^2(\mathbb{R}^d; dx)$ (dx denotes the Lebesgue measure on $\mathbb{R}^d$). For each $z \in \varrho(H)$ let G_z be the (Green's–) function satisfying

$$G_z * f = (-\Delta - z)^{-1} f \quad \forall f \in D(-\Delta).$$

Then a positive Radon measure μ on N is admissible iff $G_E * \mu \in L^2(\mathbb{R}^d; dx)$ for one and therefore all $E < 0$. Moreover for each $z \in \varrho(H)$

$$\operatorname{ran}(-\Delta'_N - \bar{z})^\perp = [G_z * \mu \mid \mu \in M(N,H)]. \qquad \blacksquare$$

Acknowledgement: I am indebted to S. Albeverio for supervising my work. It is a pleasure to thank J. Dittrich, P. Exner, R. Figari, W. Karwowski, T. Lindström, P. Šeba and A. Teta for many helpful and stimulating discussions as well as for various invitations and very pleasant visits.

References

1. Kochubei, A.N.: Funct. Anal. Appl. 16 (1982), 137–139.
2. Albeverio, S. et al.: "Non standard Methods in Stochastic Analysis and Mathematical Physics", Academic Press, 1986.
3. Albeverio, S. et al.: "Solvable models in Quantum Mechanics". Springer, 1988.
4. Albeverio, S. et al., pp. 1–42 in "Schrödinger Operators", ed. by Holden, H., Jensen, A., lecture Notes in Physics 345, Springer 1989.
5. Exner, P., Šeba, P. (Ed.): "Application of Self–Adjoint Extensions in Quantum Physics". Lecture Notes in Physics 324, 1989
6. Exner, P., Šeba, P. (Ed.): "Schrödinger Operators, Standard and Non–Standard". World Scientific, 1989.
7. Exner, P., Neidhard, H. (Ed.): "Operator Theory: Advances and Applications", Vol. 46. Birkhäuser, 1990.
8. Teta, A.: "Singular Pertubations of the Laplacian and Connection with Models of Random Media". PhD–Thesis, Trieste, 1989
9. Brasche, J.F., Teta, A.: to appear in Proc. of the Conf. "To the Memory of Raphael Høegh–Krohn".
10. Brasche, J.F., pp. 207–245 in "Stochastic Processes, Physics and Geometry", ed. by Albeverio, S. et al., World Scientific, 1990.

COHERENT STATES ASSOCIATED TO GROUP REPRESENTATIONS WHICH ARE SQUARE INTEGRABLE ON A HOMOGENEOUS SPACE

J.-P. ANTOINE

Institut de Physique Théorique, Université Catholique de Louvain

B - 1348 - Louvain-la-Neuve, Belgique

ABSTRACT

We extend the construction of coherent states to the case of a group representation which is square integrable only on a homogeneous space, using a formalism based on the notion of reproducing triple. The example of the Poincaré group in 1+1 dimensions is treated in detail.

Coherent states (CS), originally introduced by Schrödinger in the context of a harmonic oscillator, later popularized by Glauber and Klauder for the description of coherent light, have been generalized to such an extent that they find applications in every single corner of quantum theory (see Refs. 1 and 2 for the original references). Yet there are cases where the known methods fail to generate CS, for instance, the Galilei or the Poincaré groups (in 1+1 or 1+3 dimensions), and other groups of the same type. Our aim in the present paper is to treat such situations, and in fact much more general ones, by a suitable extension of the notion of coherent states . The discussion is based on joint work with S.T.Ali and J.-P.Gazeau [3-5]. First we briefly review the standard method.

1. The standard construction of coherent states

Let G be a locally compact group (not necessarily unimodular), with (left invariant) Haar measure dg, and consider a strongly continuous, irreducible, unitary representation U of G into a Hilbert space $\mathcal{H}$. Assume that the representation U is *square integrable,* i.e. there exists a vector $\eta \in \mathcal{H}$ such that :

$$c(\eta,\phi) \equiv \int_G |\langle U(g)\,\eta \mid \phi\rangle|^2 \, dg < \infty , \ \forall \phi \in \mathcal{H} \tag{1.1}$$

(equivalently, U belongs to the discrete series). Choose a fixed vector η that satisfies the admissibility condition (1.1) and normalize it by $c(\eta,\eta) = 1$. Then the orbit of η under U,

$$\mathfrak{S} = \{ \eta_g = U(g)\eta \mid g \in G \}, \tag{1.2}$$

is an overcomplete family of vectors, called *coherent states* associated to the representation U. The family $\mathfrak{S}$ determines a resolution of the identity

$$\int_G |\eta_g><\eta_g| \, dg = I. \tag{1.3}$$

More precisely, the map $W_\eta : \phi \mapsto <\eta_g \mid \phi>$ is an isometry from $\mathcal{H}$ onto a closed subspace of $L^2(G,dg)$, which is characterized by the *reproducing kernel* $K(g,g') = <\eta_g \mid \eta_{g'}>$. This is the familiar construction of coherent states [1,2]. A particularly interesting example is the affine group of the line (the "ax+b" group), in which case W_η is the *wavelet transform* [6,7] (η is then called the analyzing wavelet).

2. The Perelomov construction

In many cases, the admissibility condition (1.1) is too strong, and one has to use the generalization due to Perelomov [2]. Given η, define H as the subgroup of G that leaves η invariant up to a phase (this is obviously motivated by Quantum Mechanics, where states are defined only up to the phase) :

$$U(h)\,\eta = e^{i\alpha(h)}\,\eta. \tag{2.1}$$

Let ν be a (left) invariant measure on the coset space $X = G/H$. Then we say that the representation U is *square integrable mod H* if the following admissibility condition holds :

$$\int_{G/H} |<U(g)\,\eta \mid \phi>|^2 \, d\nu(x) < \infty, \ \forall\ \phi \in \mathcal{H} \tag{2.2}$$

(by (2.1) the integrand depends only on the coset $x = gH$). When this is the case, the construction proceeds as before. We denote again by $\mathfrak{S}$ the orbit of η under U, but now the elements of $\mathfrak{S}$ are indexed by points of the coset space X. Hence one obtains a set of generalized coherent states (GCS) :

$$\mathfrak{S} = \{ \eta_x \mid x = gH \in X = G/H \}, \tag{2.3}$$

which yields again a resolution of the identity :

$$\int_X |\eta_x><\eta_x| \, d\nu(x) = I, \tag{2.4}$$

with the same consequences as before.

Examples of this construction are plentiful ; for instance :

(i) when G is the Weyl-Heisenberg group, one gets the canonical coherent states, familiar in Quantum Mechanics [1];
(ii) G = SU(2) or a compact simple Lie group;
(iii) G = SU(1,1) or a noncompact simple Lie group, with U a representation of the discrete series.

The last two classes are treated in detail in the monograph of Perelomov [2]. It should be remarked that GCS may be defined also for a representation that is not square integrable mod H, e.g. a principal series representation, but those vectors lack many of the nice properties of the square integrable ones [2].

Now this method, however nice and powerful it may be, does not always work; for instance, it is inapplicable in the case of the Galilei group or the Poincaré groups $\mathcal{P}_+^\uparrow(1,1)$ or $\mathcal{P}_+^\uparrow(1,3)$, and more generally for a semi-direct product $G = S \wedge V$, where V is a vector space and S a semisimple group of automorphisms of V. This case has been treated recently by DeBièvre [8], using a method inspired by geometric quantization.

To get a hint towards possible generalizations, we shall look first at a simple example, namely the Poincaré group $\mathcal{P}_+^\uparrow(1,1)$ in 1+1 dimensions [3-5].

3. The case of the Poincaré group $\mathcal{P}_+^\uparrow(1,1) = SO_o(1,1) \wedge \mathbb{R}^2$

The group is parametrized in the standard way: $(a, \Lambda_p) \in \mathcal{P}_+^\uparrow(1,1)$, where

$$a = (a_o, \mathbf{a}) \equiv (a_o, \underline{a}) \in \mathbb{R}^2 \tag{3.1a}$$

$$\Lambda_p = \begin{pmatrix} \frac{p_o}{m} & \frac{\mathbf{p}}{m} \\ \frac{\mathbf{p}}{m} & \frac{p_o}{m} \end{pmatrix} \in SO_o(1,1), \text{ with } p_o = (\mathbf{p}^2 + m^2)^{1/2}, \tag{3.1b}$$

The Haar measure on $\mathcal{P}_+^\uparrow(1,1)$ is $da_o\, \mathbf{da}\, \mathbf{dp}/p_o$.

We consider the familiar Wigner representation $U_w \equiv U_w^{(m)}$ of mass m:

$$(U_w(a,\Lambda_p)\phi)(k) = e^{ik.a}\, \phi(\Lambda_p^{-1}k), \quad \phi \in \mathcal{H}_w = L^2(\mathcal{V}_m^+, \mathbf{dk}/k_o) \tag{3.2}$$

where $\mathcal{V}_m^+ = \{\, k = (k_o,\mathbf{k}),\ k_o^2 - \mathbf{k}^2 = m^2,\ k_o > 0 \,\}$ and $k.a = k_o\, a_o - \mathbf{k} \,.\, \mathbf{a}$. Then a straightforward calculation shows that this representation U_w is *not* square-integrable over $\mathcal{P}_+^\uparrow(1,1)$, since:

$$\int_{\mathcal{P}_+^\uparrow(1,1)} |\langle U_w(a,\Lambda_p)\, \phi \mid \phi \rangle|^2\, da_o\, \mathbf{da}\, \mathbf{dp}/p_o = \infty, \ \forall\, \phi \in \mathcal{H} \tag{3.3}$$

Now a natural manifold for indexing Poincaré CS is *phase space* $\Gamma = \mathcal{P}_+^{\uparrow}(1,1)/T$, where T is the time translation subgroup. But T is *not* the stability subgroup of *any* vector in $\mathcal{H}_w$, so that the method of Perelomov is inapplicable here, and we need a more general scheme for getting relativistic CS.

3.1. *A first hint*

Instead of considering the integral (3.3) over the full group $\mathcal{P}_+^{\uparrow}(1,1)$, we restrict it to the quotient space Γ. Let $(\mathbf{q}, \mathbf{p})$ denote standard coordinates on Γ; then the measure $d\mathbf{q}\, d\mathbf{p}$ is left-invariant under $\mathcal{P}_+^{\uparrow}(1,1)$. First we introduce a section $\beta_o : \Gamma \to \mathcal{P}_+^{\uparrow}(1,1)$:

$$\beta_o(\mathbf{q},\mathbf{p}) = ((0, \mathbf{q}), \Lambda_{\mathbf{p}}). \tag{3.4}$$

Then it is easily shown that a vector $\eta \in \mathcal{H}_w$ is admissible mod(T,β_o), in the sense that the following integral converges, for all $\phi \in \mathcal{H}_w$:

$$\int_\Gamma |< U_w(\beta_o(\mathbf{q},\mathbf{p}))\, \eta \mid \phi >|^2 \, d\mathbf{q}\, d\mathbf{p} < \infty, \tag{3.5}$$

if and only if η is a finite energy state, i.e. η belongs to the domain of $P_0^{1/2}$, where P_o is the energy operator on $\mathcal{H}_w$:

$$P_o\, \phi(k) = k_o\, \phi(k), \quad \phi \in \mathcal{H}_w. \tag{3.6}$$

The corresponding momentum operator reads :

$$\mathbf{P}\, \phi(k) = \mathbf{k}\, \phi(k). \tag{3.7}$$

However, the vectors $\eta_{\mathbf{q},\mathbf{p}} \equiv U_w(\beta_o(\mathbf{q},\mathbf{p}))\, \eta$ do *not* generate a resolution of the identity. Instead, one has :

$$\int_\Gamma | \eta_{\mathbf{q},\mathbf{p}} >< \eta_{\mathbf{q},\mathbf{p}} | \, d\mathbf{q}\, d\mathbf{p} = A_o \neq \lambda I, \tag{3.8}$$

where A_o is a positive bounded multiplication operator, with bounded inverse :

$$(A_o\, \phi)(k) = A_o(k)\, \phi(k), \quad \phi \in \mathcal{H}_w, \tag{3.9a}$$

$$A_o(k) = \frac{2\pi}{m} \left[\langle P_o \rangle_\eta - \frac{\mathbf{k}}{k_o} \langle \mathbf{P} \rangle_\eta \right], \tag{3.9b}$$

where $\langle X \rangle_\eta \equiv < \eta \mid X\, \eta >$ denotes the mean value of the operator X in the state η.

At this point, we have two possibilities :

(1) To impose a further restriction to the state η, namely $\langle \mathbf{P} \rangle_\eta = 0$ (this is essentially a symmetry condition). Then A_o becomes a multiple of the identity and, after a suitable normalization, we get a genuine resolution of the identity.

(2) To consider weighted CS, i.e. states of the form $\tilde{\eta}_{\mathbf{q},\mathbf{p}} = T(\mathbf{q},\mathbf{p})\, \eta_{\mathbf{q},\mathbf{p}}$, where $T(\mathbf{q},\mathbf{p})$ is a (weight) operator, corresponding essentially to $A_0^{-1/2}$, in such a way that the new states generate a resolution of the identity and the usual machinery works :

$$\int_\Gamma | \tilde{\eta}_{\mathbf{q},\mathbf{p}} >< \tilde{\eta}_{\mathbf{q},\mathbf{p}} | \, d\mathbf{q}\, d\mathbf{p} = I. \tag{3.10}$$

3.2. *A second hint*

There is another approach to CS for $\mathcal{P}_+^\uparrow(1,1)$, namely using a *contraction* procedure from the (anti-)de Sitter group $SO_0(2,1)$ as the curvature $\kappa \to 0$ (in 1+3 dimensions, the corresponding contraction is $SO_0(2,3) \to \mathcal{P}_+^\uparrow(1,3)$). Under this process, the geometry behaves as it should :

$$\begin{array}{rcl} SO_0(2,1) & \to & \mathcal{P}_+^\uparrow(1,1) \\ \text{de Sitter universe} & \to & \text{Minkowski space} \\ \text{phase space :} \quad SO_0(2,1)/SO(2)) & \to & \mathcal{P}_+^\uparrow(1,1))/T \end{array} \tag{3.11}$$

(notice that SO(2) and T are the time translation subgroups in the respective universes). As for the representations, one has the result :

$$D_+^{(E_0)} \to U_w^{(m)} \tag{3.12}$$

where $D_+^{(E_0)}$ is the discrete series representation of $SO_0(2,1)$ with minimal weight E_0 and $U_w^{(m)}$ is the Wigner representation of $\mathcal{P}_+^\uparrow(1,1)$ with mass m, the parameters behaving under contraction as $E_0 \to \infty$, $\kappa \to 0$ and $E_0\,\kappa = m$. Notice, however, that CS of $SO_0(2,1)$ do *not* lead to CS for $\mathcal{P}_+^\uparrow(1,1)$, since $U_w^{(m)}$ is not square integrable : the limit $\kappa \to 0$ is too singular [9].

Now, for performing explicitly the contraction (3.11), one has to choose a certain factorization $SO_0(2,1) = P.SO(2)$, and this choice selects a *unique* section $\beta : \Gamma \to \mathcal{P}_+^\uparrow(1,1)$. In particular, the natural Cartan factorization used in Ref.4 leads to a section β_s, which is different from β_0 :

$$\beta_s(\mathbf{q},\mathbf{p}) = (q_s, \Lambda_p), \text{ with } q_s = (\,\mathbf{q}.\mathbf{p}\, (p_0+m)^{-1}, \mathbf{q}). \tag{3.13}$$

For this new section, all the previous results hold true [4]: the vectors which are admissible mod(T,β_s) are precisely the vectors $\eta \in D(P_0^{1/2})$, the difference being that now the multiplication operator A_s has always a purely continuous spectrum, for any vector η.

At this point, a new problem arises. Choosing either one of the two sections β_0, β_s, one obtains a set of CS with all the desired properties, but the two sets are different. What is their mutual relationship ? How general is this construction ?

3.3. *A general scheme*

A detailed analysis [3,4] of the cases described so far suggests a more general scheme. Suppose one is given :

. a locally compact group G and a closed subgroup H of G such that the coset space $X = G/H$ has an invariant measure ν (a quasi-invariant measure would suffice);

. a (global) Borel section $\beta : X \to G$;

. a continuous, irreducible unitary representation $U : G \to \mathcal{H}$.

Then we tentatively define : the representation U is *square integrable mod(H,β)* if there exists a vector $\eta \in \mathcal{H}$ such that the integral

$$A_\beta = \int_X U(\beta(x)) \,|\eta\rangle \langle\eta|\, U(\beta(x))^* \, d\nu(x) \tag{3.14}$$

converges weakly to a bounded, positive, invertible operator A_β, that is,

$$\int_X |\langle U(\beta(x))\,\eta \,|\, \phi\rangle|^2 \, d\nu(x) \equiv \langle\phi \,|\, A_\beta\, \phi\rangle < \infty , \; \forall\, \phi \in \mathcal{H} \tag{3.15}$$

If this concept is the correct one, it should yield the following results :

(1) An overcomplete family of coherent states

$$\mathfrak{S}_\beta = \{ \eta_{\beta(x)} \equiv U(\beta(x))\, \eta , \; x \in X \} ; \tag{3.16}$$

(2) A map $W_K : \mathcal{H} \to \mathcal{H}_K \subset L^2(X,\nu)$: $(W_K\, \phi)(x) = \langle \eta_{\beta(x)} \,|\, \phi \rangle$, whose range $\mathcal{H}_K \equiv W_K \mathcal{H}$ is complete with respect to a suitable scalar product, for which, in addition, the map W_K is *unitary* ;

(3) $\mathcal{H}_K$ is a reproducing kernel Hilbert space, with kernel

$$K(x,y) = \langle \eta_{\beta(x)} \,|\, A_\beta^{-1}\, \eta_{\beta(y)} \rangle ; \tag{3.17}$$

(4) If the CS system $\{\eta_{\beta(x)}\}$ does not generate a resolution of the identity, one can define weighted CS $\tilde{\eta}_x = T(x)\, \eta_{\beta(x)}$ that do so :

$$\int_X |\tilde{\eta}_x\rangle \langle\tilde{\eta}_x| \, d\nu(x) = I. \tag{3.18}$$

This general scheme applies to standard CS (with $X = G$ itself), to the Perelomov setup, and to $\mathcal{P}_+^\uparrow(1,1)$ with the two sections β_o, β_s. How general is it really ? In the last case, how much does the resulting CS system depend on the section ?

It turns out, as we shall see shortly, that the situation described above is a particular case of a much more general construction that we shall explain in detail in Section 4, and which in fact makes no reference to group representations at all ! As for the dependence on β, it will be reduced to an equivalence relation among CS systems.

4. Reproducing triples

The only ingredient we need is a generalization of the resolution of the identity (2.4). Given a measure space (X,ν) and a Hilbert space $\mathcal{H}$, let A be a bounded, positive, invertible operator on $\mathcal{H}$ and $F : X \to \mathcal{L}(\mathcal{H})^+$ a ν–measurable, positive operator valued function. Then we say that $\{\mathcal{H},F,A\}$ is a *reproducing triple* if the following relation holds true :

$$\int_X F(x)\, d\nu(x) = A, \tag{4.1}$$

where the operator integral is to be understood in the weak sense. This will be the central notion in the sequel. Of course, if rank $F(x) = 1$, $\forall x \in X$, and $A = I$, we recover the usual resolution of the identity (2.4). Notice that the map $\Delta \mapsto a(\Delta) = \int_\Delta F(x)\, d\nu(x)$, where Δ runs over the Borel subsets of X, is a positive operator valued (POV) measure on X [8], with $a(X) = A$.

Let us examine the main properties of reproducing triples.

4.1. *Overcomplete family of states (OFS)*

For each $x \in X$, we denote by $\mathbb{P}(x)$ the projection on the orthogonal complement of Ker $F(x)$: $\mathbb{P}(x)\mathcal{H} = (\text{Ker } F(x))^\perp$, and define on $\mathbb{P}(x)\mathcal{H}$ the new scalar product :

$$<\phi_x \mid \psi_x>_x \; = \; <\phi_x \mid F(x)\, \psi_x>_{\mathcal{H}}, \;\; \forall\, \phi_x, \psi_x \in \mathbb{P}(x)\mathcal{H} \tag{4.2}$$

Let $\mathcal{K}_x$ be the completion of $\mathbb{P}(x)\mathcal{H}$ with respect to the corresponding norm $\|\cdot\|_x$, and $\{v_i(x),\, i = 1,\dots, d(x) \equiv \dim \mathbb{P}(x)\mathcal{H}\}$ an orthonormal basis of $\mathcal{K}_x$, contained in $\mathbb{P}(x)\mathcal{H}$. Then the set :

$$\mathfrak{S} = \{\, \eta_x^i \mid \eta_x^i = F(x) v_i(x)\,,\, i = 1,2,\dots, d(x)\, ;\, x \in X\, \}, \tag{4.3}$$

is an *overcomplete family of states*, and one has :

$$F(x) = \sum_{i=1}^{d(x)} | \eta_x^i >< \eta_x^i |, \tag{4.4}$$

$$\int_X \sum_{i=1}^{d(x)} | \eta_x^i >< \eta_x^i |\, d\nu(x) = A \text{ (weak integral).} \tag{4.5}$$

Overcompleteness means that $\mathfrak{S}$ is a total set in $\mathcal{H}$: $\mathfrak{S}^{\perp} = 0$. The states $\{\eta_x^i\}$ are the analogues of the usual coherent states; indeed, they generalize all the CS systems described previously.

4.2. *Reproducing kernel Hilbert space*

A nice feature of the usual construction of coherent states is the presence of a reproducing kernel Hilbert space. To obtain the same property here, we need first a Hilbert space of functions. For that purpose, consider the direct integral (measurability is easy to prove)

$$\tilde{\mathcal{H}} = \int_X^{\oplus} \mathcal{K}_x \, d\nu(x) \tag{4.6}$$

and the map $W_K : \mathcal{H} \to \tilde{\mathcal{H}}$, defined by

$$(W_K \phi)(x) = \sum_{i=1}^{d(x)} \langle \eta_x^i \,|\, \phi \rangle\, v_i(x) = \mathbb{P}(x)\phi \in \mathcal{K}_x \tag{4.7}$$

On $W_K[D(A^{-1})] \subset \tilde{\mathcal{H}}$, we define the new scalar product

$$\langle \Phi \,|\, \Psi \rangle_K = \langle \Phi \,|\, W_K A^{-1} W_K^{-1} \Psi \rangle_{\tilde{\mathcal{H}}} . \tag{4.8}$$

Then it is easy to show that the map W_K is an isometry from $D(A^{-1})$ onto $W_K[D(A^{-1})]$, thus it extends to a unitary map $W_K : \mathcal{H} \to \mathcal{H}_K$, where $\mathcal{H}_K$ denotes the completion of $W_K[D(A^{-1})]$ with respect to the norm $\|\cdot\|_K$. This implies that $\mathcal{H}_K = W_K \mathcal{H} \subset \tilde{\mathcal{H}}$, that is, $\mathcal{H}_K$ is a space of vector-valued functions . As a consequence, we get a new reproducing triple $\{ \mathcal{H}_K, F_K, A_K\}$, where $F_K(x) = W_K F(x) W_K^{-1}$ and A_K is the (unique) self-adjoint extension of $W_K A W_K^{-1}$, the latter being defined on $\mathrm{Ran}(W_K)$. Moreover, $\mathcal{H}_K$ possesses a *reproducing kernel* $K(x,y) : \mathcal{K}_y \to \mathcal{K}_x$, with the following properties:

. $K(x,x) > 0, \ \forall\, x \in X$ (4.9a)

. $K(x,y)^* = K(y,x)$ (4.9b)

. $\Phi(x) = \int_X K(x,y)\, \Phi(y)\, d\nu(y), \ \forall\, \Phi \in \mathcal{H}_K$, (4.9c)

the last relation being the reproducing property.

When the rank of F(x) is constant and finite : Rank $F(x) = n < \infty$, $\forall\, x \in X$, the direct integral (4.6) reduces to a space of $\mathbb{C}^n$-valued function $\tilde{\mathcal{H}} \approx L^2(X,\nu;\mathbb{C}^n)$, and the reproducing kernel becomes an n × n matrix $K(x,y) : \mathbb{C}^n \to \mathbb{C}^n$, with matrix elements

$$K_{ij}(x,y) = < \eta^i_x \mid A^{-1} \eta^j_y >_{\mathcal{H}}. \qquad (4.10)$$

4.3. *Equivalent OFS*

Consider now two reproducing triples $\{\mathcal{H}, F, A\}$ and $\{\mathcal{H}, F', A'\}$ with corresponding OFS $\mathfrak{S}$ and $\mathfrak{S}'$. Then we say that the OFS $\mathfrak{S}'$ is *weighted* with respect to the OFS $\mathfrak{S}$, $\mathfrak{S}' \prec \mathfrak{S}$, if there exists a (weakly) measurable operator valued function $T : X \to \mathcal{L}(\mathcal{H})$, such that

$$F'(x) = T(x)\, F(x)\, T(x)^* , \ \forall\, x \in X. \qquad (4.11)$$

The two OFS $\mathfrak{S}, \mathfrak{S}'$, are said to be *equivalent* if $\mathfrak{S}' \prec \mathfrak{S}$ and $\mathfrak{S} \prec \mathfrak{S}'$. In this case, we write $\mathfrak{S} \sim \mathfrak{S}'$. In particular, if the two reproducing triples have constant, finite rank n, n', respectively, it is easy to show that $n' \leq n$ implies $\mathfrak{S}' \prec \mathfrak{S}$. Thus, when n' = n, the two OFS $\mathfrak{S} = \{\eta^i_x , i = 1,...n\}$ and $\mathfrak{S}' = \{\eta'^j_x , j = 1,...n\}$ are equivalent, which means there exist two measurable families of bounded operators $\{T(x), T'(x)\}$, such that, for every $x \in X$:

$$T(x)\, T'(x)\, \mathbb{P}'(x) = \mathbb{P}'(x), \quad T'(x)\, T(x)\, \mathbb{P}(x) = \mathbb{P}(x), \qquad (4.12)$$

$$\eta'^i_x = T(x)\, \eta^i_x , \quad \eta^i_x = T'(x)\, \eta'^i_x \quad (i = 1, ...n). \qquad (4.13)$$

5. Applications

5.1. *Continuous frames*

If the triple $\{\mathcal{H}, F, A\}$ has constant, finite rank n and, in addition, A^{-1} is a bounded operator, then $\{\mathcal{H}, F, A\}$ is called a (continuous) *frame* . In that case, indeed, (4.5) leads to the "frame" condition :

$$m(A)\, \| \phi \|^2 \leq \sum_{i=1}^{n} \int_X \mid < \eta^i_x \mid \phi > \mid^2 \, d\nu(x) \leq M(A)\, \| \phi \|^2, \qquad (5.1)$$

where $m(A) = \inf \sigma(A)$ and $M(A) = \sup \sigma(A)$ (the so-called "frame bounds"). The frame is called *tight* if $m(A) = M(A)$, or equivalently if $A = \lambda I$. The terminology used here is borrowed from the theory of nonorthogonal expansions [11,12] : when X is discrete and ν is the counting measure, (4.5) reads :

$$\sum_{k=1}^{\infty} \sum_{i=1}^{n} | \eta_k^i >< \eta_k^i | = A, \tag{5.2}$$

and this is indeed the very statement that $\{\eta_k^i\ , i = 1, \ldots n, k = 1,2 \ldots\}$ is a frame in the usual sense.

Given a frame $\{\mathcal{H}, F, A\}$, we define the *dual frame* as $\{\mathcal{H}, F', A^{-1}\}$, where

$$F'(x) = A^{-1} F(x) A^{-1}, \quad x \in X, \tag{5.3}$$

is also a positive bounded operator of rank n. In other words, $\eta_x^{'i} = A^{-1}\eta_x^i$ (i=1,2,...n ; $x \in X$). Clearly, the frame bounds of the dual frame are $M(A)^{-1}$ and $m(A)^{-1}$. In the discrete case, the dual frame is a crucial ingredient for various *reconstruction formulas* [6,7,12]; in the present language, this means expanding elements of $\mathcal{H}$ in terms of those of $\mathcal{H}_K$, by a suitable approximation (truncated power expansion) of the operator W_K^{-1}.

5.2. *Coherent states*

Now we apply the general formalism just developed to the situation described at the beginning. Let G be a l.c. group and H a closed subgroup of G such that the coset space $X = G/H$ has an invariant measure ν. Choose a (global) Borel section $\beta : X \to G$. Given a representation $U : G \to \mathcal{H}$ and a finite rank operator $F \in \mathcal{L}(\mathcal{H})^+$, define the function $F_\beta : X \to \mathcal{L}(\mathcal{H})^+$ by covariance

$$F_\beta(x) = U(\beta(x))\, F\, U(\beta(x))^*. \tag{5.4}$$

In this notation, the representation U is said to be *square integrable mod(H,β)* if there exist F and $A_\beta \in \mathcal{L}(\mathcal{H})^+$, with rank F = n, such that $\{\mathcal{H}, F_\beta, A_\beta\}$ is a reproducing triple, i.e. $A_b > 0$ is invertible and the following operator integral converges weakly :

$$\int_X F_\beta(x)\, d\nu(x) = A_\beta . \tag{5.5}$$

This class of representations enjoys many properties analogous to those of the square integrable representations, in the usual sense. We describe the most interesting ones.

(i) Covariance

If U is square integrable mod(H,β), then it is also square integrable mod(H,β_g),$\forall g \in G$, where $\beta_g(x) = g\beta(g^{-1}.x)$ ($x \in X$) is the section obtained from β by the action of G on X.

(ii) Coherent states

Using the spectral decomposition $F = \sum_{i=1}^{n} \lambda_i \mid u_i >< u_i \mid$, where $\lambda_i > 0$ and $\{u_i\}_{i=1}^{n}$ is an orthonormal set, define the vectors :

$$\eta^i_{\beta(x)} = \lambda_i^{1/2}\, U(\beta(x))u_i = U(\beta(x))\eta_i \quad (i = 1, \ldots, n), \tag{5.6}$$

that is, the orbit of the family $\{\eta_i\}$.

Then, as in Section 3.1, the vectors (5.6) constitute an overcomplete family of coherent states

$$\mathfrak{S}_\beta = \{ \eta^i_{\beta(x)} , i = 1,2,\ldots,n,\ x \in X \} \tag{5.7}$$

and

$$A_\beta = \int_X \sum_{i=1}^{n} \mid \eta^i_{\beta(x)} >< \eta^i_{\beta(x)} \mid d\nu(x), \tag{5.8}$$

the integral converging weakly as usual. From there on, the whole machinery of coherent states may be developed, fundamental isometry, reproducing kernel Hilbert space, etc., but everything depends on the choice of the section β !

(iii) Weighted CS, quasi-coherent states

Let U be a representation of G, square integrable mod(H,β), and $\mathfrak{S}_\beta$ a CS system associated to the reproducing triple $\{\mathcal{H}, F_\beta, A_\beta\}$. Let $\{\mathcal{H}, F', A'\}$ be an arbitrary reproducing triple, with an OFS $\mathfrak{S}' = \{\eta'^i_x , i=1, \ldots ,n\}$. Suppose that $\mathfrak{S}' \prec \mathfrak{S}_\beta$, via the operator function $T(x) \in \mathcal{L}(\mathcal{H})$. For instance, if rank F = rank F'= n, this means :

$$\eta'^i_x = T(x)\, \eta^i_{\beta(x)} \quad (i=1, \ldots ,n\, ;\ x \in X). \tag{5.9}$$

But the two families of vectors appearing in (5.9) play different roles. The vectors $\{\eta^i_{\beta(x)}\}$ are true coherent states, that is, the orbit of the family $\{\eta_i\}$, as indicated above, whereas the vectors $\{\eta'^i_x\}$ are not. They are weighted with respect to the others, hence they will be called *weighted CS* or *quasi-coherent states.*

(iv) Equivalent families of coherent states

Let β, β' two sections $X \to G$, such that U is square integrable *both* mod(H,β) and mod(H,β'), that is, there exist operators $F, A_\beta, F', A'_{\beta'} \in \mathcal{L}(\mathcal{H})^+$, with rank F = n, rank F' = n', such that $\{\mathcal{H}, F_\beta, A_\beta\}$ and $\{\mathcal{H}, F'_{\beta'}, A'_{\beta'}\}$ are reproducing triples. Let $\mathfrak{S}_\beta$ and $\mathfrak{S}_{\beta'}$ be corresponding families of coherent states. Then $\mathfrak{S}_\beta \prec \mathfrak{S}_{\beta'}$ if $n \le n'$, and $\mathfrak{S}_\beta$ and $\mathfrak{S}_{\beta'}$ are *equivalent* if n = n' (even if F ≠ F', provided they have the same rank). In particular, if we consider the two CS systems $\mathfrak{S}_\beta$ and $\mathfrak{S}_{\beta'}$ constructed from a given F, with two different sections β, β', they are equivalent . For instance, the system $\mathfrak{S}_{\beta_g}$ is equivalent to $\mathfrak{S}_\beta$, for any $g \in G$. This equivalence means that, for every $x \in X$, there exist (non-unique) bounded operators $T_{\beta'\beta}(x)$, $T_{\beta\beta'}(x)$ such that, for all i = 1, ...n and $x \in X$:

$$\eta^j_{\beta'(x)} = T_{\beta'\beta}(x)\, \eta^i_{\beta(x)} \quad \text{and} \quad \eta^i_{\beta(x)} = T_{\beta\beta'}(x)\, \eta^j_{\beta'(x)}. \tag{5.10}$$

The simplest choice is

$$T_{\beta'\beta}(x) = U(\beta'(x))\, \mathbb{P}\, U(\beta(x))^*, \tag{5.11}$$

where $\mathbb{P}$ is the projection on $(\mathrm{Ker}\, F)^\perp$.

Thus, in this precise sense, we may say that the set of coherent states associated to the representation U does not depend on the choice of the section.

6. Back to the Poincaré group $\mathcal{P}^\uparrow_+(1,1)$

As an illustration of the general theory developed in Section 4, we shall now construct general CS systems for the Poincaré group $\mathcal{P}^\uparrow_+(1,1) = SO_o(1,1) \wedge \mathbb{R}^2$ described in Section 3 (see also Refs.3-5). Another example has been treated by the same method in a recent paper of Torrésani [13], namely the affine Weyl-Heisenberg group.

Using standard coordinates **p**, **q** on Γ, a section $\beta : \Gamma \to \mathcal{P}^\uparrow_+(1,1)$ will be parametrized as $\beta(\mathbf{q},\mathbf{p}) = (\hat{\mathbf{q}}, \Lambda_{\hat{\mathbf{p}}})$. We already know two examples [3,4], namely the Galilean section β_o and the DeSitterian section β_s. Then a general section may be written as

$$\beta(\mathbf{q},\mathbf{p}) = \beta_o(\mathbf{q},\mathbf{p})\, ((f(\mathbf{q},\mathbf{p}), \mathbf{0}), I), \tag{6.1}$$

which implies

$$\hat{q}_o = \frac{p_o}{m} f(\mathbf{q},\mathbf{p}), \quad \hat{\mathbf{q}} = \mathbf{q} + \frac{\mathbf{p}}{m} f(\mathbf{q},\mathbf{p}), \quad \hat{\mathbf{p}} = \mathbf{p}. \tag{6.2}$$

We will consider, in particular, the class $\mathcal{S}_A$ of *affine* sections defined by the two conditions :

$$f(\mathbf{q},\mathbf{p}) = \varphi(\mathbf{p}) + \mathbf{q}\,.\,\theta(\mathbf{p}), \tag{6.3}$$

$$-\frac{p_o - \mathbf{p}}{m} < \theta(\mathbf{p}) < \frac{p_o + \mathbf{p}}{m}\,. \tag{6.4}$$

The function φ in (6.3) is largely irrelevant and will usually be ignored. With $\varphi = 0$, the formulas (6.2) imply that the section coordinates $\hat{q} = (\hat{q}_o, \hat{\mathbf{q}})$ satisfy the relation :

$$\hat{q}_o = \frac{p_o\, \hat{\mathbf{q}} \,.\, \theta(\mathbf{p})}{m + \mathbf{p} \,.\, \theta(\mathbf{p})} \tag{6.5}$$

For fixed p, this means that choosing θ amounts to fixing a particular reference frame in $\hat{q}$-space. Several concrete examples will be displayed below. Notice also that, for $\varphi = 0$, the condition (6.4) is equivalent to $\hat{q}$ being spacelike, $\hat{q}_o^2 - \hat{\mathbf{q}}^2 < 0$.

Now, according to the general theory, a vector $\eta \in \mathcal{H}_w$ is admissible mod(T,β) for a section $\beta \in \mathcal{S}_A$ if

$$A_\beta = \int_{\mathbb{R}^2} U_w(\beta(\mathbf{q},\mathbf{p}))\,|\,\eta > <\eta\,|\,U_w(\beta(\mathbf{q},\mathbf{p}))^* \; d\mathbf{q}\, d\mathbf{p} \tag{6.6}$$

is a bounded, positive, invertible operator on $\mathcal{H}_w$. Then an explicit calculation yields the following results.

(1) The vector $\eta \in \mathcal{H}_w$ is admissible mod(T,β), for any section $\beta \in \mathcal{S}_A$, iff $\eta \in D(P_o^{1/2})$, where P_o is the energy operator (3.6).

(2) When η is admissible, $A_\beta \equiv A_\beta^\eta$ is a multiplication operator given by

$$(A_\beta^\eta\, \psi)(k) = A_\beta^\eta\,(k)\, \psi(k) \tag{6.7}$$

$$A_\beta^\eta(k) = \int_{\mathcal{V}_m^+} \mathcal{A}_\beta(k, p)\,|\,\eta(p)\,|^2\, \frac{dp}{p_o} \tag{6.8}$$

where $\quad \mathcal{A}_\beta(k, p) = 2\pi\, (\Lambda_k^{-1} p)_o\, [k_o - \theta(\underline{\Lambda_p^{-1} k}).\mathbf{p}]^{-1} \tag{6.9}$

Here $\underline{t} \equiv \mathbf{t}$ is another notation for the space component of the 2-vector t.

The crucial fact is that the kernel $\mathcal{A}_\beta$ satisfies the following inequalities, which result from (6.4) :

$$\frac{2\pi}{m}\,(p_o - |\,\mathbf{p}\,|) < \mathcal{A}_\beta(k, p) < \frac{2\pi}{m}\,(p_o + |\,\mathbf{p}\,|), \tag{6.10}$$

for all k, p $\in \mathcal{V}_m^+$ and any θ obeying (6.4), in particular, for any section $\beta \in \mathcal{S}_A$.

(3) As a consequence, every vector η which is admissible mod(T,β) generates a reproducing triple $\{\mathcal{H}_w, F_\beta, A_\beta^\eta\}$, of constant rank $n = 1$, and in fact a frame. The frame bounds obey the following estimates (independent of the choice of the section $\beta \in \mathcal{S}_A$) :

$$m(A^{\eta}_{\beta}) \;\geq\; m(\eta) \equiv \frac{2\pi}{m} \langle P_o - |\mathbf{P}| \rangle_{\eta} , \tag{6.11a}$$

$$M(A^{\eta}_{\beta}) \;\leq\; M(\eta) \equiv \frac{2\pi}{m} \langle P_o + |\mathbf{P}| \rangle_{\eta} . \tag{6.11b}$$

(4) It follows from the general theory developed in Section 5 that, for every affine section $\beta \in \mathcal{S}_A$, any vector η which is admissible mod(T,β) generates a family of coherent states indexed by points in phase space :

$$\mathfrak{S}_{\beta} = \{ \eta_{\beta(\mathbf{q},\mathbf{p})} \mid (\mathbf{q},\mathbf{p}) \in \Gamma \}. \tag{6.12}$$

Furthermore, all these CS systems are equivalent.

We conclude this section with some concrete examples of affine sections, specified by the form of their respective relation (6.5) :

. the Galilean section β_o, corresponding to $\hat{q}_o = 0$: here the frame is tight iff $\langle \mathbf{P} \rangle_{\eta} = 0$.

. the DeSitterian section β_S, corresponding to $\hat{q}_o = \hat{\mathbf{q}}.\mathbf{p}\,(p_o+m)^{-1}$; in this case, the frame is *never* tight; the spectrum of A^{η}_{β} is a whole interval, strictly contained in $[m(\eta),M(\eta)]$.

. the section β_{DB} of DeBièvre [8], corresponding to $\hat{q}_o = \hat{\mathbf{q}}.\mathbf{p}\,(p_o)^{-1}$; here the frame is *always* tight, for every admissible η.

7. Conclusions

The concept of reproducing triple leads to a general formalism which has all the features usually associated to coherent states : overcomplete family of states, reproducing kernel Hilbert space, quasi-coherent states, etc. Two types of applications emerge : first, the notion of continuous frame; second, a construction of CS systems associated to group representations which are square integrable only on a quotient space. In that case, CS systems corresponding to different sections are equivalent, and a genuine resolution of the identity may always be obtained with help of quasi-CS.

This construction applies to various concrete examples, where the standard approach of Perelomov fails : the Poincaré groups $\mathcal{P}^{\uparrow}_{+}(1,1)$ and $\mathcal{P}^{\uparrow}_{+}(1,3)$, the Galilei group, the Euclidean group [8], the affine Weyl-Heisenberg group [13]. Preliminary results of Ali and Goldin [14] indicate that it even extends to some infinite-dimensional groups, such as the diffeomorphism group of a manifold.

References

1. J.R.Klauder, B.S.Skagerstam, *Coherent States - Applications in Physics and Mathematical Physics*, World Scientific, Singapore 1985
2. A.Perelomov, *Generalized Coherent States and their Applications,* Springer-Verlag, Berlin 1986
3. S.T.Ali, J.-P.Antoine, Coherent states of the 1+1 dimensional Poincaré group : square integrability and a relativistic Weyl transform, *Ann. Inst. H. Poincaré* **51** (1989) 23-44
4. S.T.Ali, J.-P.Antoine, J.-P.Gazeau, De Sitter to Poincaré contraction and relativistic coherent states, *Ann. Inst. H. Poincaré* **52** (1990) 83-111
5. S.T.Ali, J.-P.Antoine, J.-P.Gazeau, Square integrability of group representations on homogeneous spaces. I. Reproducing triples and frames. II. Generalized square integrability and equivalent families of coherent states, *Ann. Inst. H. Poincaré* (1991) (to appear)
6. A.Grossmann, J.Morlet, T.Paul, Transforms associated to square integrable group representations I : General results, *J. Math. Phys.* **26** (1985) 2473-2479
7. A.Grossmann, J.Morlet, T.Paul, Transforms associated to square integrable group representations II : Examples, *Ann. Inst. H. Poincaré* **45** (1986) 293-309
8. S.DeBièvre, Coherent states over symplectic homogeneous spaces, *J. Math. Phys.* **30** (1989) 1401-1407
9. S.DeBièvre, A.El Gradechi, Quantum Mechanics and coherent states on the Anti-de Sitter space-time and their Poincaré contraction, *Preprint PAR-LPTM 90*, Université Paris VII, 1990
10. S.T.Ali, Stochastic localization, quantum mechanics on phase space and quantum space-time, *Rivista Nuovo Cim.* **8**, #11 (1985) 1-128
11. R.J.Duffin, A.C.Schaeffer, A class of nonharmonic Fourier series, *Trans. Amer. Math. Soc.* **72** (1952) 341
12. I.Daubechies, A.Grossmann, Y.Meyer, Painless nonorthogonal expansions, *J. Math. Phys.* **27** (1986) 1271-1283
13. B.Torrésani, Wavelets associated with representations of the affine Weyl-Heisenberg group, *J. Math. Phys.* (1991) (to appear)
14. S.T.Ali, G.A.Goldin, Quantization, coherent states and diffeomorphism groups, *Preprint, Concordia U.*, Montréal 1990.

SELF-DUAL YANG-MILLS FIELDS: A NEW DESCRIPTION*

PAVEL ŠŤOVÍČEK
Department of Mathematics, Faculty of Nuclear Science, Czech Technical University, Prague, Czechoslovakia
and
JINR Dubna, USSR

ABSTRACT

Takasaki proposed a new approach to SDYM fields based on a formulation of the initial value problem. The author has shown the initial condition to have a clear geometric interpretation providing a new description of the globally defined SDYM solutions.

1. Introduction

Let us first summarize some known facts about the SDYM solutions. The Penrose transformation enables to code self-dual solutions on $S^4 = \mathbb{R}^4 \cup \{ \infty \}$ into holomorphic data[1,2] and investigate in this way holomorphic vector bundles on the projective space $\mathbb{P}^3$, fulfilling some special additional properties. This is the basis for the ADHM construction[3]. Donaldson[4] has introduced the notion of framed holomorphic (instanton) bundles by distinguishing a holomorphic trivialization on a fixed (real) line in the projective space and has reduced the problem further to the investigation of framed holomorphic bundles on the projective plane $\mathbb{P}^2$, now without any additional condition. The rank r of the vector bundle correspondes to the rank of the gauge group $SU(r)$ and the topological charge (instanton number) c is equal to the second Chern class. The bundle is holomorphically trivial on almost all lines in $\mathbb{P}^2$, the exceptional lines are called jumping lines. The

*Dedicated to the memory of Jiří Blank

number of jumping lines passing through a fixed point is equal (when counted with appropriate multiplicities) to c .[5]

Takasaki[6] has introduced the notion of the initial condition in the case when the SDYM solution exists locally. The author[7] has shown that one can relate to each gauge equivalence class of SDYM solutions a special "canonical initial condition" visualized then for the globally defined solutions as a matrix-valued function on $\mathbb{P}^3$ (resp. $\mathbb{P}^2$ after the Donaldson's restriction) and called "canonical transition function".

The aim of this note is to sketch a description of the instanton moduli space $\mathrm{M}(2,c)$ (the space of global SDYM solutions modulo gauge transformations, with the fixed gauge group $SU(2)$ and the topological charge equal to c) as an algebraic set in a finite-dimensional complex space.

2. The Canonical Transition Function

First, let us introduce some notation. Let (z_0,z_1,z_2) be homogeneous coordinates on $\mathbb{P}^2$; $\xi = z_1/z_0$, $\eta = z_2/z_0$; $\mathcal{L}_0 \subset \mathbb{P}^2$ be the fixed line determined by the equation : $z_0 = 0$; $P_0, P_\infty \in \mathcal{L}_0$ be two fixed points determined by the equations - P_0: $z_0 = z_1 = 0$, P_∞: $z_0 = z_2 = 0$; $\mathcal{S}_0$, $\mathcal{S}_\infty$ be the sets of jumping lines passing through P_0, P_∞ , respectively ; $\mathcal{U}_0 = \mathbb{P}^2 \setminus \mathcal{S}_\infty$, $\mathcal{U}_\infty = \mathbb{P}^2 \setminus \mathcal{S}_0$ ($P_0 \in \mathcal{U}_0$, $P_\infty \in \mathcal{U}_\infty$); $\mathcal{O}\mathrm{M}(r,c)$ be the moduli space of framed holomorphic bundles on $\mathbb{P}^2$ identified with the instanton moduli space $\mathrm{M}(r,c)$ via the Donaldson's result.[4]

One can prolongate the distinguished trivialization on $\mathcal{L}_0$ along all the lines being not jumping and passing through P_∞ (resp. P_0) to get holomorphic trivializations $\{ s_j \}$ (resp. $\{ \hat{s}_j \}$) on $\mathcal{U}_0$ (resp. $\mathcal{U}_\infty$). The corresponding transition function

$$G = (G_{jk}) \quad , \quad s_k = \sum_j \hat{s}_j \, G_{jk} \; ,$$

is called the canonical transition function (CTF).

Proposition 1. $G(\xi,\eta)$ is unimodular and can be written in the form

$$G(\xi,\eta) = 1 + \frac{1}{p_0(\xi)\; p_\infty(\eta)}\, \mathcal{R}(\xi,\eta) ,$$

$$p_0(\xi) = \xi^c + \sum_{j=1}^{c} s_j\, \xi^{c-j} , \quad p_\infty(\eta) = \eta^c + \sum_{j=1}^{c} t_j\, \eta^{c-j} ,$$

$$\mathcal{R}(\xi,\eta) = \sum_{j,k=1}^{c} X_{jk}\, \xi^j \eta^k , \tag{1}$$

where $s_j, t_k \in \mathbb{C}$, $X_{jk} \in \mathbb{C}^{r,r}$ and the zero sets of the polynomials $p_0(\xi)$, $p_\infty(\eta)$ coincide with the sets of jumping lines $\mathcal{S}_0$, $\mathcal{S}_\infty$, respectively.

Proposition 2. The mapping relating to each point from $\mathcal{M}(r,c)$ the corresponding CTF is injective and so one gets an embedding $\mathcal{M}(r,c) \hookrightarrow \mathbb{C}^{2c+r^2c^2}$ The image is a (locally) algebraic set.

Proposition 3. Any unimodular matrix-valued function $G'(\xi,\eta)$ on $\mathbb{P}^2$ having the form from Eq.1 determines unambiguously a framed holomorphic bundle on $\mathbb{P}^2$ and G' coincides with the corresponding CTF.

Warning. One does not know a priori whether the zero set of the polynomial $p'_0(\xi)$ coincides with the set of jumping lines $\mathcal{S}'_0$; the roots are correct but the multiplicities may differ. Consequently, the topological charge c_2 need not equal to c' and must be computed.

In what follows, we restrict ourselves to the rank $r = 2$.

3. The Topological Charge

We shall suppose the polynomial $\mathcal{P}'_0(\xi)$ to have no common roots with $\mathcal{R}'_\eta(\xi) := \mathcal{R}'(\xi,\eta)$ and to have no multiple roots. The analogous assumptions hold for $\mathcal{P}'_\infty(\eta)$. Actually, this is the generic case. Hence $G'(\xi,\eta)$ can be written in the form

$$G'(\xi,\eta) = 1 + \sum_{j,k=1}^{c'} \frac{1}{(\xi-x_j)(\eta-y_k)} R_{jk} ,$$

where $x_j, y_k \in \mathbb{C}$, $R_{jk} \in \mathbb{C}^{2,2}$. We shall restrict ourselves further to the case when $\operatorname{tr} R_{jk} \neq 0$ for all j,k. Again, this condition is generically fulfilled. Now, the second Chern class equals to the sum of multiplicities $\sigma(x_j)$ of the jumping lines passing through the point P_0 ($\xi - x_j = 0$; $j = 1, \ldots, c'$). The value of $\sigma(x_j)$ can be computed in the following way. Write $G'(\xi,\eta)$ in the form

$$G'(\xi,\eta) = 1 + (\xi-x_j)^{-1} [\eta^{-1} G_1(\xi) + \eta^{-2} G_2(\xi) + \ldots] ,$$

and let $Z_0 + (\xi-x_j) Z_1 + (\xi-x_j)^2 Z_2 + \ldots$; $Z_k \in \mathbb{C}^{2c',2c'}$,

be the Taylor expansion of the matrix-valued function

$$\begin{pmatrix} (\xi-x_j)\,1, & 0, \ldots\ldots, & 0 \\ G_1 & ,(\xi-x_j)\,1, \ldots\ldots, & 0 \\ \vdots & \vdots & \vdots \\ G_{c'-1} & , G_{c'-2} \quad ,\ldots\ldots, & (\xi-x_j)\,1 \end{pmatrix} \times$$

$$\times \begin{pmatrix} G^0_{c'}, & G^0_{c'-1} & ,\ldots\ldots, & G^0_1 \\ G^0_{c'+1} & , G^0_{c'} & ,\ldots\ldots, & G^0_2 \\ \vdots & \vdots & & \vdots \\ G^0_{2c'-1} & , G^0_{2c'-2} & ,\ldots\ldots, & G^0_{c'} \end{pmatrix}$$

Here we have used the notation

$$A^{o} = tr(A)\ 1 - A \quad ,$$

for any 2×2 matrix A .

Proposition 4.

$$\sigma(x_j) = 2c' - \text{rank} \begin{pmatrix} z_0 & 0 \\ z_1 & z_0 \end{pmatrix} .$$

Proposition 5. The equality $c_2 = c'$ holds only if the necessary condition

$$R^{o}_{jk}\ R_{jl} = 0 \quad \text{for all} \quad j,k,l, \qquad (2)$$

is satisfied.

4. Concluding Remark

The Eqs.2 can be rewritten in terms of the parameters s_j, t_k, X_{jk} appearing in Eq.1. In this way one gets the desired equations determining the moduli space $\mathcal{OM}(2,c)$ as a locally algebraic set in $\mathbb{C}^{2c+4c^2}$, as claimed in Prop.2.

5. References

1. R. Ward, *Phys. Lett.* **61A** (1977) 81.
2. M. Atiyah, N. Hitchin and I. Singer, *Proc. R.Soc.Lond.* **A362** (1978) 425.
3. M. Atiyah, V. Drinfeld, N. Hitchin and Yu. Manin, *Phys. Lett.* **65A** (1978) 185.
4. S.K. Donaldson, *Commun. Math. Phys.* **93** (1984) 453.
5. Ch. Okonek, M. Schneider and H. Spindler, *Vector bundles on complex projective spaces*, (Birkhäuser, Boston, 1980).
6. K. Takasaki, *Commun. Math. Phys.* **94** (1984) 35.
7. P. Šťovíček, *Commun. Math. Phys.* (to appear).

LIST OF PARTICIPANTS

S.Albeverio	Ruhr University Bochum, Germany
A.Amann	ETH-Center Zurich, Switzerland
C.Andronache	Institute of Physics and Technology of Materials Bucharest, Roumania
V.-N.-P.Anghel	Institute of Nuclear Power Reactors Pitesti, Roumania
J.-P.Antoine	Catholic University of Louvain Louvain-la-Neuve, Belgium
E.Balslev	Aarhus University Aarhus, Denmark
M.Bednář	Institute of Physics, CSAS Prague, Czechoslovakia
C.M.Bender	Washington University St.Louis, USA
F.Bentosela	CPT-CNRS Marseille, France
S.De Bièvre	University Paris VII Paris, France
J.Bogdanowicz	Soltan Institute for Nuclear Studies Otwock-Swierk, Poland
J.Brasche	Ruhr University Bochum, Germany
R.K.Bullough	UMIST Manchester, UK
M.Chiranov	ISEH Foscani, Roumania
J.-M.Combes	University of Toulon Toulon, France

O.Cristea — University of Cluj, Cluj, Roumania

F.Defever — Institute for Theoretical Physics, Heverlee, Belgium

M.Demuth — Institute of Mathematics, Berlin, Germany

J.Dittrich — Nuclear Physics Institute, CSAS, Řež, Czechoslovakia

J.Drahoš — Charles University, Prague, Czechoslovakia

P.Duclos — University of Toulon and Var, Toulon, France

P.Exner — Nuclear Physics Institute, CSAS, Řež, Czechoslovakia

G. Ganbold — JINR, Dubna, USSR

H.N.Gheorghiu — Institute for Physics and Nuclear Engineering, Bucharest, Roumania

D.Gilbert — Sheffield City Polytechnic, Sheffield, UK

H.Grosse — University of Vienna, Vienna, Austria

M.Havlíček — Czech Technical University, Prague, Czechoslovakia

K.-E.Hellwig — Technical University, Berlin, Germany

L.Hlavatý — Institute of Physics, CSAS, Prague, Czechoslovakia

T.Hoffmann-Ostenhof — Institute for Theoretical Chemistry, Vienna, Austria

J.Hronek — Institute of Physics, CSAS, Prague, Czechoslovakia

V.Jakšić	CALTECH Pasadena, USA
A.Jensen	Aalborg University Aalborg, Denmark
G.Karner	Ruhr University Bochum, Germany
N.Khvingia	ITEP Moscow, USSR
A.N.Kochubei	Institute of Energy Systems Kiev, USSR
L.Kouba	Institute of Physics, CSAS Prague, Czechoslovakia
V.Král	Prague, Czechoslovakia
V.Ksenzov	ITEP Moscow, USSR
J.Kučera	Masaryk University Brno, Czechoslovakia
Yu.Kuperin	Leningrad State University Leningrad, USSR
P.G.L.Leach	University of Natal Durban, South Africa
M.H.Lee	University of Georgia Athens, USA
E.H.Lieb	Princeton University Princeton, USA
O.Litzman	Masaryk University Brno, Czechoslovakia
K.A.Makarov	Leningrad State University Leningrad, USSR
M.M.Malamud	Polytechnical Institute Donetsk, USSR
B.Mielnik	Research Center Mexico City, Mexico

A.Moroz	Institute of Physics, CSAS Prague, Czechoslovakia
H.Neidhardt	Institute of Mathematics Berlin, Germany
F.Nitzschner	Ruhr University Bochum, Germany
M.Pardy	Masaryk University Brno, Czechoslovakia
L.A.Pastur	Institute for Low Temperature Physics and Engineering Kharkov, USSR
B.S.Pavlov	Leningrad State University Leningrad, USSR
I.Yu.Popov	Leningrad Institute of Precise Mechanics and Optics Leningrad, USSR
E.Prokhorenko	Yerevan Physics Institute Yerevan, USSR
B.Rudolf	Czech Technical University Prague, Czechoslovakia
A.E.Ryzhkov	Leningrad Institute of Precise Mechanics and Optics Leningrad, USSR
K.Sandler	Nuclear Physics Institute, CSAS Řež, Czechoslovakia
R.Seiler	Technical University Berlin, Germany
M.Z.Solomyak	Leningrad State University Leningrad, USSR
V.R.Struletskaia	MIEM Moscow, USSR

P.Šeba — Ruhr University
Bochum, Germany
and
Nuclear Physics Institute, CSAS
Řež, Czechoslovakia

P.Šťovíček — Czech Technical University
Prague, Czechoslovakia

M.Tater — Nuclear Physics Institute, CSAS
Řež, Czechoslovakia

A.Teta — Ruhr University
Bochum, Germany

M.Thoma — Czech Technical University
Prague, Czechoslovakia

M.Vymazal — Nuclear Physics Institute, CSAS
Řež, Czechoslovakia

K.Yajima — University of Tokyo
Tokyo, Japan

V.A.Zagrebnov — JINR
Dubna, USSR

B.N.Zakhariev — JINR
Dubna, USSR

G.M.Zhislin — Radiophysical Research Institute
Nizhnii Novgorod, USSR

M.Znojil — Nuclear Physics Institute, CSAS
Řež, Czechoslovakia

LIST OF UNPUBLISHED CONTRIBUTIONS

Lectures

J.-M.Combes: Dynamical behaviour of some multi-well Schrödinger operators

L.A.Pastur: Large-time asymptotics of some Wiener integrals and problems of the theory of disordered systems

B.S.Pavlov: Waveguide effects in many-body lattice problems

R.Seiler: The integer Hall conductivity in two dimensions: The Bellisard-Connes index

Short contributions

J.Dittrich: Dirac-Coulomb Hamiltonian with contact shell interaction

J.Hronek: Spectral theory of the Dirac operator with periodic potential

A.N.Kochubei: Pseudo-differential operators and Markov processes

K.A.Makarov: Scattering on fractal measures

I.Yu.Popov: Localization of resonances in the problem of scattering on a region with small aperture

B.N.Zakhariev: Schrödinger equation for lattices and periodic potentials

Posters

M.Chiranov: Minimum uncertainty supersymmetric states

O.Cristea: Dissipative quantum dynamics on pseudo-riemannian manifold

F.Defever: Superconformal algebras and Clifford algebras

G.Ganbold: The improvement of variational approach to the polaron energy

K.-E.Hellwig: On the notion and representations of classical convex structures

G.Karner: The dynamics of periodically kicked quantum particles

V.Ksenzov: Calculation of effective Lagrangian of vector, pseudovector, pseudoscalar mesons and dilaton in σ-model

O.Litzman: Rigorous formulae for the dynamical theory of diffraction of neutrons on a semisimple crystal

A.Moroz: Horn shaped regions - summability method and strong asymptotic conditions with some applications

E.Prokhorenko: Dispersion law in charged matter considering elastic properties

V.R.Struletskaia: Quasi-free boson and fermion dynamical systems. Stochastic evolutions and stochastic quantization.

www.ingramcontent.com/pod-product-compliance
Ingram Content Group UK Ltd.
Pitfield, Milton Keynes, MK11 3LW, UK
UKHW020359150926
8624IPUK00004B/250

9 789810 205614